AF368963

# Machining: State-of-the-Art 2022

# Machining: State-of-the-Art 2022

Editors

**Francisco J. G. Silva**
**Filipe Daniel Fernandes**
**Vitor Fernando Crespim Sousa**

Basel • Beijing • Wuhan • Barcelona • Belgrade • Novi Sad • Cluj • Manchester

*Editors*

Francisco J. G. Silva
ISEP—School of Engineering,
Polytechnic of Porto
Porto, Portugal

Filipe Daniel Fernandes
ISEP—Scool of Engineering,
Polytechnic of Porto
Porto, Portugal

Vitor Fernando Crespim
Sousa
ISEP—School of Engineering,
Polytechnic of Porto
Porto, Portugal

*Editorial Office*
MDPI
St. Alban-Anlage 66
4052 Basel, Switzerland

This is a reprint of articles from the Special Issue published online in the open access journal *Metals* (ISSN 2075-4701) (available at: https://www.mdpi.com/journal/metals/special_issues/machining2021).

For citation purposes, cite each article independently as indicated on the article page online and as indicated below:

Lastname, A.A.; Lastname, B.B. Article Title. *Journal Name* **Year**, *Volume Number*, Page Range.

ISBN 978-3-0365-9128-5 (Hbk)
ISBN 978-3-0365-9129-2 (PDF)
doi.org/10.3390/books978-3-0365-9129-2

Cover image courtesy of Francisco José Gomes da Silva

# Contents

# About the Editors

**Francisco J. G. Silva**

Francisco J. G. Silva is a Dr.Habil with a PhD, MSc and BSc in Mechanical Engineering from FEUP and ISEP (Portugal). He is a postgraduate in Materials and Manufacturing. He is currently the Head of the Mechanical Engineering Research Center at ISEP, Polytechnic of Porto. He was also the Head of the Master's Degree in Mechanical Engineering of ISEP (2014–2022) and Head of the Bachelor's Degree in Mechanical Engineering at ESEIG, Polytechnic of Porto (2003 to 2006). He has supervised more than 10 PhD students at FEUP (Portugal), as well as more than 200 MSc students at ISEP, and has co-supervised more than 40 MSc students at ISEP and FEUP. He has published more than 280 papers (ISI+SCOPUS) and 12 international books. He has reviewed more than 800 papers, and has been the Associate Editor and an Editorial Board Member of more than 10 indexed international journals.

**Filipe Daniel Fernandes**

Filipe Daniel Fernandes is presently an Assistant Professor at the School of Engineering at the Polytechnic of Porto, and a collaborator at the University of Coimbra and Colaborator at IPN where he is collaborating in the planning, preparation and submission of research project proposals. He is dedicated to the surface engineering field, and to efforts to extend the lifetime of components working under high temperature solicitations. He has won different prizes and awards for which he was recently highlighted at the "2023 Young scientist awards" for his outstanding contribution towards research, as recognized at the Indo-European conference of Advanced Manufacturing and Materials Processing, IECAdvaMAP-2023. He has published 73 manuscripts in peer-reviewed journals (h-index 18) and participated in more than 56 conferences, 5 as a keynote and 3 as an invited speaker. He has been the PI of several projects, having attracted EUR 2 million in industrial research projects and EUR 500,000 in fundamental research projects. He is a member of the advisor committee/scientific commission of conferences, a member of a doctoral program, a member of the evaluation comities of research projects, a member of scientific societies, a member of the Editorial Board of scientific journals and an organizer of seminars for high school students and graduate researchers. He supervises postdocs, PhDs and Master's students, and has participated on the juries of PhD and Master's theses and research contracts.

**Vitor Fernando Crispim Sousa**

Vitor Fernando Crespim Sousa, born in 1994 in Porto, Portugal, is a researcher at the Institute of Science and Innovation in Mechanical Engineering (INEGI), Porto, Portugal. His main research interests are focused on machining processes, mostly of hard-to-machine materials, using coated tools. He has cosupervised around six MSc dissertations. He is currently focused on the study of the machinability of nickel-based alloys. He has participated, as an author and coauthor, in a total of 26 scientific articles already published in international indexed journals. He has also participated in the writing of three published conference articles and three book chapters. Vitor Fernando Crispim Sousa has presented at a total of seven international conferences. He has also reviewed more than 50 papers.

# Preface

Although research efforts around manufacturing processes are increasingly concentrated on additive manufacturing, conventional processes still have a very high margin of progression, which deserves to be investigated. Furthermore, machining presents levels of surface quality not yet attainable by most additive manufacturing processes, which is why some more advanced equipment even incorporates both technologies. Given that machining continues to be a topic around which there is substantial research, it was considered appropriate to develop this Special Issue that brings together a batch of twelve excellent works, some of which are reviews. Thus, this Special Issue is intended to be extremely useful to researchers starting their work on these themes. The other articles also provide new data in terms of research, which may be useful to anyone who researches or teaches in this field of knowledge.

After a brief contextualization at the beginning of this Special Issue, some broad-spectrum review articles are presented. The first addresses hybrid manufacturing, which, as previously mentioned, is beginning to assume increasing importance in the metalworking industry and is also a hot topic of research. A retrospective of the most recent developments is carried out, allowing those new in the topic to evolve quickly by reading this review. Next, another review is presented, focusing on the difficulties presented in the machining of fiber metal laminates, a multimaterial that is beginning to gain considerable importance, mainly in the aeronautical and aerospace industries. Another review then follows, this one now focused on the conventional and unconventional machining of Inconel alloys, a nickel alloy used more and more frequently in very demanding applications and one that is traditionally demanding in terms of tools in machining operations. In fact, this alloy is also the main focus of the following two articles, the first studying the wear behavior of milling tools coated with TiN/TiAlN through PVD, using HiPIMS technology. Concerning the machining of Inconel alloys, the behavior of turning tools is also studied in the next paper, using PCBN tools. The Special Issue continues with a paper studying the wear behavior of coated tools in UNS S32101 Duplex Stainless Steel milling operations. This duplex stainless steel is traditionally a material that is difficult to cut due to its mechanical strength and elevated ductility, with the chip tending to adhere to the tools. Drilling, and the respective consequences of this operation when applied to fiber metal laminates, is also covered in this Special Issue, through a study focused on the phenomena of delamination between the different layers of these laminates when they are exposed to the characteristic forces of the drilling process.

Environmental sustainability in machining processes is a constant concern, mainly due to the lubricants and coolants used. This Special Issue presents two studies, one focused on the use of $CO_2$ as a coolant in machining operations, and another review focused on the environment and on how to make machining more environmentally friendly using specific techniques and products.

Finally, the economic aspect of machining is addressed, through two different approaches to estimating the time and cost of machining operations. One of the approaches is more conventional, while the other uses artificial intelligence algorithms to calculate the estimated machining time in certain operations. The results obtained allow the prediction that the models will converge to increasingly reliable results that are closer to reality.

With a wide range of subjects, all of them sharply focused on current machining concerns, this Special Issue is expected to be of great usefulness to all those looking to deepen their knowledge in this important area of metalworking production.

**Francisco J. G. Silva, Filipe Daniel Fernandes, and Vitor Fernando Crespim Sousa**
*Editors*

*Editorial*

# Machining: State-of-the-Art 2022

**Francisco J. G. Silva *, Filipe Fernandes and Vitor F. C. Sousa**

ISEP, Polytechnic of Porto, Rua Dr. António Bernardino de Almeida, 4249-015 Porto, Portugal;
fid@isep.ipp.pt (F.F.); vcris@isep.ipp.pt (V.F.C.S.)
* Correspondence: fgs@isep.ipp.pt

## 1. Introduction and Scope

Although additive manufacturing is gaining prominence in the market, many applications require very high levels of precision, which are currently not attainable by additive manufacturing [1]. Consequently, research in the field of machining continues to be highly vigorous [2,3], as evolution can take place towards the use of hybrid manufacturing processes, which complement additive processes with subtractive processes [4]. Given that machining processes themselves have already been explored in terms of the phenomena to which they are related, as well as machining trajectories, the aspects related to sustainability, the tools used in advanced materials and the study of their wear phenomena have recently been deeply explored [5,6]. In fact, increasing tool life is a matter of economic and environmental sustainability, as it increases the useful machining time of each piece of equipment and reduces the ecological footprint, due to the lower rate of energy consumed and the reduced need to recycle worn-out tools. This Special Issue intends to bring together some of the best works conducted on machining published in 2022 and 2023. Specifically, it aims to provide comprehensive and in-depth reviews that enable new researchers to quickly and efficiently familiarize themselves with specific subjects. By compiling and presenting the results of numerous prior publications in a condensed and well-structured manner, this initiative facilitates access to a substantial amount of information within a short timeframe. However, the investigation of new coatings for tools and even the study of models for the quick budgeting of machining processes have also been taken into account in this Special Issue, which aims to constitute a very interesting piece of work for all those who investigate machining processes.

## 2. Contributions

This Special Issue has eleven contributions, five of which are reviews of current issues that deserve particular attention. Two of the articles are essentially focused on models for the management of machining processes, with a view to quickly organizing budgeting processes. The remaining articles are essentially experimental studies, with a view to determining better machining parameters, extending the useful life of tools or studying the wear phenomena that affect tools.

The first review focuses on an extremely important issue: the sustainability of tool-cooling processes during machining [7]. $CO_2$ is often used for cooling tools during machining processes and is included in cryogenic cooling machining processes. Given that $CO_2$ does not constitute an added threat to the environment, its use can significantly reduce the wear of tools whose wear mechanisms are particularly associated with the temperature developed during the process. It is clear that cryogenic machining has added advantages over other tool-cooling processes, such as flood cooling, MQL (minimum quantity lubrication) or ADL (aerosol dry lubrication). This review allows readers to gain a quick understanding of the advantages of using $CO_2$ in the cryogenic machining of different materials, as well as in establishing the best manufacturing parameters when this cooling technique is adopted.

The issue of sustainability associated with the lubrication and cooling of tools and material to be machined is also addressed in the second review presented in this Special

**Citation:** Silva, F.J.G.; Fernandes, F.; Sousa, V.F.C. Machining: State-of-the-Art 2022. *Metals* **2023**, *13*, 1036. https://doi.org/10.3390/met13061036

Received: 17 May 2023
Accepted: 19 May 2023
Published: 29 May 2023

Issue [8]. Additionally, the review considers the surface finish quality achieved through the machining process. By examining these aspects, the review contributes to a more comprehensive understanding of sustainable machining practices, emphasizing the need to balance environmental concerns with achieving the desired production quality. Indeed, the drawbacks usually brought about by the use of mineral-based oils are well known, and there are already solutions based on vegetable oils that can be used as an alternative. However, these solutions still present considerable limitations, namely the amount of fluid that is required, the quality of the surface obtained using the process and the tribological aspects related to tool wear. These limitations are conveniently described and analyzed, allowing readers to understand that this solution still does not present a higher performance when compared to mineral-based fluids. However, this review also dissects nanofluids, which have very interesting properties in thermal and physical terms; however, research efforts will be necessary to make this solution more economical.

The need for increasingly demanding properties on the part of the market has led to the development of new solutions in terms of materials, which now have a new range, called fiber metal laminates (FMLs) [9], with concerns in sustainability. The market's growing demand for materials with increasingly stringent properties has driven the development of new solutions. One such solution is the emergence of fiber metal laminates (FMLs) [9], which offer a novel range of properties.

The review work presented in this field is quite comprehensive, reviewing the manufacturing processes of these multi-materials, and subsequently analyzing the challenges presented by the machining processes, when applied to this specific case. Indeed, the combination of different materials leads to the existence of interfaces and the need for tools to be prepared to jointly machine materials with significantly different properties. These factors pose challenges that are difficult to overcome, a fact that has given rise to numerous studies with a view to understanding the scale of these challenges and finding increasingly better solutions to overcome them. In that work, the parameters that best correspond to the requirements imposed by this type of material are also studied, as well as a SWOT analysis that aimed to bring together the trends in terms of the strengths, weaknesses, opportunities and threats found in the different analyzed works.

This Special Issue also includes a review article that compiles previously dictated work on the integration of additive processes with subtractive processes [10]. Effectively, there are still problems in obtaining surface finishes in additive manufacturing that are compatible with the needs of contact and relative movement existing in many applications. Thus, the so-called hybrid manufacturing processes have gained prominence in the market, starting from additive manufacturing, and ending with subtractive manufacturing (machining or other similar techniques), with a view to obtaining the necessary final precision and finish quality. In an attempt to reduce the need to use multiple pieces of equipment and transport parts between these same pieces of equipment (internal logistics), there is a strong tendency for additive and subtractive operations to be performed using the same piece of equipment. The work develops each of the technologies, describing the advantages and limitations of each one and the integration of both processes.

Another review included in this Special Issue is related to the machining of IN-CONEL [11]. In fact, the problems related to the machining of INCONEL are known, due to its high mechanical resistance and problems of conduction of the heat generated in the contact. This review, based on many consulted articles, covers what has been recently published on this subject, presenting the data in a structured way, as well as some SWOT analyses that allow the readers to quickly and concisely understand the main challenges imposed on the machining processes of nickel alloys. Moreover, this study also extends the analysis to non-conventional machining processes, allowing a wider perspective of the difficulties and current solutions in machining some hard-to-cut metallic alloys.

Within this Special Issue, two additional research articles focused on the INCONEL alloy are included. These articles delve into the study of tool life, investigating the wear mechanisms involved and identifying parameters that contribute to a prolonged tool lifespan and enhanced efficiency when machining this alloy. One of the papers [12] studies the wear mechanisms of tools coated with TiN/TiAlN via PVD, using the HiPIMS technology, which provides greater adhesion of the coatings to the substrate. It is a double coating, with a first layer of TiN to ensure optimal adhesion to the substrate, and a complementary layer of TiAlN, which has shown remarkable tribological properties in terms of wear resistance in machining operations. The study was conducted based on milling operations. The main objective of the study was to analyze the failure mechanisms of these coatings, concluding that there is abrasion, material adhesion, cratering and adhesive wear. Another work, not using solid coated tools, but polycrystalline cubic boron nitride (PCBN) inserts, studied the influence of two different binder phases (TiN and TiC) on the performance of tools in turning operations of INCONEL 718 alloys [13]. The work revealed that the turning process is sensitive to the cutting speed established for the process, also revealing that the different phases induce different wear mechanisms in the tools. It was also proved that the TiN phase presents a better performance than the TiC phase, also presenting a greater consistency in terms of the final result achieved in the machining process.

Another work presented in this Special Issue [14] focuses on the study of wear phenomena in coated tools, specifically when machining UNS S32101 duplex stainless steel. These alloys, having different machining characteristics from INCONEL alloys, are also classified as difficult to machine, and have a high number of applications where machining is required. Therefore, these alloys, despite having been previously studied, continue to deserve the attention of several research groups. In this study, tools with two or four flutes, as well as TiAlN, TiAlSiN and AlCrN coatings produced via PVD, were tested. The observed wear mechanisms were duly dissected and analyzed for different cutting lengths. It was reported that the TiAlSiN coating presented the best wear behavior, considering the parameters used in the conducted milling operations.

Returning to fiber metal laminates (FMLs), this Special Issue also presents a work on drilling operations, with the focus on studying the delamination occurring between different layers, as well as the fracture mechanisms involved [15]. A modeling study was also carried out, based on experimental results. In fact, FMLs are of particular relevance for the aeronautical industry, due to the fact that they have a lower density, combined with high mechanical resistance, significant impact resistance and high ductility. However, the application of these multi-materials in aircraft presupposes that they are easy to drill, which is not the case, due to problems of decohesion between layers, because of the shear forces developed on the material in general, which is particularly felt in the interfaces. The study revealed that the use of specific step tools with a secondary cutting edge significantly improves the machining performance. The developed model, despite being simplistic, proved to be efficient in predicting the delamination that can be induced in the interfaces of the FMLs used in this work.

Given that machining is widely used in industrial terms, predicting the time of machining operations is extremely useful, and is also the subject of in-depth studies aimed at developing models capable of reliably estimating the time required for certain machining operations, thus minimizing the response time in the budgeting of this type of operation, since outsourcing is a very common practice in this type of industry [16].

Machining plays a crucial role in industrial applications, making the accurate prediction of the machining time highly valuable. Consequently, in-depth studies are dedicated to developing models capable of reliably estimating the time required for specific machining operations. Such models enable the minimization of the response time in budgeting these operations, which is particularly important considering that outsourcing is a prevalent practice in the industry.

The papers included in this Special Issue include two different approaches, with one being a more conventional approach [17], which can be used by any company that regularly uses MS Excel spreadsheets. This model presents different degrees of complexity in machining, distributing these degrees of difficulty by levels in different aspects, and building a model that can be easily adjusted by users, with a view to improving its operational reliability. The model presented in the second work of this nature in this Special Issue uses more advanced techniques, namely artificial neural networks (ANNs) [18]. In this case, a group of drawings of parts was considered, to "teach" the system, followed by a smaller set for testing the accuracy, in terms of the time estimates offered by the model. The maximum precision achieved, in this case, was 2.52%, which is a very acceptable value for this sector. This value is significantly better than the average estimation accuracy found in the traditional model, which was around 14% [17]. However, these models can be fine-tuned by increasing the volume of data considered for learning, improving the accuracy of the estimate offered by the model.

### 3. Conclusions and Outlook

Machining continues to be a field of strong investment by researchers, looking for new solutions that increase the competitiveness and sustainability of this type of subtractive process. Despite the strong increase in number that additive processes have undergone in the last two decades, the quality of the finish exhibited by machining processes is difficult to match using additive processes. Hence, hybrid processes and equipment are emerging, with a view to responding more effectively to market needs. More recently developed alloys also lead to new research needs, as well as new materials, such as fiber metal laminates.

(The emergence of recently developed alloys and new materials, such as fiber metal laminates (FMLs), creates new research requirements and opportunities.)

Environmental sustainability, in addition to economic sustainability, has also been a factor to be considered in the area of machining. The development of more durable tools, as well as the establishment of cutting conditions that promote a longer tool life, has received increasing attention from researchers. The development of tools capable of assisting professionals in budgeting operations is also another concern of researchers, who have been developing models that aim to increase the accuracy of estimates. It can thus be seen that the machining area remains particularly active in terms of research and development, which is extremely beneficial for all those who are particularly attracted to these technologies. All of this can be found in this Special Issue.

**Conflicts of Interest:** The authors declare no conflict of interest.

## References

1. Teixeira, Ó.; Silva, F.J.G.; Atzeni, E. Residual stresses and heat treatments of Inconel 718 parts manufactured via metal laser beam powder bed fusion: An overview. *Int. J. Adv. Manuf. Technol.* **2021**, *113*, 3130–3162. [CrossRef]
2. Sousa, V.F.C.; Silva, F.J.G. Recent Advances on Coated Milling Tool Technology—A Comprehensive Review. *Coatings* **2021**, *10*, 235. [CrossRef]
3. Sousa, V.F.C.; Silva, F.J.G. Recent Advances in Turning Processes Using Coated Tools—A Comprehensive Review. *Metals* **2020**, *10*, 270. [CrossRef]
4. Pragana, J.P.M.; Sampaio, R.F.V.; Bragança, I.M.F.; Silva, C.M.A.; Martins, P.A.F. Hybrid metal additive manufacturing: A state–of–the-art review. *Adv. Ind. Manuf. Eng.* **2021**, *2*, 100032. [CrossRef]
5. Martinho, R.P.; Silva, F.J.G.; Martins, C.; Lopes, H. Comparative study of PVD and CVD cutting tools performance in milling of duplex stainless steel. *Int. J. Adv. Manuf. Technol.* **2019**, *102*, 2423–2439. [CrossRef]
6. Silva, F.J.G.; Gouveia, R.M. *Cleaner Production—Toward a Better Future*; Springer Nature: Cham, Switzerland, 2020.
7. Proud, L.; Tapoglou, N.; Slatter, T. A Review of $CO_2$ Coolants for Sustainable Machining. *Metals* **2022**, *12*, 283. [CrossRef]
8. Khan, M.A.; Hussain, M.; Lodhi, S.K.; Zazoum, B.; Asad, M.; Afzal, A. Green metalworking fluids for sustainable machining operations and other sustainable systems: A review. *Metals* **2022**, *12*, 1466. [CrossRef]
9. Costa, R.D.F.S.; Sales-Contini, R.C.M.; Silva, F.J.G.; Sebbe, N.; Jesus, A.M.P. A Critical Review on Fiber Metal Laminates (FML): From Manufacturing to Sustainable Processing. *Metals* **2023**, *13*, 638. [CrossRef]
10. Sebbe, N.P.V.; Fernandes, F.; Sousa, V.F.C.; Silva, F.J.G. Hybrid Manufacturing Processes Used in the Production of Complex Parts: A Comprehensive Review. *Metals* **2022**, *12*, 1874. [CrossRef]
11. Pedroso, A.F.V.; Sousa, V.F.C.; Sebbe, N.P.V.; Silva, F.J.G.; Campilho, R.D.S.G.; Sales-Contini, R.C.M.; Jesus, A.M.P. A Comprehensive Review on the Conventional and Non-Conventional Machining and Tool-Wear Mechanisms of INCONEL®. *Metals* **2023**, *13*, 585. [CrossRef]
12. Sousa, V.F.C.; Fernandes, F.; Silva, F.J.G.; Costa, R.D.F.S.; Sebbe, N.; Sales-Contini, R.C.M. Wear Behavior Phenomena of TiN/TiAlN HiPIMS PVD-Coated Tools on Milling Inconel 718. *Metals* **2023**, *13*, 684. [CrossRef]

13. Matos, F.; Silva, T.E.F.; Sousa, V.F.C.; Marques, F.; Figueiredo, D.; Silva, F.J.G.; de Jesus, A.M.P. On the Influence of Binder Material in PCBN Cutting Tools for Turning Operations of Inconel 718. *Metals* **2023**, *13*, 934. [CrossRef]
14. Sousa, V.F.C.; Silva, F.J.G.; Alexandre, R.; Pinto, G.; Baptista, A.; Fecheira, J.S. Investigations on the Wear Performance of Coated Toolsin Machining UNS S32101 Duplex Stainless Steels. *Metals* **2022**, *12*, 896. [CrossRef]
15. Marques, F.; Silva, F.G.A.; Silva, T.E.F.; Rosa, P.A.R.; Marques, A.T.; de Jesus, A.M.P. Delamination of Fibre Metal Laminates Due to Drilling: Experimental Study and Fracture Mechanics-Based Modelling. *Metals* **2022**, *12*, 1262. [CrossRef]
16. Ferreira, V.; Silva, F.J.G.; Martinho, R.P.; Pimentel, C.; Godina, R.; Pinto, B. A comprehensive supplier classification model for SME outsourcing. *Procedia Manuf.* **2019**, *38*, 1461–1472. [CrossRef]
17. Silva, F.J.G.; Sousa, V.F.C.; Pinto, A.G.P.; Ferreira, L.P.F.; Pereira, T. Build Up an Economical Tool for Machining Operations Cost Estimation. *Metals* **2022**, *12*, 1205. [CrossRef]
18. Rodrigues, A.; Silva, F.J.G.; Sousa, V.F.C.; Pinto, A.G.; Ferreira, L.P.; Pereira, T. Using an Artificial Neural Network Approach to Predict Machining Time. *Metals* **2022**, *12*, 1709. [CrossRef]

 **_metals_**

*Review*

# Hybrid Manufacturing Processes Used in the Production of Complex Parts: A Comprehensive Review

Naiara P. V. Sebbe [1], Filipe Fernandes [1,2,*], Vitor F. C. Sousa [1] and Francisco J. G. Silva [1,3]

[1]  ISEP—School of Engineering, Polytechnic of Porto, Rua Dr. António Bernardino de Almeida 431, 4200-072 Porto, Portugal

[2]  CEMMPRE—Centre for Mechanical Engineering Materials and Processes, Department of Mechanical Engineering, University of Coimbra, Rua Luís Reis Santos, 3030-788 Coimbra, Portugal

[3]  INEGI—Driving Science and Innovation, Rua Dr. Roberto Frias 400, 4200-465 Porto, Portugal

*  Correspondence: fid@isep.ipp.pt

**Abstract:** Additive manufacturing is defined as a process based on the superposition of layers of materials in order to obtain 3D parts; however, the process does not allow achieve the adequate and necessary surface finishing. In addition, with the development of new materials with superior properties, some of them acquire high hardness and strength, consequently decreasing their ability to be machined. To overcome this shortcoming, a new technology assembling additive and subtractive processes, was developed and implemented. In this process, the additive methods are integrated into a single machine with subtractive processes, often called hybrid manufacturing. The additive manufacturing process is used to produce the part with high efficiency and flexibility, whilst machining is then triggered to give a good surface finishing and dimensional accuracy. With this, and without the need to transport the part from one machine to another, the manufacturing time of the part is reduced, as well as the production costs, since the waste of material is minimized, with the additive–subtractive integration. This work aimed to carry out an extensive literature review regarding additive manufacturing methods, such as binder blasting, directed energy deposition, material extrusion, material jetting, powder bed fusion, sheet laminating and vat polymerization, as well as machining processes, studying the additive-subtractive integration, in order to analyze recent developments in this area, the techniques used, and the results obtained. To perform this review, ScienceDirect, Web of Knowledge and Google Scholar were used as the main source of information because they are powerful search engines in science information. Specialized books have been also used, as well as several websites. The main keywords used in searching information were: "CNC machining", "hybrid machining", "hybrid manufacturing", "additive manufacturing", "high-speed machining" and "post-processing". The conjunction of these keywords was crucial to filter the huge information currently available about additive manufacturing. The search was mainly focused on publications of the current century. The work intends to provide structured information on the research carried out about each one of the two considered processes (additive manufacturing and machining), and on how these developments can be taken into consideration in studies about hybrid machining, helping researchers to increase their knowledge in this field in a faster way. An outlook about the integration of these processes is also performed. Additionally, a SWOT analysis is also provided for additive manufacturing, machining and hybrid manufacturing processes, observing the aspects inherent to these technologies.

**Keywords:** hybrid manufacturing; CNC machining; additive manufacturing; 5-axis machining; manufacturing processes

**Citation:** Sebbe, N.P.V.; Fernandes, F.; Sousa, V.F.C.; Silva, F.J.G. Hybrid Manufacturing Processes Used in the Production of Complex Parts: A Comprehensive Review. *Metals* **2022**, *12*, 1874. https://doi.org/10.3390/met12111874

Academic Editors: Manoj Gupta and Pavel Krakhmalev

Received: 11 September 2022
Accepted: 27 October 2022
Published: 2 November 2022

**Publisher's Note:** MDPI stays neutral with regard to jurisdictional claims in published maps and institutional affiliations.



## 1. Introduction

Nowadays, with the advances in materials for several industrial sectors, such as the aerospace, biomedical and automotive sectors, new methods/processes are needed to fulfil the production demand and required quality [1]. Standard methods of manufacturing have thus started to be put aside, whilst more advanced methods of manufacturing raised, so that customer needs are met without affecting the profitability of the companies [2].

It is in this way that the concept of hybrid manufacturing arises. The idea behind hybrid manufacturing is joining different processes on the same setup in order to achieve the effect known as "1 + 1 = 3" [3,4]. The term "hybrid manufacturing" is directly linked to the integration of different processes, and its development is related to the requirements and complexity of new parts [5].

Thus, the objective of hybrid manufacturing is the joining of two or more distinct processes in a single piece of equipment, observing the unique advantages of each one, while minimizing the limitations of the process [6]. It is observed that there are several types of hybrid manufacturing, although the most common is the type that combines laser additive manufacturing and 5-axis machining processes [7].

Additive manufacturing allows the production of 3D geometric parts through layer overlay [8,9], being a suitable method for efficient production of parts; however, the cost is high and requires a high financial capital, which makes the technique not yet widely used [10]. 5-axis machining is a process more and more used in the industry [11], allowing a single configuration to machine five sides of the part [12] and providing high accuracy and surface quality [13].

The present work aims to carry out a broad bibliographic review about hybrid manufacturing regarding additive and subtractive processes in a single equipment. Additive manufacturing and its manufacturing processes, such as binder blasting, directed energy deposition, material extrusion, material jetting, powder bed fusion, sheet laminating and vat polymerization, and their respective subclassifications were thoroughly and properly analyzed and described. An approach to machining is also taken, emphasizing 5-axis machining. From this, hybrid manufacturing is discussed through a broad review of the work carried out and the integration between additive and subtractive processes, as well as the challenges faced and future opportunities.

## 2. Additive Manufacturing

Additive manufacturing is the process of manufacturing parts layer by layer, enabling the production of parts with more robust and complex geometries [14]. This technology is of great relevance and represents a real challenge for today's industries, given its flexibility and ability to provide differentiated products and parts [15–18]. Additive manufacturing is popularly known as "3D printing", and it is a process which takes place under digital control. In this process, the raw material is placed in the equipment in the form of wire or powder, and the part produced according to the geometry contained in a CAD project (computer-aided design) [19]. Figure 1 exposes the basic principle of additive manufacturing of a part grown layer by layer [20].

**Figure 1.** Basic principle of additive manufacturing. Reproduced from [20], 2012, Elsevier.

The classification of additive manufacturing processes, in the past, was based on the criterion of separating materials into a liquid basis, solid basis or powder basis [21]. However, in 2010, the American Society for Testing and Materials (ASTM) published the standard "ASTM F42—Additive Manufacturing", which considers additive manufacturing in seven categories [22]:

(i) Binder blasting;

(ii) Directed energy deposition—DED (comprised of processes such as direct deposition of metal), such as wire arc additive manufacturing (WAAM);

(iii)  Material extrusion (which includes fuse deposition modeling (FDM));

(iv)  Material jetting;

(v)  Powder bed fusion (which includes processes such as direct metal laser sintering (DMLS) and selective laser sintering (SLS));

(vi)  Sheet laminating (ultrasonic additive manufacturing (UAM) and laminated object manufacturing (LOM));

(vii)  Vat polymerization.

In binder blasting, an ink jet is selectively deposited onto bond powder materials, typically plaster or starch, and creates three-dimensional objects, consisting of placing thin layers of this powder, with the print head ejecting and depositing drops of binder [23]. In the study carried out by Chen et al. [24], alumina powder was used as a material processed in different ways to obtain granules with different properties, and aiming at the final quality of the part in terms of compressive strength and density. With this, it was verified that the alumina samples were produced properly and with density and compressive strength within the expected ranges.

The directed energy deposition (DED) process uses thermal energy to build the 3D product, layer by layer, with very good properties. The volumetric density of the product can be practically 100%, and its use with hybrid systems is very frequent, given the ability to deposit heterogeneous materials on the substrate with adequate characteristics [25–27]. Figure 2 illustrates the DED process.

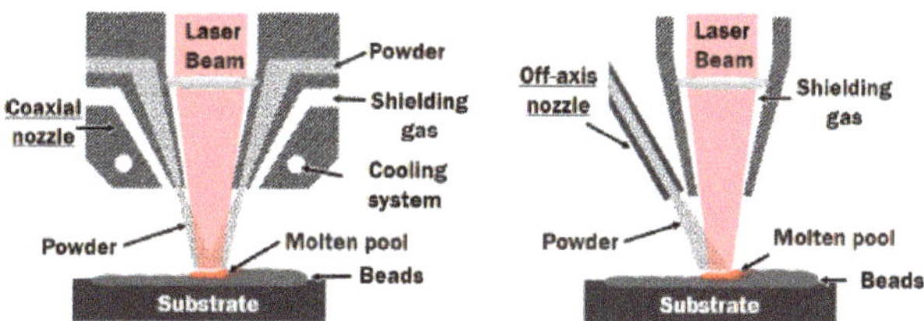


**Figure 2.** DED process illustration. Reproduced from [27]. Creative Commons CC BY license: https://creativecommons.org/licenses/by/4.0/; accessed on 21 October 2022.

In the work by Chen et al. [28], it was observed that changes in temperature directly influence the melting and cooling of the material, and consequently also influence the final microstructure and the resulting hardness. Thus, it was possible to predict the hardness distribution of a part by monitoring the temperature changes that occurred. For this purpose, some key temperature features (KTFs) were defined, according to the temperature field present in the DED process, and the resulting hardness predictions were made, showing that predictions were consistent with the real hardness trends.

Tekumalla et al. [29] studied high-vanadium high-speed steels (HVHSSs), which are considered difficult to machine materials due to their high hardness. Thus, the DED method was used, with two alloy compositions: Fe-10V-4.5Cr-2.5C and Fe-15V-13Cr-4.5C, which are highly wear-resistant alloys due to their vanadium and carbon content, and after processing, high hardness of the products was confirmed through a comparison performed regarding these materials after conventional heat treatment. In the work by Radhakrishnan et al. [30], the laser-directed energy deposition method was used to deposit titanium/titanium carbide (Ti/TiC) with 20%, 40% and 60% TiC. The formation of the non-stoichiometric compound TiC0.55 was observed, and hardness increased with TiC content increasing.

Still regarding the DED process, wire feeding processes are divided into three types: (i) wire and arc additive manufacturing (WAAM) processes, which combine an arc, working as a thermal energy source, and a wire, which is the raw material, providing energy efficiency less than 90%; (ii) additive manufacturing of wire and laser (WLAM), which in turn uses the concept of laser cladding and welding in order to produce metal parts that do not have porosity; and finally, (iii) additive manufacturing of wire and electron beam (WEAM), which, as the name implies, uses a beam of electrons as a source of energy [27].

Grossi et al. [31] analyzed the dynamic behavior of a NACA 9403 airfoil produced using the WAAM technique, exploring by finite element analysis (FE) the subsequent machining process in order to predict the dynamics of the part during that machining process. Thus, the system was able to simulate the material removal process and the dynamic behavior of the part previously generated by WAAM, updating its variable geometry.

On the other hand, using another technology known as PAW (plasma arc welding), and in order to analyze the deposition process based on process characteristics and mechanical and microstructural properties, Artaza et al. [32] produced two Ti6Al4V walls with a high deposition rate (2 kg/h) and in an inert argon atmosphere, with no significant differences in microstructural or mechanical properties after thermal heat treatments. Still in relation to PAW technology and Ti6Al4V alloy, Veiga et al. [33] analyzed the production, as well as the quality, of the alloy produced using this technology with subsequent milling, and observed that there were no large deviations in the mechanical properties of the samples in different positions and orientations, and the up-milling showed torque values that were slightly larger. However, the quality of the final surface was superior.

The principle of material extrusion is a technique widely used popularly in domestic environments. This process uses a wire which is extruded and gradually forms the product. A layer of material is deposited horizontally through a nozzle, and then with increments in the vertical direction, until the desired shape is reached [20,34]. There are several technologies for this process, for example the FDM (fused deposition modeling) process, the most common being associated with the extrusion of thermoplastics. However, there are research and investigation opportunities for other processes, including the development of new materials and new technologies, within this same process [35].

Awasthi et al. [36] addressed the challenges faced by the FDM technique, given the wide variety of new materials emerging, emphasizing thermoplastic elastomers (TPE) that are compatible with the process. However, the process is limited due to the difficulty of printing, as well as the generation of defects in the produced parts. In the work by Jin et al. [37], a mathematical model was made and analyzed to verify the surface of the product manufactured through the FDM technique, indicating that the precision of the upper surface is determined through the ratio between the flow of the molten material and the feed rate of the nozzle, whilst the lateral surface is related to the thickness of the process layer and the bedding angle. Thus, it is suggested that the aforementioned ratio be properly used and the thickness of the layer is kept as thin as possible.

Thus, in that work, the process parameters were split into two groups: the first being the pre-processing parameters and the second the manufacturing parameters. Seven different proportions between Q (filament flowrate) and f (feed rate) were selected. It was observed that with increasing the ratio between Q and f, the surface quality became more consistent, but worsened when the ratio reached a value greater than the reference value of 0.585.

Yap et al. [38] used the additive manufacturing technique of material jetting to create a methodology that analyzes the capability of this process regarding the ideal quality and the dimensional accuracy of the manufactured part, utilizing three specific benchmarks. On the other hand, Tyagi et al. [39] made a thorough analysis of the material jetting technique, investigating the fundamentals of the process, as well as the characteristics and properties of the produced parts, especially the influences on mechanical properties, emphasizing that the orientation used is crucial to obtain better mechanical properties.

In relation to the powder bed fusion (PBF) technique, a heat source is used to melt the material in powder form, and thus giving rise to three-dimensional objects layer by layer, and with highly complex geometries [40–42]. This technology is often used to produce parts for the aerospace industry [43]. The selective laser sintering (SLS) process is the main technique behind this process [20], and the most suitable when it is desired to produce on a large scale [44]. Based on this, Nar et al. [45] characterized the surface topography of the LS PA12 specimens. Compared to the injection molding technique, SLS produces surfaces with high roughness, due to the nature of the powder used, and this can affect their performance in its various features.

Another technique of additive manufacturing is called sheet laminating, and as the name implies, metal sheets are joined to form the product. UAM (ultrasonic additive manufacturing), a hybrid technique, is included in this class [46]. Figure 3 exemplifies this process [47].

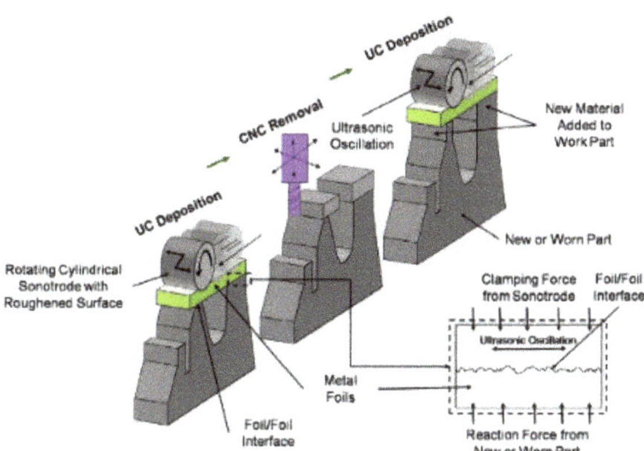


**Figure 3.** Representative image of the UAM process. Reproduced from [47]. Creative Commons CC BY license: https://creativecommons.org/licenses/by-nc-nd/4.0/; accessed on 21 October 2022.

Hehr et al. [48], in their work, thoroughly analyzed the entire history of the UAM process, in the case of the technological developments achieved and the areas in which it is most used. Knowing that the technique uses ultrasonic energy with the function of bonding the metal layers, the authors explained in a chronological way all the advances of the technique, as well as the characterization methods and how they influence the mechanical properties obtained.

Still in relation to additive manufacturing, another technique used is called vat polymerization. In this technique, UV or laser light is used in the photopolymerization of a liquid photopolymer, which solidifies, layer by layer, forming the product in 3D [49,50]. Revilla-León et al. [51], analyzed the influence of the design (solid, alveolar, and hollow) of the mold base with two different thicknesses (1 and 2 mm), in order to evaluate the accuracy of the obtained mold. For this, a digital mold of each design was made, and the honeycomb and hollow designs were further subdivided into the two thicknesses mentioned above. Additionally, a comparison of the mold obtained via printing with the digital mold was made. It was observed that the vat polymerization process can provide a good complement to the extrusion material process, being used for the manufacture of surgical instruments and in the dental field as well [52].

Based on this, it is known that nowadays, additive manufacturing is an industrial area with a last-generation of metal or composite printing machinery, used in an under-development sector that often uses cheap plastic filament printers to reduce production costs [14]. However, the continuous growth and studies around additive manufacturing demonstrate that it may have a significant place in the future of the industry, given the possibility of manufacturing lightweight and resistant materials and products, thus being able to be widely used in the aerospace as well as in the automotive industries, which requires high dimensional accuracy. In addition, in medicine, there has been a great advance when using this technology due to the possibility of producing bone prototypes and tailor-made protheses. Furthermore, in 2012, at least four additional significant technologies were observed in the additive manufacturing sector, which confirms its great growth and development [53].

As examples of recent advances, there is, according to the work of Khondoker et al. [54], an extruder that can carry out the deposition of mechanically interlocked extrudates composed of two immiscible polymers, which prints two filaments through a single nozzle, and with this, a reduction in adhesion failures between the filaments is observed. Furthermore, Cresswell-Boyes et al. [55] developed three-dimensional (3D) precise artificial teeth with two different materials (polylactic acid and thermoplastic elastomer) through tomography files generated from scans obtained by X-ray microtomography (XMT), using several open-source programs. The next approach will be to replicate the properties of natural teeth. In turn, Park et al. [56] demonstrated for the first time the multifunctional integration of various semiconductor devices produced through additive manufacturing, with the production of high-performance polymeric photodetectors and integrated multifunctional devices. Figure 4 illustrates this work.

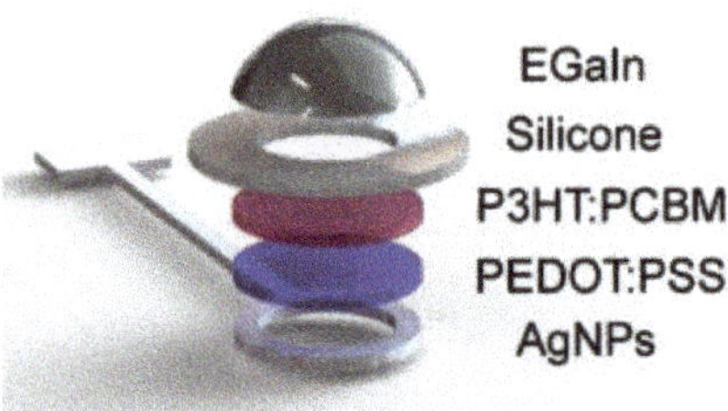


**Figure 4.** Structure of photodetector. Reproduced from [56], 2018, Wiley.

Therefore, additive manufacturing, according to Tofail et al. [57], is at a crossroads between a promising but unproven process for producing functional parts. Therefore, as observed by Pragana et al. [58], new applications have been developed in the field of additive manufacturing to overcoming the limitations of the several developed processes. Based on the factors that permeate this technology in development and growth, a brief diagnosis of this context was carried out through a SWOT analysis/diagram, as shown in Figure 5.

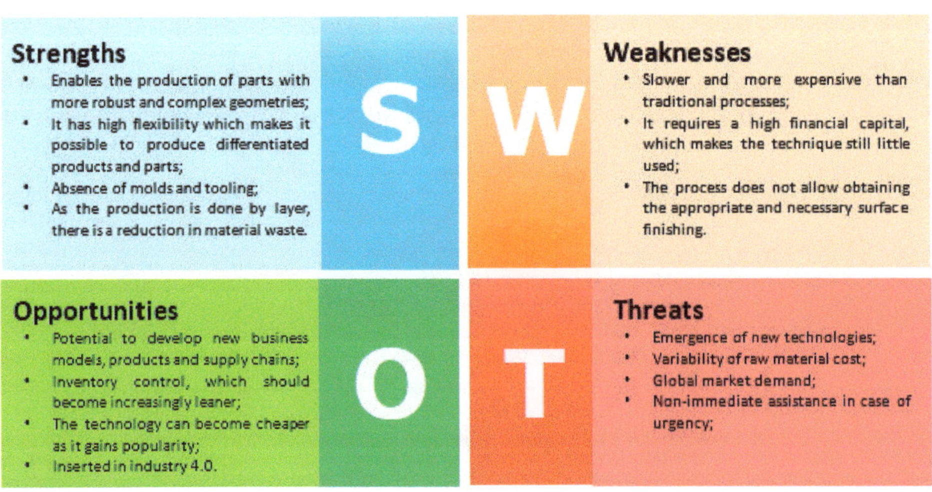


**Figure 5.** SWOT analysis regarding additive manufacturing.

## 3. CNC Machining

With the increase of industrial competition and energetic costs, it is very important for companies to reduce the production cycle and product prices, while ensuring the good quality of the parts. This is one of the reasons why companies should use CNC machines, since they can achieve high accuracy and short times in production/processing [59]. A CNC machine can be defined as a manufacturing process that is controlled through a computer with CAD and CAM systems, with the entire operation being carried out on the machine [60,61]. Figure 6 illustrates this process.

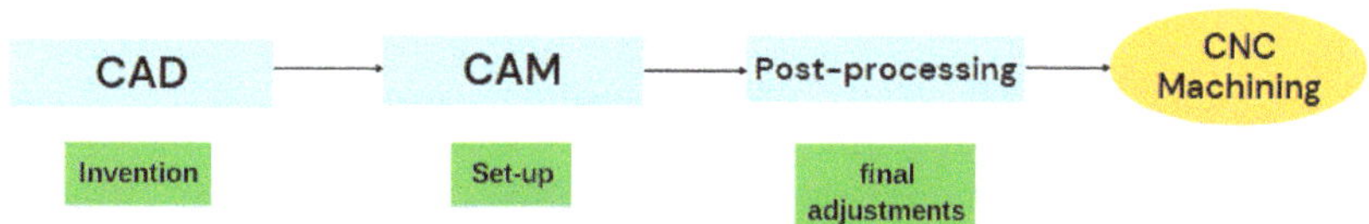


**Figure 6.** Representation of the CAM-CAD process on a CNC machine.

As in any process, some advantages and limitations can be observed. The advantages are good surface quality and dimensional tolerance, being ideal for high-end applications, in addition to the ability to process complex profiles, which would be hard to produce using conventional methods. In terms of limitations, there are the high cost of acquisition of the equipment and the need of specialized operators [62–65].

Gray et al. [66] has compared 3-axis machining and 5-axis machining, concluding that a 3-axis machine provided with an additional rotary/tilt table would improve the surface finish. On the other hand, Sheen et al. [67] developed a method to reliably determine the characteristics of the part to be machined between the upper and lower profiles. These characteristics have been used to program the subsequent manufacturing sequence that can be organized automatically with several generated cutting tool paths.

In the work by Senatore et al. [68], the recent search for cutting toolpaths on freeform surfaces was performed. The authors aimed at increasing productivity through a parallel plane milling strategy, determining an indicator to assist in choosing the best cutter suitable for end milling free-form surfaces. Furthermore, in the work by D'Souza et al. [69], an algorithm was designed to find the sequence of tools with the lowest cost in a 3-axis milling machine for machining free pockets, and further, the method can be adjusted to suit the set of tools available for milling.

4-axis machining has one more rotation for the X axis, called the A axis, which can cause the part to rotate, thus being able to be machined on 5 sides, and being more economical than 3-axis machining [70]. Axinte et al. [71] carried out a theoretical and experimental analysis regarding the functional feasibility of a 4-axis MMT (miniature multi-axis machine tools), reaching the conclusion that the micro-equipment can process small parts with satisfactory precision. Ding et al. [72] based his work on the contour EDM machining of free-form surfaces, thus developing a method capable of imposing tool paths for rough milling 4-axis contour EDM with a cylindrical electrode.

Using a 3-axis CNC machining center and incorporating a 4th axis via a horizontally oscillating rotary table, Ligten et al. [73] focused their work on producing aspherical surfaces, which were produced on this machining center in less than 15% of the time needed in the optical industries. Furthermore, it was observed in the work by Tang et al. [74] that an obstacle occurs when trying to maximize the number of machined surfaces in a 4-axis numerical control (NC) machine, and with that, the authors proposed a time algorithm $O((E+Iwb)2N)$, in order to minimize the number of setups needed to machine a part on a 4-axis NC machine.

Furthermore, post-processing technology is the crucial point for CNC automatic programming technology, as it is directly influenced by the processing quality and the production efficiency. Thus, Magambo et al. [75] created a post-processor for CNC systems, indicating that their use generates better production efficiency and reliability. On the other hand, Baghi et al. [76] compared the effects of post-processing machining and annealing at 850 °C on titanium (Ti64) parts manufactured via selective laser melting (SLM). It is worth noting that each post-processing method acts in a certain way in relation to orientation, whether vertical or horizontal, for example. The fish scale defect was verified in samples that were only machined, being considered a failure of vertical samples, or even anisotropy of elongation between the vertical and horizontal samples that reduced by 125% on machined samples to 36% on annealed samples. Still in relation to post-processing, Chen et al. [77] analyzed hypoid gears because their manufacturing process results in large errors, since a simplified model of blades of cutting systems is used, and with this, a generic approach to programming and CNC post-processing was proposed that can be applied directly in the manufacture of these gears with great efficiency and superior quality.

*Machining with 5 or More Axis*

5-axis machining, in addition to the X, Y, and Z axes, has two rotary axes, which greatly improves the machining efficiency and accuracy. This process is commonly used to machine blades, rotors, dies, molds and propellers, among others [78–80]. As a result, it is considered the most important piece of equipment related to cutting in the industries, which allows the production of final components with much more complex shapes, and which would be impossible to obtain in a 3- or 4-axis machining [81]. Figure 7 exemplifies a 5-axis tool with table tilt A-C and X, Y, and Z axis, showing the flexibility presented by these systems.

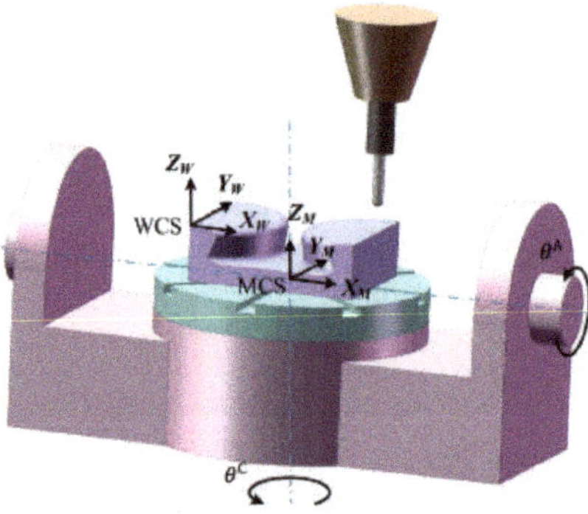


**Figure 7.** 5-axis tool with table tilt A-C and X, Y, and Z axis. Adapted from [78], 2021, Elsevier.

In the study by My et al. [82], a mathematical model was proposed that would be able to analyze and compare the kinematic performance of six configurations of 5-axis machines. Xu et al. [83] emphasized the great need to avoid changing the tool orientation in 5-axis machining. Thus, a smoothing method oriented to kinematic performance was proposed in the work.

Using a recent methodology called double-flank milling, Bizzarri et al. [84] investigated the fabrication of screw rotors, since the methodology was applied in 5-axis flank machining, and it was observed that for symmetrical profiles, double-flank milling is possible, and it works as a designed tool. Using this same methodology, Bo et al. [85] demonstrated a customized tool for machining narrow and curved regions with high precision, and the created algorithm was validated using commercial software.

Prabha et al. [86] used the Unigraphics NX6 CAD/CAM software, which is integrated into 5-axis CNC machines, to machine steam turbine blades, and then make the measurement through 3D coordinates, observing the great efficiency of the 5-axis machine in relation to dimensional accuracy. On the other hand, Huang et al. [87] emphasized the geometric errors occurring in a 5-axis machine, despite the precision of the machine, and defined two models: "Rotary axis component displacement" and "Rotary axis line displacement", which, analyzed through the machining of five axes, has presented discrepant results. Therefore, the accuracy of 5-axis machining is the determining factor for success in processing and manufacturing parts and products [88]. Furthermore, it has a great advantage in terms of its flexibility and efficiency [89].

In 6-axis machining, one more rotation is added. Indeed, adding one more axis expands the variety of movements and transitions, reducing cutting time. Therefore, the 6-axle machine is used for machining very complex geometries, such as turbines or engine blocks [90]. Moriya et al. [91], through 6-axis control non-rotating cutting tools, processed curved V-shaped microgrooves on a curved surface. Due to the high difficulty of machining sharp corners using 3- to 5-axis machining, Japitana et al. [92] created a method for this process through 6-axis machining with ultrasonic vibration cutting, thus being able to create sharp corners on a protruding surface. Krimpenis and Noeas [93] also referred to the advantages of the machining process associated with additive manufacturing in the microfabrication, providing an insight into how these processes can be integrated.

Yuanfei et al. [94] designed a 6-axis CNC system through open architecture based on an industrial personal computer (IPC) and digital motion controller (DMC), thus analyzing the CNC design as well as the developed software. In turn, Carpiuc-Prisacari et al. [95] used a 6-axis machine to perform an analysis on the propagation of mixed-mode cracks; for this it was necessary to initialize a crack, and its propagation through rotation, traction and shear was obtained by the 6-axis machine. Based on robotic machining with six degrees of freedom, Huynh et al. [96] modeled several flexible industrial robots with six axes to be used with milling operations, with torsion springs and dampers positioned on each axis to assist in flexibility, since the lack of joint stiffness is a possible limitation. Milling operations on aluminum and steel corners were simulated, the latter being unsatisfactory because the results did not correspond to the experimental results.

5-axis CNC machining becomes even more critical when machining difficult-to-machine materials, where the vibrations generated and the tool moving simultaneously in different axes can cause stiffness problems, causing imperfections in the machined surface. Indeed, the machinability of the material to be machined strongly impacts the process results and machine behavior [97,98].

Thus, it is observed that the machining of parts has a very broad market, but with some limitations, such as waste of material, for example. Figure 8 illustrates the SWOT analysis for this process, emphasizing that the tool and CNC market represents a large share of the global market, and this aspect is of great importance for the process.

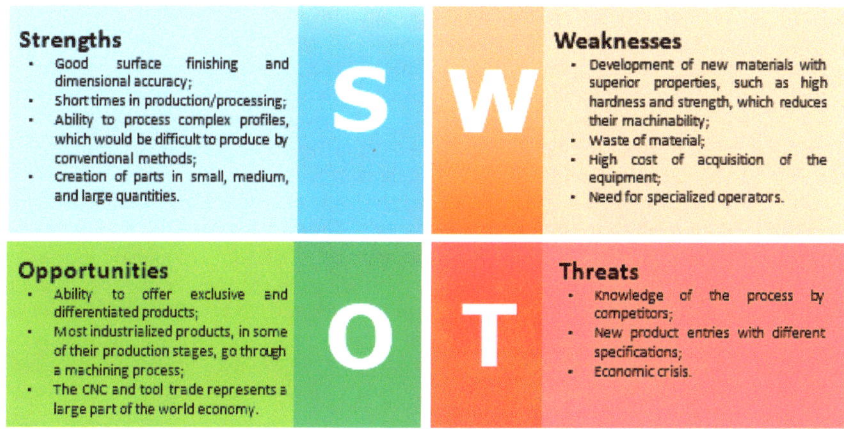


**Figure 8.** SWOT analysis regarding machining.

## 4. Hybrid Manufacturing Processes

Nowadays, due to the development of new materials with better mechanical properties, less specific weight and better performance, the machining industry is faced with an enormous challenge to machine those materials, and thus, new processes had to be coupled and applied to satisfy this need. This is the case in hybrid manufacturing processes [1,99]. The concept of hybrid manufacturing is basically defined as the manufacturing process that joins two or even more manufacturing methods in a single piece of equipment (Figure 9), which has a great effect on the overall performance of the process [1]. The typical example of hybrid manufacturing is the combination of additive manufacturing and machining, using powder bed fusion (PBF) or directed energy deposition (DED) as an additive process with 5-axis machining [100]. This process is especially important when it comes to materials that are difficult to machine. However, many studies still need to be carried out, especially when dealing with complex geometries in 5-axis machining [101].

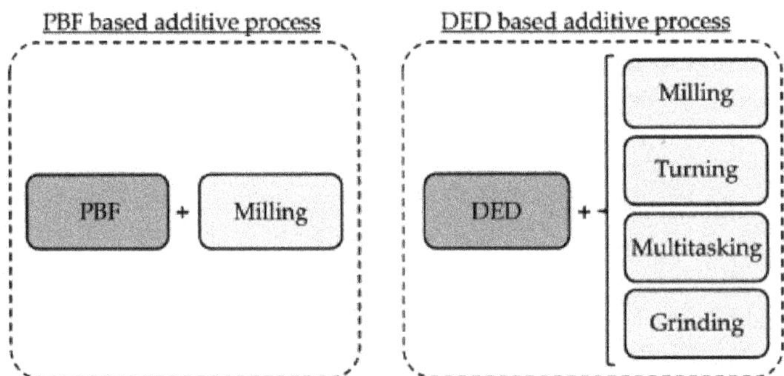


**Figure 9.** Different combinations of processes that can be performed using hybrid manufacturing, with powder bed fusion (PBF) or directed energy deposition (DED). Reproduced from [100]. MDPI Open Access Information and Policy: https://www.mdpi.com/openaccess; accessed on 21 October 2022.

Soshi et al. [102], in their work, manufactured a prototype of a mold normally produced by injection through a hybrid process involving DED technology as an additive process, followed by milling for surface finish and subtractive process. What could be observed in relation to the cooling performance of the mold is the uniformity of this cooling in relation to the traditional process and more stable temperature throughout the cycle, thus improving the cooling performance. In turn, Chen et al. [103] proposed an algorithm capable of calculating the machinability of the material and the product, determining as well the minimum number of alternations between additive and subtractive processes to form a highly complex part and guarantee an ideal cutting tool path in the subtractive operation. The algorithm was called Top-Down_Sequential_Maximization; however, only 3-axis machining was used by default.

Flynn et al. [104] addressed the issue of additive and subtractive hybrid machines through a careful review of the literature, considering the DED method in conjunction with CNC machining, with this being proposed in a future vision that forms a closed circuit. Li et al. [105] developed a 6-axis hybrid additive–subtractive manufacturing process, using a robotic arm with six degrees of freedom, together with a platform for manufacturing parts, and observed an improvement in the surface quality of the part, reduction of material waste and production time. In addition, this enables the need for a support, given the flexibility found in the six axes. Figure 10 illustrates the configuration used.

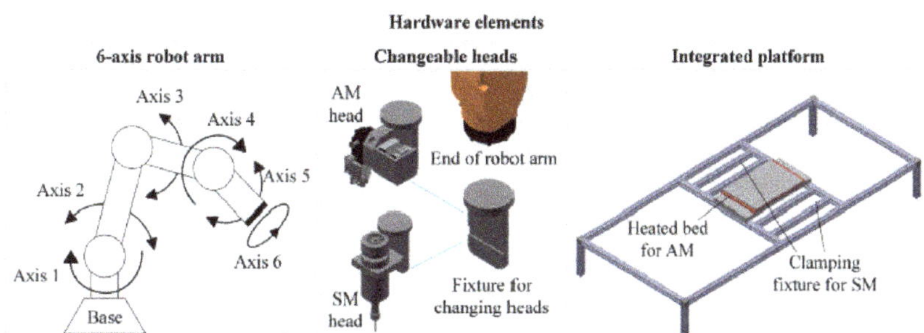



**Figure 10.** Configuration used in the study for the hybrid additive–subtractive manufacturing process. Reproduced from [105], 2018, Elsevier.

Following this lead, Yamazaki et al. [10] developed a hybrid manufacture system integrating the additive method LMD (laser metal deposition), with turning and milling tools, through the Mazak Hybrid multi-tasking machine and exposed application examples for this process, such as the oil industry. With this, it was possible to notice a huge advantage over the standard manufacturing process.

Chen et al. [106] implemented planning algorithms capable of combining additive technology with subtractive technology. In this case, in the first stage a subtractive technology was used to create the beginning of the product's geometry, followed by additive manufacturing, and ending with a machining process for surface finishing. The additive process used was the DED method and a 5-axis machine, more specifically the 3+2-axis machine. Newman et al. [107] worked with a structure called iAtractive, based on which a system called Re-Plan was created for process planning, and with it analyzed the capacity for the integration between additive and subtractive processes through case studies. With Re-Plan, it was possible to perceive that according to the geometry and complexity of the part, the material is added or removed, and even the material of an existing part can be reused to form a product with a new identity.

From a perspective considering the energy expenditure of the process and environmental damage, Yang et al. [108] analyzed the energy consumption performance when using a 6-axis robotic arm as a subtractive process, after manufacturing via additive manufacturing using the fused deposition modeling process. Several case studies and scenarios were used, and the result obtained was quite complex. Regarding future work, it is intended to carry out a mathematical and numerical analysis between the configuration of the robot arm and the energy consumption.

Liou et al. [109] combined the laser deposition process and a 5-axis CNC milling system through process planning and visualization, aiming at integrating all the processes. Additionally, Grzesik [110] highlighted additive–subtractive hybrid manufacturing through the techniques of LMD (laser metal deposition) and multi-axis CNC machining. Furthermore, Ren et al. [111] used the hybrid machining technique with DED and 5-axis machining to repair dies, targeting the corroded and worn surface. In addition, several other authors have based their studies on hybrid manufacturing, and Table 1 summarizes some pertinent works.

**Table 1.** Summary of studies about hybrid manufacturing, coupling addictive processes and machining.

| Author/Year | Technique | Material | Description |
|---|---|---|---|
| Cococcetta et al. (2021) [112] | 3D printed + Dry, MQL and Cryogenic machining | CFRP | Analysis of printing and machining parameters and cooling/lubrication conditions in the production of printed CFRP thermoplastic compounds, from the comparison of three post-processing methods: dry, minimum quantity lubrication (MQL) and cryogenic machining. |
| Tapoglou et al. (2020) [113] | DED + Milling | 316L Steel | Production of 316L stainless steel parts using the DED technique with subsequent machining, from defining the best parameters for material deposition, to analyzing the machinability of the deposited material. |
| Hällgren et al. (2016) [114] | PBF + HSM | Aluminium | Analysis of the cost of serial production of parts if print speed increases, if machine cost or part mass is reduced, using GMP techniques such as additive manufacturing and high-speed machining (HSM). |
| Kaynak et al. (2018) [115] | SLM + FM/VSF/DF | 316L Steel | In order to improve the surface quality of the parts manufactured via selective laser melting (SLM), three post-processing techniques were performed: finishing machining (FM), vibrating machining surface finishing operations (VSF) and drag finishing (DF). The surface was analyzed to meet quality requirements, and it was verified that the DF method resulted in a less rough and more consistent surface finish. |
| Bai et al. (2020) [116] | SLM + Milling | 6511 Steel | Analysis of the production of 6511 martensitic stainless steel parts through selective laser melting (SLM) and end milling, optimizing the process parameters, and taking into account the residual stresses arising from the phase transformation of the martensitic steel. |
| Kaynak et al. (2018) [117] | SLM + Machining | Inconel 718 | Study of surface finish through machining, including surface roughness, microhardness and XRD analysis of Inconel 718 alloy, produced via SLM, verifying that the roughness had a decrease of 90% after post-processing with machining. |
| Salonitis et al. (2015) [118] | Laser Cladding + HSM | Steel | Quality verification of a steel tube manufactured via laser cladding after the high speed machining process, which reduced residual stresses and distortion from the additive manufacturing process. |
| Pal et al. (2016) [119] | DMLS + Machining | SS PH1 | Characterization of the mechanical properties of the PH1 stainless steel product produced via direct metal laser sintering (DMLS) in conjunction with machining process parameters and post-processing. |
| Careri et al. (2021) [120] | DED + Machining | Inconel 718 | Evaluation of surface finish, microstructure, microhardness and residual stresses from the manufacture of parts in Inconel 718 by DED, followed by two possible post-processing: machining + heat treatment or heat treatment + machining, verifying that the best production strategy was AD + M + DA. |
| Heigel et al. (2018) [121] | PBF + Machining | Stainless steel | Production of stainless steel cylinders through the laser powder bed fusion process, followed by machining, and analysis of residual stresses from the processes. |
| Speidel et al. (2021) [122] | PBF + EJM | Ti-6Al-4V | Due to the poor surface quality due to unmelted powder from the powder bed fusion process, this study analyzed the application of electrochemical jet machining in order to produce a good surface finish and increase the functionality of the part. |
| W. Grzesik [123] | General | General | This work provides a recent and brief overview of hybrid machining. |

As noted, hybrid manufacturing brings numerous advantages and benefits. However, some challenges are still encountered, as depicted in Figure 11 [100].

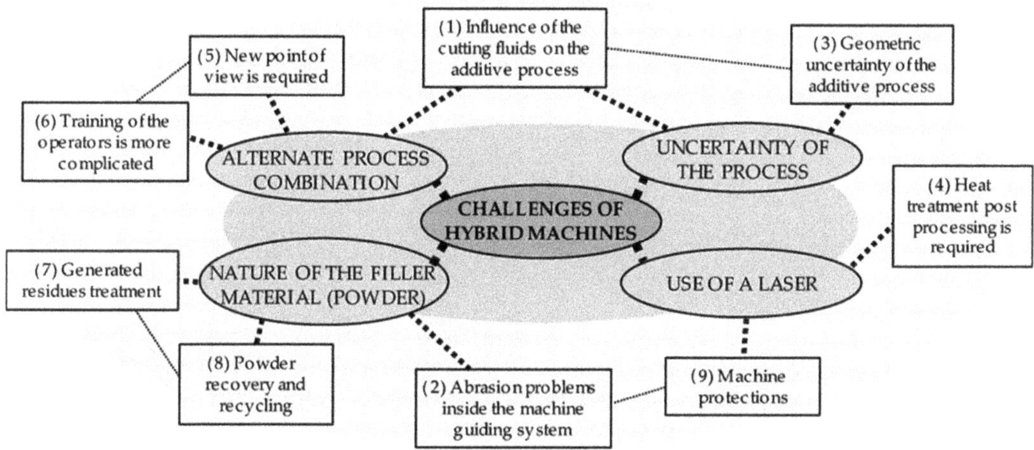



**Figure 11.** Schematic figure illustrating the challenges of hybrid machining. Reproduced from [100]. MDPI Open Access Information and Policy: https://www.mdpi.com/openaccess; accessed on 21 October 2022.

As a challenge to be faced, the recycling of metallic powder can be considered, since it can be harmful to human health, especially nickel or cobalt compounds, and must be removed from the machine after the addition operations are carried out, so that they do not affect machine components in order to avoid damage and breakdown [7]. In addition, the protection of the machine, safety and reliability of the process must be considered, since due to the heat generated during the additive process, it can result in the melting of some areas, which must have adequate protection [98]. Additionally, when talking about the challenges faced when implementing hybrid production, it is important to emphasize the training of qualified technicians so that they are able to understand and apply both techniques, and the correct disposal of waste, with a focus on handling the waste, dust and recycling liquid waste, such as lubricating oils and cutting fluids. Another problem faced is precisely in the use of these cutting fluids, since laser technology was developed regarding a fluid-free environment, and on the other hand, in machining this is widely used and practically essential or mandatory for some materials to be machined [7,98].

As hybrid manufacturing is a process that is still under development and expansion, many challenges are encountered and need to be resolved, both in relation to scientific and research parameters, as well as the technical parameters for its implementation on an industrial scale [124]. In this sense, some questions must be answered, such as, for example, what would be the best processing sequence, how the microstructure and properties of the part would behave after the cycles, or even how the software used would be able to detect collisions after adding more material [125].

However, according to Dilberoglu et al. [126], it is expected that in the future the use of hybrid production systems will be increased with additional improvements, as the implementation of this technology has attracted the attention of researchers in modern industries. Thus, based on the current scenario, and with the analysis carried out in this work, the following SWOT matrix for hybrid manufacturing resulted (Figure 12).

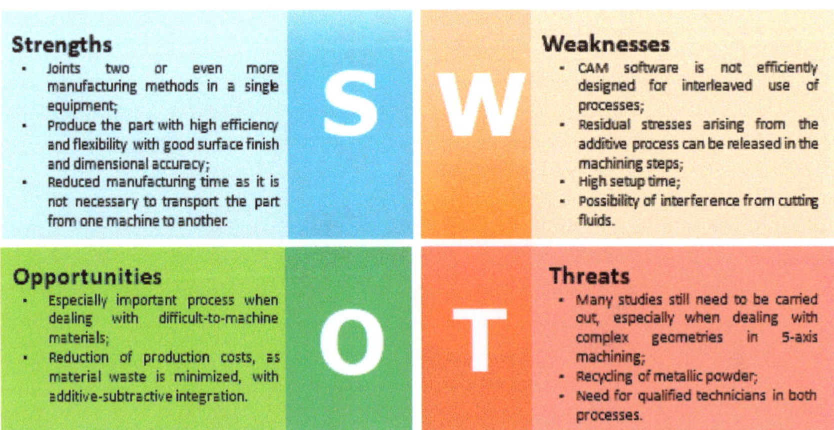



**Figure 12.** SWOT analysis corresponding to hybrid manufacturing.

Through this review work, it was possible to verify that there is a lot of work to be done in the field of hybrid manufacturing, which incorporates some developments carried out in each of the integrated processes (additive manufacturing and machining). This was evident through this study, which seeks to provide a structured, broad and integrated view of the contributions that have been made lately by several researchers in each of the areas. Additionally, it was possible to see how other researchers have taken advantage of these developments, and taking the knowledge acquired in each one of the considered processes, to use it in the integration of the two processes.

There is still a lot of work to be developed in the area of hybrid production. The interaction of the additive and subtractive processes does not have to be sequential and carried out in this order, although the additive process needs to be the first, but it can be reused later if there are problems with accessing the tools in certain areas. Indeed, the parts tend to have an increasingly complex shape, which also requires increasingly refined techniques for their manufacture. CAM software has not yet been developed efficiently for an interleaved use of processes, and this interleaved use may be absolutely necessary to realize different geometries. This is certainly a subject that will be of interest to researchers in the near future. Another issue to be aware of is the level of internal stresses left on the parts after machining. Although the volume of chips to be removed is generally small, finishing operations generally raise the temperature, which may influence the mechanical properties of the material being worked. Thus, there may be overlapping stresses in parts of the parts, which can lead to deformations that may interfere with the functionality and expected lifetime of the machined part. Thus, the induction of stresses and their level are also subjects that will certainly be investigated in the near future.

## 5. Conclusions

Additive manufacturing is undoubtedly an effective and emerging method for the manufacture of complex and difficult parts to be produced by other methods; however, a major limitation of the processes involved is the poor surface quality and dimensional accuracy, requiring a subtractive method to achieve the required surface quality and shape tolerance.

Thus, hybrid manufacturing has gained great prominence, bringing many benefits to modern industry. It guaranteed, through the integration between additive and subtractive processes, the production of parts that would previously be difficult to be produced with only one of the individual techniques, either due to the high hardness of the material, or because of the high complexity and dimensional tolerance.

One of the great advantages of this process is the high flexibility and versatility, as well as the reduction of waste, a problem highly faced by conventional machining. In addition, the possibility of using only one machine during the entire process eliminates transport times and process inventories. As a limitation, we can highlight the residual stresses present in the additive manufacturing process, which can result in distortions in the part, requiring further heat treatment.

In addition, in terms of challenges for this technology to be implemented, we can mention the recycling process of metallic powder, reliability and safety in the process, and training. Indeed, the proper disposal of waste, as well as the use of cutting fluids in machining remains an environmental issue, since the additive manufacturing technology-using laser was developed applying environmentally friendly means, free of these fluids.

Based on this, this work has shown through SWOT analysis and a broad bibliographic review the studies and developments, as well as the great growth of this process, thus contributing to the study and analysis of the development of new design of parts, which previously would be impractical, and today are already a promising reality. Moreover, the combination of both technologies helps in overcoming the traditional issues related to the surface finishing of additively manufactured parts, providing the surface quality usually required in many applications. In this way, it is expected that in the future there will be a greater integration of hybrid machines in production lines, so that they can produce parts that are closer to the final product shape without the need for post-processing.

In the future, it is expected that the efforts to integrate the two technologies will evolve positively in the sense of increasing the rigidity of the equipment, the versatility of the transition from one technology to the other, and the complementarity in terms of work in both technologies. In fact, it is possible to take advantage of the knowledge acquired in terms of surface quality left by the additive manufacturing and try to minimize surface defects, thus minimizing the machining work to be carried out later. Furthermore, 5-axis machining makes it possible to take better advantage of topological optimization, making more complex shapes in a single fixture, thus increasing the accuracy obtained. The training needs are evident, as the requirements at this level start with the design of the parts, and end with obtaining the final parts in the equipment. Thus, there will have to be a greater integration of knowledge between those who develop the products and those who make them possible through the programming and operation of hybrid equipment. The improvement of these factors and a natural reduction of investment costs in the near future, will be decisive for the rapid growth of this technology in the market.

**Author Contributions:** Conceptualization, N.P.V.S. and F.J.G.S.; methodology, N.P.V.S.; validation, F.J.G.S., F.F. and V.F.C.S.; writing—original draft and designed, N.P.V.S.; writing—review and editing, F.J.G.S., F.F. and V.F.C.S.; Supervision, F.J.G.S.; Funding acquisition, F.J.G.S. and F.F. All authors have read and agreed to the published version of the manuscript.

**Funding:** The present work was done and funded under the scope of the projects ON-SURF (ANI | P2020 | POCI-01-0247-FEDER-024521 and MCTool21 "Manufacturing of cutting tools for the 21st century: from nano-scale material design to numerical process simulation" (ref.: "POCI-01-0247-FEDER-045940") co-funded by Portugal 2020 and FEDER, through COMPETE 2020-Operational Programme for Competitiveness and Internationalisation. This work is also sponsored by FEDER National funds FCT under the project CEMMPRE ref. "UIDB/00285/2020". F.J.G. Silva also thanks INEGI-Instituto de Ciência e Inovação em Engenharia Mecânica e Engenharia Indústria due to its support.



## References

1. Bhattacharyya, B.; Doloi, B. *Modern Machining Technology—Advanced, Hybrid, Micro Machining and Super Finishing Technology;* Academic Press: San Diego, CA, USA, 2020; pp. 461–591, ISBN 9780128128947.
2. Berman, B. 3-D printing: The new industrial revolution. *Bus. Horiz.* **2012**, *55*, 155–162. [CrossRef]

3. Luo, X.; Cai, Y.; Chavoshi, S.Z. *Hybrid Machining-Theory, Methods, and Case Studies*; Academic Press: London, UK, 2018; pp. 1–15, ISBN 9780128130599.

4. Schuh, G.; Kreysa, J.; Orilski, S. Roadmap Hybride Produktion: Wie 1 + 1 = 3-Effekte in der Produktion maximiert werden können. *Z. Wirtsch. Fabr.* **2009**, *104*, 385–391. [CrossRef]

5. Saxena, K.K.; Bellotti, M.; Qian, J.; Reynaerts, D.; Lauwers, B.; Luo, X. *Chapter 2—Overview of Hybrid Machining Processes, Hybrid Machining*; Academic Press: London, UK, 2018; pp. 21–41, ISBN 9780128130599.

6. Karunakaran, K.P.; Suryakumar, S.; Pushpa, V.; Akula, S. Low cost integration of additive and subtractive processes for hybrid layered manufacturing. *Robot. Comput.-Integr. Manuf.* **2010**, *26*, 490–499. [CrossRef]

7. Cortina, M.; Ruiz, J.E.; Villarón, I.; Arrizubieta, J.I.; Borgiattino, H.; Dünky, A.; Martinez, J.; Baine, S. Máquina Híbridas: A Integração de Processos. Available online: https://www.intermetal.pt/Artigos/265587-Maquinas-hibridas-integrando-processos.html (accessed on 13 January 2022).

8. Matias, E.; Rao, B. 3D printing: On its historical evolution and the implications for business. In Proceedings of the PICMET'15: Management of the Technology Age, Hilton Portland, OR, USA, 2–6 August 2015; pp. 551–558. [CrossRef]

9. Diegel, O. 10.02—Additive Manufacturing: An Overview. In *Comprehensive Materials Processing*; Hashmi, S., Batalha, G.F., Van Tyne, C.J., Yilbas, B., Eds.; Elsevier: Amsterdam, The Netherlands, 2014; pp. 3–18, ISBN 9780080965338.

10. Yamazaki, T. Development of A Hybrid Multi-Tasking Machine Tool: Integration of Additive Manufacturing Technology with CNC Machining. *Procedia CIRP* **2016**, *42*, 81–86. [CrossRef]

11. Ashby, M.F.; Jones, D.R.H. *Engineering Materials 2: An Introduction to Microstructures and Processing*; Butterworth-Heinemann: Oxford, UK, 2013; pp. 280–296, ISBN 9780080966687.

12. Mazak. 5-Axis. What Is 5-Axis Machining? Available online: https://www.mazakusa.com/machines/process/5-axis/ (accessed on 13 January 2022).

13. Saxer, M.; Dimitrov, D.; de Beer, N. High-speed 5-axis machining for tooling applications. *S. Afr. J. Ind. Eng.* **2012**, *23*, 144–153. [CrossRef]

14. Frazier, W.E. Metal Additive Manufacturing: A Review. *J. Mater. Eng. Perform.* **2014**, *23*, 1917–1928. [CrossRef]

15. Bogers, M.; Hadar, R.; Bilberg, A. Additive manufacturing for consumer-centric business models: Implications for supply chains in consumer goods manufacturing. *Technol. Forecast. Soc. Chang.* **2015**, *102*, 225–239. [CrossRef]

16. Phaal, R.; O'Sullivan, E.; Routley, M.; Ford, S.; Probert, D. A framework for mapping industrial emergence. *Technol. Forecast. Soc. Chang.* **2011**, *78*, 217–230. [CrossRef]

17. Afuah, A. *Business Model Innovation: Concepts, Analysis and Cases*; Routledge: New York, NY, USA, 2018; ISBN 9781138330528.

18. Zott, C.; Amit, R.; Massa, L. The business model: Recent developments and future research. *J. Manag.* **2011**, *37*, 1019–1042. [CrossRef]

19. Savolainen, J.; Collan, M. Additive manufacturing technology and business model change—A review of literature. *Addit. Manuf.* **2020**, *32*, 101070. [CrossRef]

20. Tempelman, E.; Shercliff, H.; van Eyben, B.N. *Manufacturing and Design—Understanding the Principles of How Things Are Made*; Butterworth-Heinemann: Oxford, UK, 2014; pp. 187–200, ISBN 9780080999227.

21. Alghamdi, S.S.; John, S.; Choudhury, N.R.; Dutta, N.K. Additive Manufacturing of Polymer Materials: Progress, Promise and Challenges. *Polymers* **2021**, *13*, 753. [CrossRef] [PubMed]

22. Loughborough University. The 7 Categories of Additive Manufacturing—Additive Manufacturing Research Group. Available online: https://www.lboro.ac.uk/research/amrg/about/the7categoriesofadditivemanufacturing/ (accessed on 13 January 2022).

23. Sachs, E.M.; Haggerty, J.S.; Cima, M.J.; Williams, P.A. Three-Dimensional Printing Techniques. Massachusetts Institute of Technology. U.S. Patent No. 5,204,055, 20 April 1993.

24. Chen, Q.; Juste, E.; Lasgorceix, M.; Petit, F.; Leriche, A. Binder jetting process with ceramic powders: Influence of powder properties and printing parameters. *Open Ceram.* **2022**, *9*, 100218. [CrossRef]

25. Lee, K.; Kim, H.; Ahn, D.; Lee, H. Thermo-mechanical characteristics of inconel 718 layer deposited on AISI 1045 steel substrate using a directed energy deposition process. *J. Mater. Res. Technol.* **2022**, *17*, 293–309. [CrossRef]

26. Ahn, D.G. Direct metal additive manufacturing processes and their sustainable applications for green technology: A review. *Int. J. Precis. Eng. Manuf. Green Technol.* **2016**, *3*, 381–395. [CrossRef]

27. Ahn, D.G. Directed Energy Deposition (DED) Process: State of the Art. *Int. J. Precis. Eng. Manuf. Green Technol.* **2021**, *8*, 703–742. [CrossRef]

28. Chen, Z.; Guo, X.; Jing, S. Hardness Prediction and Verification Based on Key Temperature Features During the Directed Energy Deposition Process. *Int. J. Precis. Eng. Manuf. Green Technol.* **2020**, *8*, 453–469. [CrossRef]

29. Tekumalla, S.; Tosi, R.; Tan, X.; Seita, M. Directed energy deposition and characterization of high-speed steels with high vanadium content. *Addit. Manuf. Lett.* **2022**, *2*, 100029. [CrossRef]

30. Radhakrishnan, M.; Hassan, M.; Long, B.; Otazu, D.; Lienert, T.; Anderoglu, O. Microstructures and properties of Ti/TiC composites fabricated by laser-directed energy deposition. *Addit. Manuf.* **2021**, *46*, 102198. [CrossRef]

31. Grossi, N.; Scippa, A.; Venturini, G.; Campatelli, G. Process Parameters Optimization of Thin-Wall Machining for Wire Arc Additive Manufactured Parts. *Appl. Sci.* **2020**, *10*, 7575. [CrossRef]

32. Artaza, T.; Suárez, A.; Veiga, F.; Braceras, I.; Tabernero, I.; Larrañaga, O.; Lamikiz, A. Wire arc additive manufacturing Ti6Al4V aeronautical parts using plasma arc welding: Analysis of heat-treatment processes in different atmospheres. *J. Mater. Res. Technol.* **2020**, *9*, 15454–15466. [CrossRef]
33. Veiga, F.; Gil Del Val, A.; Suárez, A.; Alonso, U. Analysis of the Machining Process of Titanium Ti6Al-4V Parts Manufactured by Wire Arc Additive Manufacturing (WAAM). *Materials* **2020**, *13*, 766. [CrossRef] [PubMed]
34. Loughborough University. About Additive Manufacturing. Available online: https://www.lboro.ac.uk/research/amrg/about/the7categoriesofadditivemanufacturing/materialextrusion/ (accessed on 13 January 2022).
35. Rosen, D. Design for Additive Manufacturing: Past, Present, and Future Directions. *J. Mech. Des.* **2014**, *136*, 090301. [CrossRef]
36. Awasthi, P.; Banerjee, S.S. Fused deposition modeling of thermoplastic elastomeric materials: Challenges and opportunities. *Addit. Manuf.* **2021**, *46*, 102177. [CrossRef]
37. Jin, Y.-A.; Li, H.; He, Y.; Fu, J.-Z. Quantitative analysis of surface profile in fused deposition modeling. *Addit. Manuf.* **2015**, *8*, 142–148. [CrossRef]
38. Yap, Y.L.; Wang, C.; Sing, S.L.; Dikshit, V.; Yeong, W.Y.; Wei, J. Material jetting additive manufacturing: An experimental study using designed metrological benchmarks. *Precis. Eng.* **2017**, *50*, 275–285. [CrossRef]
39. Tyagi, S.; Yadav, A.; Deshmukh, S. Review on mechanical characterization of 3D printed parts created using material jetting process. *Mater. Today Proc.* **2022**, *51*, 1012–1016. [CrossRef]
40. Thompson, M.K.; Moroni, G.; Vaneker, T.; Fadel, G.; Campbell, R.I.; Gibson, I.; Bernard, A.; Schulz, J.; Graf, P.; Ahuja, B.; et al. Design for Additive Manufacturing: Trends, opportunities, considerations, and constraints. *CIRP Ann.* **2016**, *65*, 737–760. [CrossRef]
41. Goodridge, R.; Ziegelmeier, S. *Laser Additive Manufacturing—Materials, Design, Technologies, and Applications*; Brandt, M., Ed.; Woodhead Publishing: Sawston, UK, 2017; pp. 181–204, ISBN 9780081004333.
42. Zhang, Y.; Jarosinski, W.; Jung, Y.-G.; Zhang, J. *Additive Manufacturing: Materials, Processes, Quantifications and Applications*; Butterworth-Heinemann: Cambridge, MA, USA, 2018; pp. 39–51, ISBN 978-0-12-812155-9.
43. Najmon, J.C.; Raeisi, S.; Tovar, A. *Additive Manufacturing for the Aerospace Industry*; Froes, F., Boyer, R., Eds.; Elsevier: Chennai, India, 2019; pp. 7–31, ISBN 9780128140628.
44. Gan, X.; Fei, G.; Wang, J.; Wang, Z.; Lavorgna, M.; Xia, H. *Structure and Properties of Additive Manufactured Polymer Components*; Friedrich, K., Walter, R., Soutis, C., Suresh, G.A., Habil, I., Fiedler, B., Eds.; Woodhead Publishing: Sawston, UK, 2020; pp. 149–185, ISBN 9780128195352.
45. Nar, K.; Majewski, C.; Lewis, R. A comprehensive characterisation of Laser Sintered Polyamide-12 surfaces. *Polym. Test.* **2022**, *106*, 107450. [CrossRef]
46. Friel, R.J. *Power Ultrasonics—Applications of High-Intensity Ultrasound*; Juan, A., Juárez, G., Karl, F.G., Eds.; Woodhead Publishing: Sawston, UK, 2015; pp. 313–335, ISBN 978-1-78242-028-6.
47. Schwope, L.-A.; Friel, R.J.; Johnson, K.; Harris, R.A. Field repair and replacement part fabrication of military components using ultrasonic consolidation cold metal deposition. In Proceedings of the RTO-MP-AVT-163—Additive Technology for Repair of Military Hardware, Bonn, Germany, 19–22 October 2009; p. 22.
48. Hehr, A.; Norfolk, M. A comprehensive review of ultrasonic additive manufacturing. *Rapid Prototyp. J.* **2019**, *26*, 445–458. [CrossRef]
49. Chartrain, N.A.; Williams, C.B.; Whittington, A.R. A review on fabricating tissue scaffolds using vat photopolymerization. *Acta Biomater.* **2018**, *74*, 90–111. [CrossRef] [PubMed]
50. Al Rashid, A.; Ahmed, W.; Khalid, M.Y.; Koç, M. Vat photopolymerization of polymers and polymer composites: Processes and applications. *Addit. Manuf.* **2021**, *47*, 102279. [CrossRef]
51. Revilla-León, M.; Piedra-Cascón, W.; Aragoneses, R.; Sadeghpour, M.; Barmak, B.A.; Zandinejad, A.; Raigrodski, A.J. Influence of base design on the manufacturing accuracy of vat-polymerized diagnostic casts: An in vitro study. *J. Prosthet. Dent.* **2021**. [CrossRef]
52. Leary, M. *Design for Additive Manufacturing*; Elsevier: Amsterdam, The Netherlands, 2020; pp. 283–293, ISBN 978-0-12-816721-2.
53. Wong, K.V.; Hernandez, A. A Review of Additive Manufacturing. International Scholarly Research Network. *ISRN Mech. Eng.* **2012**, *2012*, 208760. [CrossRef]
54. Khondoker, M.A.H.; Asad, A.; Sameoto, D. Printing with mechanically interlocked extrudates using a custom bi-extruder for fused deposition modelling. *Rapid Prototyp. J.* **2018**, *24*, 921–934. [CrossRef]
55. Cresswell-Boyes, A.J.; Barber, A.H.; Mills, D.; Tatla, A.; Davis, G.R. Approaches to 3D printing teeth from X-ray microtomography. *J. Microsc.* **2018**, *272*, 207–212. [CrossRef]
56. Park, S.H.; Su, R.; Jeong, J.; Guo, S.Z.; Qiu, K.; Joung, D.; Meng, F.; McAlpine, M.C. 3D printed polymer photodetectors. *Adv. Mater.* **2018**, *30*, e1803980. [CrossRef]
57. Tofail, S.A.M.; Koumoulos, E.P.; Bandyopadhyay, A.; Bose, S.; O'Donoghue, L.; Charitidis, C. Additive manufacturing: Scientific and technological challenges, market uptake and opportunities. *Mater. Today* **2017**, *21*, 22–37. [CrossRef]
58. Pragana, J.P.M.; Sampaio, R.F.V.; Bragança, I.M.F.; Silva, C.M.A.; Martins, P.A.F. Hybrid metal additive manufacturing: A state–of–the-art review. *Adv. Ind. Manuf. Eng.* **2021**, *2*, 100032. [CrossRef]
59. Zębala, W.; Plaza, M. Comparative study of 3- and 5-axis CNC centers for free-form machining of difficult-to-cut material. *Int. J. Prod. Econ.* **2014**, *158*, 345–358. [CrossRef]

60. Baksi, S. CNC Machine: Types, Parts, Advantages, Disadvantages, Applications, and Specifications. Available online: https://learnmechanical.com/cnc-machine/#What_is_a_CNC_Machine (accessed on 20 January 2022).

61. Hay, A. Ultimate Guide to CNC Machining. Available online: https://jiga.io/resource-center/cnc-machining/what-is-cnc-machining-guide/ (accessed on 22 January 2022).

62. Varotsis, A.B. Introduction to CNC Machining. Available online: https://www.hubs.com/knowledge-base/cnc-machining-manufacturing-technology-explained/#pros-cons (accessed on 22 January 2022).

63. Li, M. What Is CNC Milling, Advantages and Disadvantages. Available online: https://www.3qmachining.com/what-is-cnc-milling-advantages-and-disadvantages/ (accessed on 22 January 2022).

64. Cloud, N.C. What's the Difference between 3-Axis, 4-Axis & 5-Axis Milling? Available online: https://cloudnc.com/cnc-best-practices-3-whats-the-difference-between-3-axis-4-axis-5-axis-milling/ (accessed on 22 January 2022).

65. Yu, Z.; Zhi-Tong, C.; Tao, N.; Ru-Feng, X. Tool orientation optimization for 3 + 2-axis CNC machining of sculptured surface. *Comput. Aided Des.* **2016**, *77*, 60–72. [CrossRef]

66. Gray, P.; Bedi, S.; Ismail, F.; Rao, N.; Morphy, G. Comparison of 5-Axis and 3-Axis Finish Machining of Hydroforming Die Inserts. *Int. J. Adv. Manuf. Technol.* **2001**, *17*, 562–569. [CrossRef]

67. Sheen, B.T.; You, C.F. Machining feature recognition and tool-path generation for 3-axis CNC milling. *Comput. Aided Des.* **2006**, *38*, 553–562. [CrossRef]

68. Senatore, J.; Segonds, S.; Rubio, W.; Dessein, G. Correlation between machining direction, cutter geometry and step-over distance in 3-axis milling: Application to milling by zones. *Comput. Aided Des.* **2012**, *44*, 1151–1160. [CrossRef]

69. D'Souza, R.M.; Sequin, C.; Wright, P.K. Automated tool sequence selection for 3-axis machining of free-form pockets. *Comput. Aided Des.* **2004**, *36*, 595–605. [CrossRef]

70. Huang, L. What Is 4-Axis and 5-Axis CNC Machining? Available online: https://www.rapiddirect.com/blog/4-axis-and-5-axis-cnc-machining/ (accessed on 22 January 2022).

71. Axinte, D.A.; Shukor, S.A.; Bozdana, A.T. An analysis of the functional capability of an in-house developed miniature 4-axis machine tool. *Int. J. Mach. Tools Manuf.* **2010**, *50*, 191–203. [CrossRef]

72. Ding, S.; Jiang, R. Tool path generation for 4-axis contour EDM rough machining. *Int. J. Mach. Tools Manuf.* **2004**, *44*, 1493–1502. [CrossRef]

73. Van Ligten, R.F.; Venkatesh, V.C. Diamond Grinding of Aspheric Surfaces on a CNC 4-Axis Machining Centre. *CIRP Ann.* **1985**, *34*, 295–298. [CrossRef]

74. Tang, K.; Chen, L.; Chou, S.Y. Optimal workpiece setups for 4-axis numerical control machining based on machinability. *Comput. Ind.* **1998**, *37*, 27–41. [CrossRef]

75. Magambo, S.; Ying, L. The NC Machining Post-Processing Technology Based on UG. *Int. J. Sci. Res.* **2013**, *2*, 131–134.

76. Baghi, D.A.; Nafisi, S.; Hashemi, R.; Ebendorff-Heidepriem, H.; Ghomashchi, R. Effective post processing of SLM fabricated Ti-6Al-4 V alloy: Machining vs thermal treatment. *J. Manuf. Process.* **2021**, *68*, 1031–1046. [CrossRef]

77. Chen, Z.C.; Wasif, M. A generic and theoretical approach to programming and post-processing for hypoid gear machining on multi-axis CNC face-milling machines. *Int. J. Adv. Manuf. Technol.* **2015**, *81*, 135–148. [CrossRef]

78. Wang, Y.; Xu, J.; Sun, Y. Tool orientation adjustment for improving the kinematics performance of 5-axis ball-end machining via CPM method. *Robot. Comput. Integr. Manuf.* **2021**, *68*, 102070. [CrossRef]

79. Xu, J.; Ji, Y.; Sun, Y.; Lee, Y. Spiral Tool Path Generation Method on Mesh Surfaces Guided by Radial Curves. *J. Manuf. Sci. Eng.* **2018**, *140*, 071016. [CrossRef]

80. Xu, K.; Li, Y. Region based five-axis tool path generation for freeform surface machining via image representation. *Robot. Comput. Integr. Manuf.* **2019**, *57*, 230–240. [CrossRef]

81. Bologa, O.; Breaz, R.E.; Racz, S.G.; Crenganiş, M. Decision-making Tool for Moving from 3-axes to 5-axes CNC Machine-tool. *Procedia Comput. Sci.* **2016**, *91*, 184–192. [CrossRef]

82. My, C.A.; Bohez, E.L.J. A novel differential kinematics model to compare the kinematic performances of 5-axis CNC machines. *Int. J. Mech. Sci.* **2019**, *163*, 105117. [CrossRef]

83. Xu, J.; Zhang, D.; Sun, Y. Kinematics performance oriented smoothing method to plan tool orientations for 5-axis ball-end CNC machining. *Int. J. Mech. Sci.* **2019**, *157–158*, 293–303. [CrossRef]

84. Bizzarri, M.; Bartoň, M. Manufacturing of Screw Rotors Via 5-axis Double-Flank CNC Machining. *Comput. Aided Des.* **2020**, *132*, 102960. [CrossRef]

85. Bo, P.; González, H.; Calleja, A.; de Lacalle, L.N.L.; Bartoň, M. 5-axis double-flank CNC machining of spiral bevel gears via custom-shaped milling tools—Part I: Modeling and simulation. *Precis. Eng.* **2020**, *62*, 204–212. [CrossRef]

86. Prabha, K.A.; Prasad, B.S. Machining of Steam Turbine Blade on 5-Axis CNC Machine. *Mater. Today Proc.* **2019**, *18*, 3001–3007. [CrossRef]

87. Huang, N.; Bi, Q.; Wang, Y. Identification of two different geometric error definitions for the rotary axis of the 5-axis machine tools. *Int. J. Mach. Tools Manuf.* **2015**, *91*, 109–114. [CrossRef]

88. Wiessner, M.; Blaser, P.; Böhl, S.; Mayr, J.; Knapp, W.; Wegener, K. Thermal test piece for 5-axis machine tools. *Precis. Eng.* **2018**, *52*, S0141635917305767. [CrossRef]

89. Chen, Q.; Li, W.; Jiang, C.; Zhou, Z.; Min, S. Separation and compensation of geometric errors of rotary axis in 5-axis ultra-precision machine tool by empirical mode decomposition method. *J. Manuf. Process.* **2021**, *68*, 1509–1523. [CrossRef]

90. Ahmarn, M.; Stanislav, S.M. Optimization of rotations for six-axis machining. *Int. J. Adv. Manuf. Technol.* **2011**, *53*, 435–451. [CrossRef]

91. Moriya, T.; Nakamoto, K.; Ishida, T.; Takeuchi, Y. Creation of V-shaped microgrooves with flat-ends by 6-axis control ultraprecision machining. *CIRP Ann. Manuf. Technol.* **2010**, *59*, 61–66. [CrossRef]

92. Japitana, F.H.; Morishige, K.; Yasuda, S.; Takeuchi, Y. Manufacture of Overhanging Sharp Corner by Means of 6-Axis Control Machining with the Application of Ultrasonic Vibrations. *JSME Int. J. Ser. C* **2003**, *46*, 306–313. [CrossRef]

93. Krimpenis, A.A.; Noeas, G.D. Application of Hybrid Manufacturing processes in microfabrication. *J. Manuf. Process.* **2022**, *80*, 328–346. [CrossRef]

94. Qin, Y.; Xiao, J.; Wang, G. The Open Architecture CNC System Based on 6-axis Flame Pipe Cutting Machine. In Proceedings of the 2011 Third International Conference on Measuring Technology and Mechatronics Automation, Shanghai, China, 6–7 January 2011. [CrossRef]

95. Carpiuc-Prisacari, A.; Poncelet, M.; Kazymyrenko, K.; Leclerc, H.; Hild, F. A complex mixed-mode crack propagation test performed with a 6-axis testing machine and full-field measurements. *Eng. Fract. Mech.* **2017**, *176*, 1–22. [CrossRef]

96. Huynh, H.N.; Riviere-Lorphevre, E.; Verlinden, O. Multibody modelling of a flexible 6-axis robot dedicated to robotic machining. In Proceedings of the 5th Joint International Conference on Multibody System Dynamics, Lisbon, Portugal, 24–28 June 2018.

97. Duplak, J.; Hatala, M.; Duplakova, D.; Steranka, J. Comprehensive analysis and study of the machinability of a high strength aluminum alloy (EN AW-AlZn5.5MgCu) in the high-feed milling. *Adv. Prod. Eng. Manag.* **2018**, *13*, 455–465. [CrossRef]

98. Zajac, J.; Duplak, J.; Duplakova, D.; Cizmar, P.; Olexa, I.; Bittner, A. Prediction of Cutting Material Durability by T = f(vc) Dependence for Turning Processes. *Processes* **2020**, *8*, 789. [CrossRef]

99. Priyadarshi, S. *Hybrid Machining Processes, Short Term Course on Advanced Machining Processes, 1–5 July 2013*; Mechanical Engineering Department, MNNIT Allahabad: Uttar Pradesh, India, 2013.

100. Cortina, M.; Arrizubieta, J.; Ruiz, J.; Ukar, E.; Lamikiz, A. Latest Developments in Industrial Hybrid Machine Tools that Combine Additive and Subtractive Operations. *Materials* **2018**, *11*, 2583. [CrossRef] [PubMed]

101. Lin, Z.; Fu, J.; Shen, H.; Gan, W. A generic uniform scallop tool path generation method for five-axis machining of freeform surface. *Comput. Aided Des.* **2014**, *56*, 120–132. [CrossRef]

102. Soshi, M.; Ring, J.; Young, C.; Oda, Y.; Mori, M. Innovative grid molding and cooling using an additive and subtractive hybrid CNC machine tool. *CIRP Ann. Manuf. Technol.* **2017**, *66*, 401–404. [CrossRef]

103. Chen, L.; Xu, K.; Tang, K. Optimized sequence planning for multi-axis hybrid machining of complex geometries. *Comput. Graph.* **2017**, *70*, 176–187. [CrossRef]

104. Flynn, J.M.; Shokrani, A.; Newman, S.T.; Dhokia, V. Hybrid Additive and Subtractive Machine Tools-Research and Industrial Developments. *Int. J. Mach. Tools Manuf.* **2016**, *101*, 79–101. [CrossRef]

105. Li, L.; Haghighi, A.; Yang, Y. A novel 6-axis hybrid additive-subtractive manufacturing process: Design and case studies. *J. Manuf. Process.* **2018**, *33*, 150–160. [CrossRef]

106. Chen, N.; Frank, M. Process planning for hybrid additive and subtractive manufacturing to integrate machining and directed energy deposition. *Procedia Manuf.* **2019**, *34*, 205–213. [CrossRef]

107. Newman, S.T.; Zhu, Z.; Dhokia, V.; Shokrani, A. Process planning for additive and subtractive manufacturing technologies. *CIRP Ann. Manuf. Technol.* **2015**, *64*, 467–470. [CrossRef]

108. Yang, Y.; Haghighi, A.; Li, L. Energy Consumption Study for 6-Axis Hybrid Additive-Subtractive Manufacturing Process. In Proceedings of the 2018 Industrial and Systems Engineering Conference, Orlando, FL, USA, 19–22 May 2018.

109. Liou, F.W.; Choi, J.; Landers, R.G.; Janardhan, V.; Balakrishnan, S.N.; Agarwal, S. Research and development of a hybrid rapid manufacturing process. In Proceedings of the Twelfth Annual Solid Freeform Fabrication Symposium, Austin, TX, USA, 6–8 August 2001.

110. Grzesik, W. Hybrid manufacturing of metallic parts integrated additive and subtractive processes. *Mechanik* **2018**, *91*, 468–475. [CrossRef]

111. Ren, L.; Padathu, A.P.; Ruan, J.; Sparks, T.; Liou, F.W. Three dimensional die repair using a hybrid manufacturing system. In Proceedings of the 17th Solid Freeform Fabrication Symposium, Austin, TX, USA, 14–16 August 2006; pp. 51–59.

112. Cococcetta, N.; Jahan, M.P.; Schoop, J.; Ma, J.; Pearl, D.; Hassan, M. Post-processing of 3D printed thermoplastic CFRP composites using cryogenic machining. *J. Manuf. Process.* **2021**, *68*, 332–346. [CrossRef]

113. Tapoglou, N.; Clulow, J. Investigation of hybrid manufacturing of stainless steel 316L components using direct energy deposition. *Proc. Inst. Mech. Eng. Part B J. Eng. Manuf.* **2020**, *235*, 1633–1643. [CrossRef]

114. Hällgren, S.; Pejryd, L.; Ekengren, J. Additive Manufacturing and High Speed Machining—Cost Comparison of short Lead Time Manufacturing Methods. *Procedia CIRP* **2016**, *50*, 384–389. [CrossRef]

115. Kaynak, Y.; Kitay, O. The effect of post-processing operations on surface characteristics of 316L stainless steel produced by selective laser melting. *Addit. Manuf.* **2018**, *26*, 84–93. [CrossRef]

116. Bai, Q.; Wu, B.; Qiu, X.; Zhang, B.; Chean, J. Experimental study on additive/subtractive hybrid manufacturing of 6511 steel: Process optimization and machining characteristics. *Int. J. Adv. Manuf. Technol.* **2020**, *108*, 1–10. [CrossRef]

117. Kaynak, Y.; Tascioglu, E. Finish machining-induced surface roughness, microhardness and XRD analysis of selective laser melted Inconel 718 alloy. *Procedia CIRP* **2018**, *71*, 500–504. [CrossRef]

118. Salonitis, K.; D'Alvise, L.; Schoinochoritis, B.; Chantzis, D. Additive manufacturing and post-processing simulation: Laser cladding followed by high speed machining. *Int. J. Adv. Manuf. Technol.* **2015**, *85*, 2401–2411. [CrossRef]
119. Pal, S.; Tiyyagura, H.R.; Drstvenšek, I.; Kumar, C.S. The effect of post-processing and machining process parameters on properties of Stainless Steel PH1 product produced by Direct Metal Laser Sintering. *Procedia Eng.* **2016**, *149*, 359–365. [CrossRef]
120. Careri, F.; Imbrogno, S.; Umbrello, D.; Attallah, M.M.; Outeiro, J.; Batista, A.C. Machining and heat treatment as post-processing strategies for Ni-superalloys structures fabricated using direct energy deposition. *J. Manuf. Process.* **2021**, *61*, 236–244. [CrossRef]
121. Heigel, J.C.; Phan, T.K.; Fox, J.C.; Gnaupel-Herold, T.H. Experimental Investigation of Residual Stress and its Impact on Machining in Hybrid Additive/Subtractive Manufacturing. *Procedia Manuf.* **2018**, *26*, 929–940. [CrossRef]
122. Speidel, A.; Sélo, R.; Bisterov, I.; Mitchell-Smith, J.; Clare, A.T. Post processing of additively manufactured parts using electrochemical jet machining. *Mater. Lett.* **2021**, *292*, 129671. [CrossRef]
123. Grzesik, W. Hybrid additive and subtractive manufacturing processes and systems: A review. *J. Mach. Eng.* **2018**, *18*, 5–24. [CrossRef]
124. Altıparmak, S.C.; Yardley, V.A.; Shi, Z.; Lin, J. Challenges in additive manufacturing of high-strength aluminium alloys and current developments in hybrid additive manufacturing. *Int. J. Lightweight Mater. Manuf.* **2021**, *4*, 246–261. [CrossRef]
125. Simpson, T.W. Combining Additive and Subtractive Processes for Hybrid Machining. Available online: https://www.mmsonline.com/articles/combining-additive-and-subtractive-processesfor-hybrid-machining (accessed on 16 January 2022).
126. Dilberoglu, U.M.; Gharehpapagh, B.; Yaman, U.; Dolen, M. Current trends and research opportunities in hybrid additive manufacturing. *Int. J. Adv. Manuf. Technol.* **2021**, *113*, 623–648. [CrossRef]

*Review*

# A Critical Review on Fiber Metal Laminates (FML): From Manufacturing to Sustainable Processing

Rúben D. F. S. Costa [1,2], Rita C. M. Sales-Contini [3], Francisco J. G. Silva [2,4,*], Naiara Sebbe [4] and Abílio M. P. Jesus [1,2]

[1]   FEUP, Faculty of Engineering, University of Porto, Rua Dr. Roberto Frias 400, 4200-465 Porto, Portugal
[2]   Associate Laboratory for Energy, Transports and Aerospace (LAETA-INEGI), Rua Dr. Roberto Frias 400, 4200-465 Porto, Portugal
[3]   Aeronautical Structures Laboratory, Faculdade de Tecnologia de São José dos Campos Prof. Jessen Vidal, Centro Paula Souza, São José dos Campos, 1350 Distrito Eugênio de Melo, São Paulo 12247-014, SP, Brazil
[4]   ISEP, Polytechnic of Porto, Rua Dr. António Bernardino de Almeida, 4249-015 Porto, Portugal
*   Correspondence: fgs@isep.ipp.pt; Tel.: +351-228340500

**Abstract:** Composite materials such as Fiber Metal Laminates (FMLs) have attracted the interest of the aerospace and automotive industries due to their high strength to weight ratio, but to use them as structures it is necessary to master the manufacturing and wiring techniques of these materials. Therefore, this paper aims to address and summarize the drilling and milling processes in FMLs based on a literature review of papers published from 2000 to 2023. Parameters used in multi-material manufacturing and machining such as drilling and milling, tool geometry, tool coating, lubricants and coolants published by researchers were analyzed, compared and discussed. Machining process parameters related to sustainability were also analyzed. A SWOT analysis was carried out and discussed to identify opportunities for improvement in the machining process. There are opportunities to develop the surface treatment of aluminum alloys, such as testing other combinations than those already used, testing non-traditional surface treatments and manufacturing modes, and developing sustainable techniques during the FML manufacturing process. In the area of tooling, the opportunities are mainly related to coatings for tools and changing machining parameters to achieve an optimum finished part. Finally, to improve the sustainability of the process, it is necessary to test coated drills under cryogenic conditions to reduce the use of lubricants during the machining process.

**Keywords:** fiber metal laminate; drilling; milling; coolant; tool wear; tool geometry; sustainability

**Citation:** Costa, R.D.F.S.; Sales-Contini, R.C.M.; Silva, F.J.G.; Sebbe, N.; Jesus, A.M.P. A Critical Review on Fiber Metal Laminates (FML): From Manufacturing to Sustainable Processing. *Metals* **2023**, *13*, 638. https://doi.org/10.3390/met13040638

Academic Editor: Francesco Colangelo

Received: 16 February 2023
Revised: 12 March 2023
Accepted: 17 March 2023
Published: 23 March 2023

## 1. Introduction

Composite materials have been gaining visibility for the last few decades due to their advanced properties in comparison to the other material families, specifically in the aeronautic industry, but also with a growing interest in the automobile industry. This tendency began after the Second World War with innovations for military applications, where these materials brought a significant weight reduction in structural design and presented excellent fatigue properties and corrosion resistance as well [1,2].

Fiber-reinforced polymers (FRPs) are heterogeneous composite materials which combine lightweight, stiff and brittle reinforcing fibers, such as aramid, glass and carbon (known, respectively, as AFRPs, GFRPs and CFRPs) bound together by a polymeric matrix (thermoplastic or thermoset) [2]. The fibers, the reinforcing phase, contribute to the improvement of the mechanical properties of the laminated composite, whereas the matrix transfers the load to the inner fibers and at the same time it protects them from external damage and provides the composite material with its high fracture toughness [3].

In spite of this, there was a need to improve even further the properties of these materials so they could sustain the harsher conditions existent in aircraft, a requisite satisfied with the appearance of the fiber-metal laminates (FMLs) [4]. These are, by definition, hybrid structures which combine alternate phases of FRPs in the form of plies with thin sheets of a metal alloy, usually aluminum (FRP/Al) or titanium (FRP/Ti) [5,6]. An example of a FML with the fibers incorporated in epoxy resin is observable in Figure 1, which presents a three/two lay-up, since three metal layers alternate with two reinforcement ones.

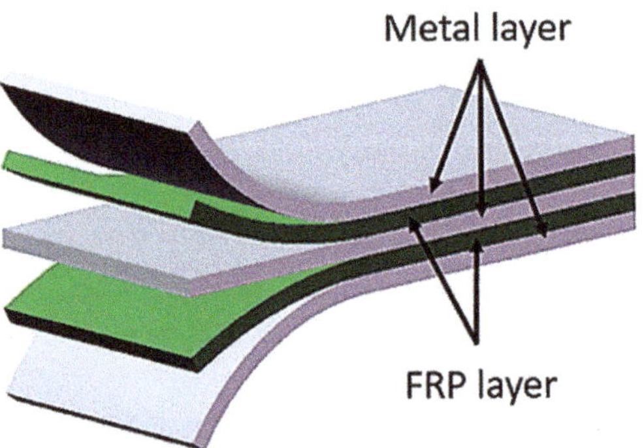


**Figure 1.** Fiber metal laminate configuration (adapted from [7]).

Nowadays, several types of FMLs exist, depending on the chosen metal to use or even the polymer, based on the intended application for the new material. For example, if these are composed of elastomer interlayers, their designation becomes FMEL, for fiber-metal-elastomer laminates [8]. On the other hand, considering just the metal counterpart, the most common FMLs are the CARALL (Carbon Reinforced Aluminum Laminate), GLARE (Glass Reinforced Aluminum Laminate) and ARALL (Aramid Reinforced Aluminum Laminate) [9–12]. Apart from the metal and reinforcement constituents, as well as the lay-up configuration in which the metal layers can be outside or inside the multi-material, the direction of the laminate must also be considered, namely, if it is a unidirectional hybrid laminate or a cross ply (woven), as shown in Figure 2.

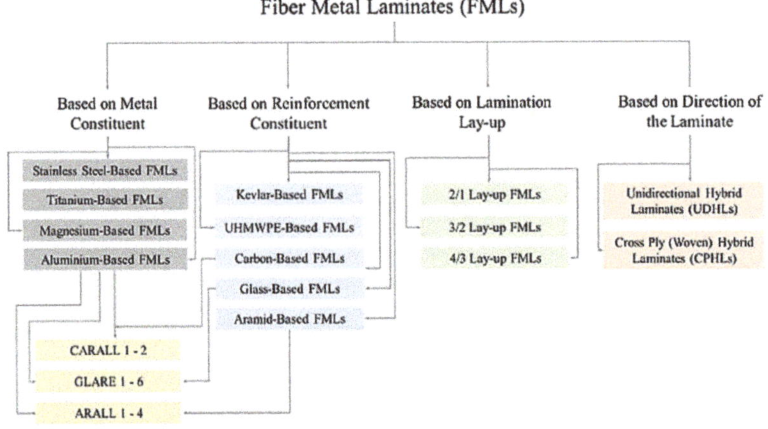


**Figure 2.** Classification of FMLs [13].

Through this union, FMLs combine the durability and easiness of fabrication associated with metal alloys with the outstanding specific properties and excellent fracture and fatigue resistance of high-performance composite materials [14,15]. Furthermore, metals lack fatigue strength and corrosion resistance, whereas composites have a low bearing strength and impact strength, in addition to their problem of reparability [16]. Hence, the combination of both materials overcomes the existing negative issues individually. The result is a high-strength, lightweight material with improved thermal, mechanical and tribological properties [17].

Regarding the manufacturing of multi-materials, this is a difficult process, due to the different properties of the materials to be joined and relatively weak adhesion between them [18–21]. Surface treatments can also be executed to enhance the adhesion at the metal-composite interface [22–25]. An adequate surface treatment of the metallic layer is indispensable to guarantee a good mechanical and adhesive bond between the composite laminates and metal surfaces. This treatment can be mechanical, through an abrasion to produce a macro-roughened surface and to remove undesirable oxide layers; chemical, with the immersion of the substrate in an acid solution; or electrochemical, with a coupling agent or dry surface, such as plasma-sprayed coating or ion-beam-enhanced deposition [1]. After the treatment, metal sheets may also be annealed, in order to relieve mechanical and thermal stresses, another step to facilitate the adhesion [26].

Several methods can be employed to achieve the intended configuration, with the most prominent being the fusion, chemical bonding (bond dual materials using structural adhesive) [27–29], physical and mechanical bonding, such as screwing or riveting (SPR) [21] and friction-based joining processes [30]. Figure 3 presents some of the processing methods used in the fabrication of Carbon Fiber Reinforced Metal Matrix Composites (CF/MMCs).

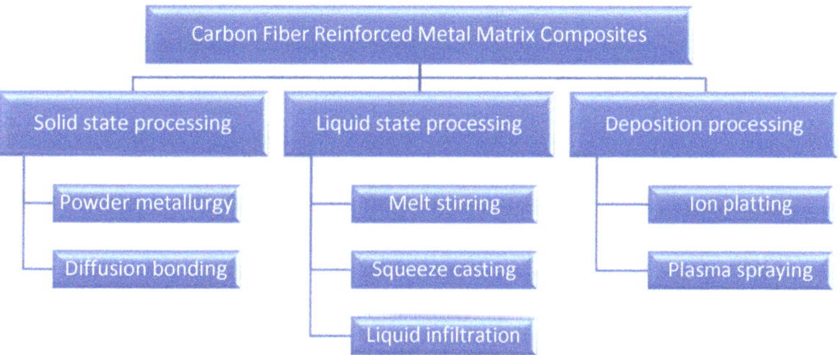


**Figure 3.** Processing methods for fabrication of CF/MMCs (adapted from [17]).

The shown techniques can be of diverse natures. In the solid-state processing, the composites are formed as a result of bonding between the metallic matrix and carbon fiber due to mutual diffusion occurring between the two in solid states at high temperature/pressure. The liquid-based metallurgy methods include processes such as casting and gravity or vacuum infiltration. This technique has as benefits short processing times, high fiber contents and low cost; however, the carbon fibers are likely to separate and float on the surface due to the significant density differences. Finally, the deposition processing consists of depositing the matrix on the fibers through various methods, followed by consolidation of the final product [17].

Alternatively, other fabrication methods also exist, such as forming techniques [31,32], where the cured fiber and resin matrix layers are deformed elastically, and the metal layers are deformed plastically by deep drawing, with a combination of optimized inner glass fiber patches and non-cured FMLs [33,34] or vacuum infusion, where there is no need of an autoclave or press [35,36].

In the case of CFRP/Al and CFRP/Ti stacks, firstly, the metal surfaces are treated for an optimal adhesion between the alloy and epoxy resin. After that, they are cured in a hot press, in order to achieve their final configuration, with the adhesive impregnated fibers prepregs successfully joined with the metallic layers [1].

The multi-material is of the utmost importance to the construction of aircraft due to its enhanced properties, surpassing the other families of materials in specific applications [37]. The CFRP composites are applied in the fabrication of major structural members of aircraft, such as floor beams, frame panels and a significant portion of the tail sections [38]. Moreover, 25% of the Airbus A380 airframe consists of composites, from which 22% are carbon- or glass-fiber-reinforced plastics and 3% are GLARE [39]. The application of GLARE in the upper fuselage shell of the Airbus A380 resulted in 15–30% weight savings over aluminum panels alone with significant improvement in fatigue properties [40]. Considering the Boeing Commercial Airplanes, 30% of the Boeing 767's outer structure is made from composites, and the Boeing B-787 contains almost 57% of its primary structure built only from composite materials, which confers it its wide range of flexibility. In the same way, the blades used in the cooler section of compressor and fan cases, usually employed in aero engines, are manufactured from CFRP, which reduced the assembly total weight by 180 kg and operational costs by 20% [41,42]. Furthermore, CARALL is also used in helicopter struts to absorb shocks, as well as plane seats. This FML shows an extreme fatigue fracture development resistance and outstanding pressure intensity and energy absorption [43]. Its high stiffness and strength with good impact properties also give CARALL laminates a great advantage for space applications [1].

Not only does the aeronautic industry benefit from the use of these materials, but the automobile sector is also adopting the same practice [44,45]. The reasons behind this are mostly the reduction in fuel consumption and the $CO_2$ emissions of future car generations. The combination of high-strength steel alloys and CFRP prepregs in an FML is a promising approach to producing lightweight car structures with a high stiffness-to-weight ratio [33]. In an automotive chassis, for example, it is beneficial to use CFRP where the chassis needs to be light and stiff aluminum where the deformation needs to be controlled [46]. Due to the high damping and good vibration resistance of FRP, metal-FRP hybrid components can also be used to improve the noise, vibration, and harshness (NVH) performance of automotive parts, which enhances the driving experience [47].

Based on a literature survey between 2000 and 2023, the main contribution of this manuscript is to provide information in a structured way about composites and manufacturing and machining operations of various materials, mainly in the field of drilling, and how the process can be performed in a sustainable way. The major novelty brought by this study is to group this information to perform a SWOT study, performing the critical analysis of the results already existing in the literature to prospect opportunities for improvements in the manufacturing of FML, in the machining process making it as sustainable as possible. Thus, in a structured and organized manner, it is intended to facilitate reader access and guide to what can still be studied in depth to contribute to creating a robust database on the topic in question.

## 2. Methodology

### 2.1. Literature Review

To develop this review, scientific databases such as Clivarate/Web of Science and Scopus were searched for selected papers related to FML and their production steps. In both scientific databases, the advanced search by title, abstract and keywords was used. The chosen search range was between 2000 and 2023. The selection of publications started with the first level of keywords used: Fiber Metal Laminates (FML), process, fabrication, and machining. The following keywords were associated with topics related to manufacturing steps and material characterization, such as CFRP/Al stacks, CFRP/Ti stacks, cooling, damage, defects, drilling, milling, surface treatment, tools, microscopy, tomography and their comparison.

Using the Web of Science search tools, the publication selection started with the combination of the keywords FML and process (6504 papers found), FML and manufacturing (3014 papers found), FML and fabrication (2439 papers found) and FML and machining (856 papers found). The number of papers decreases when a specific keyword is used. If the keyword FML is changed to "CFRP/Al stacks and machining", the number of papers decreases drastically to 37. Using the Scopus search tools, the combination of the keywords CFRP/Al stacks and processing or CFRP/Al stacks and machining yielded 4445 papers and 3104 papers, respectively, indicating that this research platform has more specific papers related to the topic discussed here. As mentioned above, the number of papers decreases when a specific keyword is used. In this case, when the keyword "CFRP/Al stacks" is changed to "FML and machinability", the number of papers decreases drastically to 24. The diagram presents the route followed to reach the best literature information about the FML machining process (Figure 4).

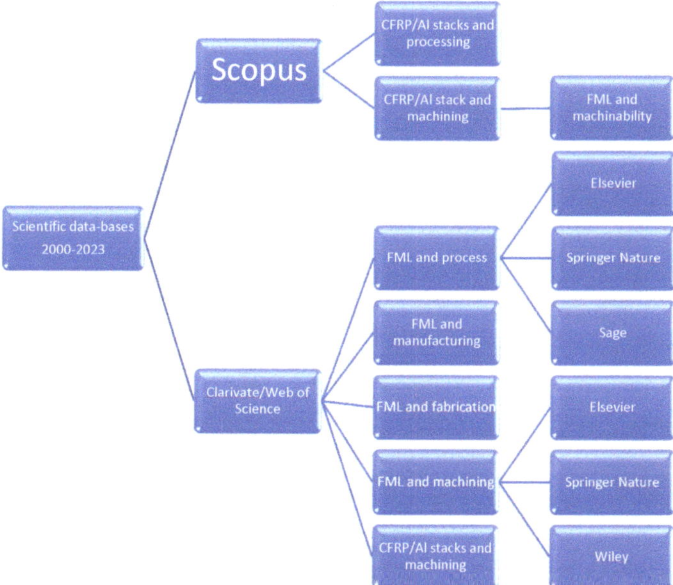


**Figure 4.** Route to reach best literature information about FML machining process.

Using the keyword "FML and machining", the first three publishers with more publications on the subject are Elsevier, Springer Nature and Sage, and using the keyword "FML and process", the first three publishers with more publications on the subject are Elsevier, Springer Nature, and Wiley, as can be seen in Figure 5a,b.

Another relevant piece of information extracted from the database was related to the countries that most published on the subject of FML machining, and it can be seen in Figure 6 that China leads the publications, followed by the USA and India, with Brazil being in the 13th position.

The literature review is divided into three main sections: (i) multi-material, (ii) machining tools and (iii) process sustainability. The first section discusses the multi-material manufacturing process, focusing on CRFP/Al stacks and CRFP/Ti stacks, followed by the main machining processes: drilling and milling, addressing the most used coatings and substrates for use in multi-material drilling/milling processes. The same section will also cover process parameters and the main lubrication methods used in multi-material machining. Finally, the main process failures according to quality levels and acceptance criteria will be presented, as well as how to analyze these process failures. The second section examines the types of tools used in the machining processes in terms of geometry, the presence or absence of coatings, and the tool wear. Finally, in the third section, the importance of studying the sustainability of the process is presented regarding the recycling or reuse of the chips produced for the different thermoplastic and thermosetting composite materials, and regarding the products used for cooling during the machining process, considering the gains in operational efficiency.

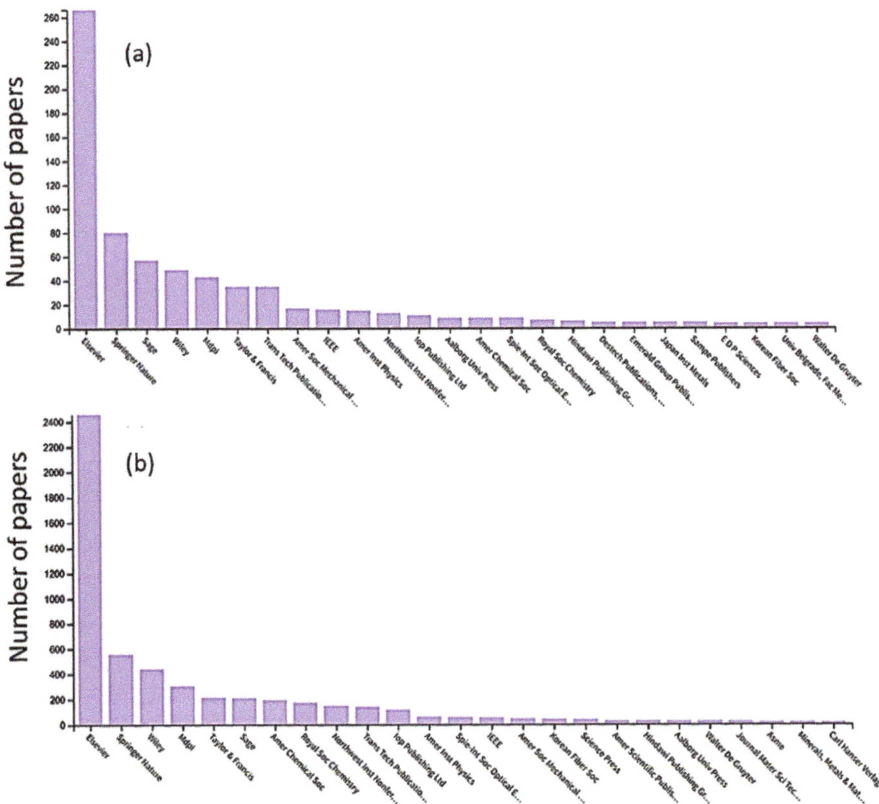

**Figure 5.** Number of papers produced per publisher using the keyword (**a**) publishers with more publications on the subject and (**b**) FML and process.

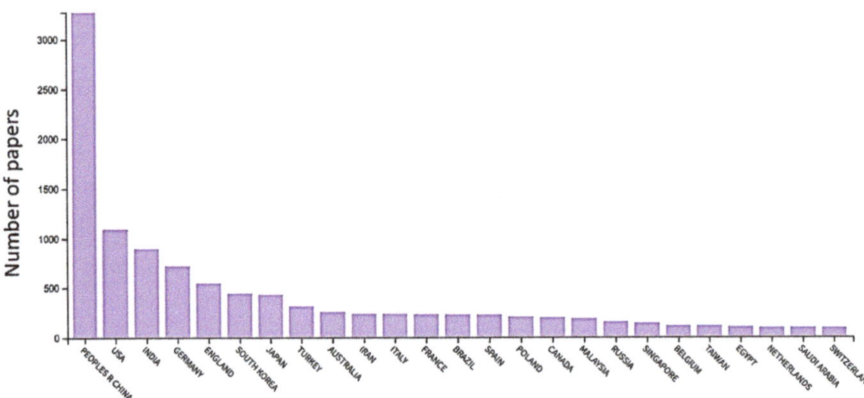

**Figure 6.** Number of papers by country produced on the FML and machining area.

*2.2. SWOT Analysis*

The SWOT analysis reinforces the vision of the strategic context, helping in the knowledge of characteristics on the theme that may not be apparent, in addition to also helping to understand its importance in the market. It is one of the tools that allows a broad view of the theme under study, considering the internal analysis of the process (Strengths and Weaknesses), which can be controlled, and the external analysis (Opportunities and Threats) are elements that cannot be controlled, but their understanding is essential to create the strategic plan of the process [48,49]. The SWOT analysis is carried out to identify where the multi-material machining process turns the threats into opportunities.

The information to carry out the SWOT analysis was extracted from the bibliographic review of the multi-material machining process and corresponding critical analysis. From the information collected, a list of strengths and weaknesses, opportunities, and threats was created by comparing the different sources of information. After compiling the lists, the most relevant items were prioritized, for an effective analysis of about 3 to 5 items per category that were highlighted as relevant. The items were crossed to check the relationships between them and were rated from 0 to 3, where 0 is no relationship, 1 is a weak relationship, 2 is a moderate relationship and 3 is a strong relationship. Based on the SWOT matrix, improvement actions will be established for the topic, identifying opportunities, weaknesses, sustainability and threats to the multi-material machining process.

## 3. Multi-Material Machining Processes

### 3.1. Multi-Material Process Preparation

The manufacturing process for FMLs is similar to that for polymer composites, using an autoclave process. To produce an FML, six steps must be followed: (i) sheet metal surface preparation, (ii) material deposition: prepreg and metal layers, (iii) mold cleaning and vacuum bag preparation, (iv) curing process, (v) stretching process, and (vi) inspection, usually imaging techniques, and mechanical testing [1].

All these steps must be carefully followed to obtain an FML with excellent mechanical quality, which depends on the correct choice and execution of the material preparation methods. A weak interfacial interaction between the prepreg and the metal can lead to a delamination process in composite structures, which is a major defect. The correct choice of surface treatment is then required to improve the bonding energy between the materials [47].

Adhesive bonding processes such as chemical, coupling agent, dry, electrochemical and mechanical surface treatments can be used as single or combined methods to produce an FML. All these surface treatments change the surface topography and increase the surface roughness, as well as the interaction between the metal and the bonding site.

Some surface treatments were used as a single method. Kwon et al. [21] used a single method, sanding, a mechanical surface treatment, using three types of sandpaper and different sanding times. They concluded that longer sanding times increased surface energy and improved mechanical adhesion between metal and composite. Drozdziel-Jurkiewicz et al. [22] investigated some aluminum and titanium surface treatments using mechanical, chemical and electrochemical methods to achieve high adhesion at the metal-composite interface. These surface treatments improve the interfacial fracture toughness of the FML, increase the bond strength and provide high adhesion at the FML interface. The study showed that chromic acid anodization is still the most effective in improving the expected bond strength. In their studies, Thirukumaran et al. [20] used the first step, mechanical abrasion, to create a roughened surface at the macro level and remove an unwanted oxide layer. They then tested the chemical process, also known as acid etching, using two chemical solutions: potassium dichromate and ferric sulfate. The treatment with ferrous sulfate, which contains $SO_4^{2e}$ ions, had a better performance, promoting pitting, which increased the roughness and favored good adhesion between the substrate surfaces, as indicated by the higher tensile strength values.

Laser and plasma techniques are also used for the surface treatment of aluminum sheets to produce FML. Dieckhof et al. [50] treated AA2024 and AA5028 sheets and found that both techniques allowed the formation of a structured anodic oxide layer, which improved corrosion protection and interfacial adhesion. Park et al. [51] prepared the metal surface by mechanical abrasion, followed by phosphoric acid anodization and alkaline acid etching techniques. They also observed that the rough substrates were essential to improve the bond strength of the metal sheet-prepreg interface. In addition, they understood that selecting different autoclave pressures can influence the quality of the FML, reducing the void content and avoiding premature failure of the FML. Furthermore, Cheng et al. [52] optimized the production of pores on an aluminum substrate. They chemically treated the metal surfaces with three different electrolytes, all containing $SO_4^{2e}$ ions combined with oxalic acid and/or ferrous sulfate. To improve the interfacial adhesion between the composite and the metal sheet, they also used CNTs for interfacial bridging and final bond strength. They concluded that a rougher substrate surface potentially helps to improve wetting, contact area, and mechanical occlusion of the bonded joint.

Analyzing the surface treatment process in aluminum alloy found in the literature, the combination of mechanical and electrochemical methods is the best used to reach excellent surface preparation. All the papers analyzed showed that a good surface preparation favors an adequate roughness of the sheet metal surface, contributing to an optimal mechanical interlock and excellent interfacial adhesion, thus guaranteeing an FML of sufficient quality so that it can be machined without suffering delamination processes.

### 3.2. Numerical Simulation Applied to FML Machining

The numerical simulation technique is an economical option used during the development of a product or process. In the numerical simulation, the cost of equipment and man-hours is lower compared to the equipment used in lubrication processes, finishing, and the qualification of the finished product, as it includes the cost of acquisition, maintenance, and man-hours per machine. In this way, numerical simulation helps to reduce costs if it is carried out before experimental tests, reducing the number of tests, and making the choice of parameters more reliable. In addition, numerical simulation can be used to help understand the mechanism of the machining process by predicting cutting forces and temperature effects.

In the aeronautical field, the FML machining process requires a proper design of the processing parameters to avoid the occurrence of various problems and to minimize defects dependent on the process temperature variation, which can be detrimental to the integrity of the parts. These aspects become more critical in dry machining, such as matrix burning, fiber extraction and delamination.

In their work, Parodo et al. [53] monitor the temperature measured on the tool and the workpiece during dry drilling of Al/GFRP (GLARE) and Al/CFRP (CARALL). The influence of cutting speed on the temperature trends was analyzed. In addition, a numerical model was developed to analyze the process of temperature evolution during drilling. The numerical simulation also indicated that the temperature fields are dictated by the thermal properties of the carbon and glass fibers: temperature profiles within the CARALL were found to be smoother than those observed in GLARE. Giasin et al. [40] studied the machinability of GLARE laminate by experimental techniques and analytical simulation. The chosen parameters were the effect of feed rate and spindle speed on the cutting forces and hole quality. A 3D FE model was also developed to help understand the mechanism of drilling GLARE. To evaluate the numerical model's efficiency, the collected thrust force and torque data were compared, and it was shown that the FE drilling model can predict the cutting forces within acceptable levels. It was the first contribution to the simulation of the drilling process of FMLs. Zitoune et al. [54] investigated the effect of machining parameters and tool geometry on cutting forces, hole quality, and CFRP/Al interface using experimental and numerical simulation. From the experimental study, it was found that the tool-enhanced geometry induced less thrust force compared to the standard one. The numerical analysis was based on the linear fracture mechanics of the CFRP and the plastic behavior of the aluminum with isotropic hardening. The results showed the critical thrust force responsible for the delamination of the last layer as a function of the aluminum thickness and, on the other hand, the maximum shear force responsible for the separation of the CFRP/Al interface was predicted as a function of the aluminum thickness.

Kim et al. [41] presented a method for predicting a cutting force model to optimize the feed direction. In this study, they used a CFRP with six different absolute fiber orientation angles. The fiber cutting angle significantly affects the cutting characteristics of the material and can be easily changed by varying the feed direction. By changing the feed direction angle, which is a simple adjustment in the milling process, this method effectively reduces the cutting force in material milling. In addition, since a predictive cutting force model is used, it is possible to derive the optimum feed direction under different cutting conditions with minimal experimentation.

Numerical simulation of the drilling process is relatively new; therefore, it is an area in the machining process that can be explored. It was observed that many of the authors investigate the influence of tool geometry on cutting forces and the delamination process of the material. However, a smaller number of authors study the temperature change in the profiles and how this can affect the properties of the material, the tool or can even be useful to study new types of lubricants that are more sustainable.

### 3.3. Multi-Material Machining Processes

Conventional machining processes such as drilling, milling, turning, and water jet cutting can be applied to composites, as long as the correct tool design and operating conditions are used. An FML is difficult to machine due to its anisotropic properties, inhomogeneous structure, and high abrasiveness of its constituents. These operations typically result in damage to the laminate such as fiber debonding, spalling, matrix cracking, fiber failure or pullout, and very fast wear of the cutting tool [55,56].

In the aerospace industry, the FML is used because of its light weight and stability at elevated temperatures. In these composite structures, large numbers of cut-outs and holes need to be produced. As mentioned above, FMLs are basically composed of a sheet of metal, aluminum, or titanium, and composite, carbon or glass fiber reinforcement, and a thermoset or thermoplastic matrix. Composites require high speed and low feed, drilling titanium requires low speed and high feed, while drilling aluminum requires a balance between speed and feed. Accordingly, the great challenge to drilling the FML is the right choice of parameters processes because of the materials' constituent properties that also vary with the environmental conditions [57].

Kumar et al. [58] compared the machinability of conventional drilling of hybrid titanium/carbon-fiber-reinforced polymer/titanium (Ti/CFRP/Ti) stack laminates in a single shot under dry and cryogenic conditions. The results indicate that the low temperature affects the hole quality but increases the thrust force due to the increase in the hardness of the Ti sheet at low temperatures. Azwan et al. [59] investigated the effect of different drilling parameters on FMLs, such as the drilling speed, feed rate and thickness of the FMLs on the strength of composite materials. They concluded that drilling at a lower spindle speed and a lower feed rate generates higher workload than at a higher speed, for the same feed rate. The thicker FML induces an higher workload compared to the one with less thickness. Zitoune et al. [60] studied the influence of cutting variables on thrust, torque, hole quality and chip during the drilling of a CFRP/Al stack. They observed that the magnitude of thrust force and torque during the drilling of Al compared to CFRP is doubled at a low feed rate. Hole circularity and surface roughness increase with increasing feed rate. The aluminum layer also has a better finish compared to the CFRP one.

Another machining process to produce holes commonly used in aircraft part finishing is milling. The milling process uses rotary cutters to remove material by advancing a cutter into a part. This method is used as an alternative to drilling these joints with conventional twist drills, named after helical milling, where a milling tool rotates in a helical path and creates the hole [61]. Helical milling has also been investigated for making holes in FMLs, with several advantages such as lower cutting forces, lower heat generation and easy chip evacuation [62]. The main difference between drilling and helical milling is that in the former, the hole diameter is determined by the tool diameter, whereas in the latter, the hole is defined by a combination of the tool diameter and the helical path, resulting in greater flexibility in the hole diameter [63].

One of the compounds used in the aerospace industry that is difficult to machine is a unidirectional CFRP and Ti6Al4V, due to the hardness characteristics of titanium sheets and the abrasive characteristics of CFRP. They observed diameter variations, which may be due to the different Young's moduli of titanium and CFRP, as well as variations in surface roughness caused by material-specific chip formation mechanisms [61]. Hemant et al. [64] evaluated the helical hole milling process in GLARE laminates. Although the material and parameters are different from those used by [61], similar process dissimilarities were observed, with variations in hole diameter depending on the material layer, production of discontinuous powdery chips and surface roughness. Therefore, to obtain an excellent quality finished product, the milling process parameters need to be adjusted to produce holes of uniform diameter throughout their depth, continuous chips in the metal zone, low surface roughness and a composite layer without delamination.

### 3.3.1. Comparison of Drilling and Milling Processes Parameters

The different properties of the constituent materials in FMLs are a challenging task when hole-making is needed. Improper selection of the process and process parameters can result in poor surface quality, inadequate dimensional quality, dimensional inaccuracy, or even component failure [65].

However, there are several interrelated factors. The most crucial factors which affect the machinability of the material are cutting forces, tool geometry, materials, coatings, chip formation, analysis of tool wear, hole metrics such as the hole size and circularity error, surface roughness and burrs formation [66]. Bolar et al. [65] compared two hole-making techniques used in CARALL: drilling and milling. In their study, the two hole-making techniques were evaluated using various performance measures including thrust, radial force, chip morphology, surface roughness, machining temperature, hole diameter accuracy and burr size. Comparing the results, the authors found some advantages of the helical hole milling process in terms of lower thrust and radial force. The intermittent cutting and convenient chip evacuation and heat dissipation helped to lower the temperature and prevent some material damage. The discontinuous aluminum chips produced by the helical milling process proved to be beneficial, with the holes showing a superior surface finish. However, excessive axial feed in helical milling resulted in tool deformation and chatter, leading to surface quality degradation. Further analysis of the machined surface using microscopic inspection showed that the delamination process occurred when conventional drilling was used. On the other hand, the helically milled hole was free of such defects and showed no signs of delamination. They also found that the formation of oversized holes after the drilling process was incredibly significant. Finally, the exit burr's height was much lower with helical milling due to the lower thermal impact and lower thrust. Considering all the aspects presented here, it is safe to say that helical milling is a suitable alternative for machining holes in FMLs.

Another study comparing drilling and milling was carried out by Barman et al. [63]. In their work, they evaluated the two-hole machining process in titanium alloy Ti6Al4V material. They carried out the machining tests considering the thrust force, surface roughness, hole diameter and machining temperature. The morphology of the chips produced and burr formation were also investigated. The magnitude of the force components (thrust and radial force) and the cutting temperature was lower in the milling process, which produced discontinuous powdery chips that evacuated easily without damaging the machined hole surface. The final quality of the holes and the diameter are better in the helical milling process, which produced burr-free holes, contrary to drilling. A similar study using helical milling and drilling techniques to machine AISI D2 tool steel was previously carried out by Iyer et al. [67]. They found that helical milling produced H7-quality holes with good surface roughness compared to drilling. Another comparative study on helical milling of a larger thickness CFRP/Ti stack and its individual layers was conducted by Wang et al. [5]. Their experimental results showed that as the number of holes increased, the cutting forces increased due to tool wear and its dependence on the material type. Indeed, the abrasive nature of the CFRP resulted in an increase in cutting force. The hole edge quality is good while machining the titanium alloy, and low delamination is registered at the tool entry and exit points in the material, indicating that if quality problems of the holes appear, it refers to the milling of CFRP. Hole size was inversely proportional when comparing a single layer to a stack. In the drilling process, with CFRP, smaller hole sizes can be achieved compared to the titanium plate; however, oversized CFRP holes and undersized titanium alloy holes are observed when helically milling stacks.

Therefore, there is the agreement in the literature when comparing the drilling and milling processes. All results showed similarity regardless of the material used, Ti/CRFP stacks, Al/CRFP stacks, Ti metal alloy, CRFP or steel, indicating that the best process and the one that promotes less material damage is the helical milling process.

### 3.3.2. Lubrication Processes during Machining

During the machining process, a large amount of cutting heat and friction heat can be generated due to the abrasive nature of the material and tool wear, which usually makes the temperature elevate rapidly. Temperature is one of the most important factors affecting the machined hole surface quality, especially for the CFRPs. When drilling CFRPs, the temperature can reach about 150~250 °C, leading to a risk of thermal degradation of matrix resin, carbonization of thermosetting matrices, the fusion of thermoplastic matrices, and sometimes, burning of the carbon fibers [68]. In addition, the thermal effects can affect the quality of machined holes. These, however, can be minimized using coolants supplied either directly or indirectly to the cutting tool-workpiece interaction zone, to remove part of the generated heat. Nevertheless, the use of coolants adds extra costs for handling, disposal, and environmental impact [69].

The coolants generally used in machining processes are water-oil emulsion, a mixture of a soluble oil cutting fluid and mineral oil lubricant [70], micro-lubrication, also known as minimum quantity lubrication (MQL) [71], cryogenic coolants $CO_2$ [72] and liquid nitrogen ($LN_2$) [73,74], air cooling [62] and vegetable oil [75], among others.

Shyha et al. [70] analyzed the hole quality/integrity after drilling of titanium/CFRP/aluminum stacks under flood cutting fluid water/oil emulsion of 7–8% volume solution and spray mist, a mixture of soluble oil cutting fluid and mineral oil lubricant. Some significant improvements in the machining process were observed. Burr height was generally less than 500 μm, except when spray mist was used. Delamination of the CFRP laminates was significantly reduced due to the support provided by the Al and Ti layers. Surface roughness was significantly lower when using through-spindle cutting fluid compared to spray mist application, especially on the Al section. Spiral-shaped continuous aluminum chips were prevalent, while both short and long helical chips were found with the titanium material when cutting under wet regime. In contrast, the CFRP layer typically produced dusty black composite particles suspended in the soluble oil of the coolant emulsion.

Kumar and Gururaja [76] investigated the cryogenic cooling effects during the drilling of Ti/CFRP/Ti stacks. The parameters such as thrust, torque, delamination, burr height and surface roughness were considered to investigate the effects of liquid nitrogen ($LN_2$) as a coolant. The results are compared with those obtained from dry drilling of Ti/CFRP/Ti stacks, indicating that torque, top surface of metal composite interface, exit burr height and surface roughness decreased when drilling was performed under a cryogenic environment. Nonetheless, thrust force and damage to the lower surface of the metal composite interface are increased under $LN_2$ cooling conditions. Biermann and Hartmann [72] analyzed a cryogenic process cooling with $CO_2$ and found that the chosen coolant improved burr formation in drilling compared to dry machining and resulted in higher sustainability comparing to machining with cooling lubricant. Many authors have investigated the influence of the application of cryogenic liquid nitrogen cooling and minimum quantity lubrication (MQL) during GLARE machining. Giasin et al. [77] observed that the use of MQL and cryogenic liquid nitrogen cooling increased the cutting forces; however, both reduced the surface roughness of machined holes, adhesion and built-up edge formation on the cutting tool compared to dry drilling. Examination of the microhardness of the top and bottom aluminum sheets near the hole edges after machining showed that it increased when both coolants were used. Pereira et al. [62] reported some benefits of air-cooling application. It was used to cool the cutting point and to remove the chips, obtaining better results in terms of cutting forces and temperature. The cutting temperature reduction in helical milling was 30%. When drilling the CFRP, the cutting heat becomes lower by increasing the revolution speed. Chips pulverized into small sizes reduce cutting temperature by absorbing heat on the cutting of CFRP composite plate. However, the powder-like chip is bad for the machining center, even promoting corrosion and wear of part of the equipment.

Lubricants and coolants play an important role in the machining process to prevent delamination damage in the material during the drilling process, especially in FMLs, because the difference in modulus of elasticity between composite and metal causes different machining deformations. The main defects observed after machining and their discussion are presented below.

### 3.3.3. Machined FML Defects and Analysis Techniques

Drilling holes in both CRFP and FML can result in material damage as the drill passes through the material. Several types of defects are associated with drilling operations, both at the entry and exit of the hole: dimensional defects, surface roughness and surface integrity problems of the hole wall. Defects induced by machining process conditions include matrix cracking, fiber fracture, debonding, delamination and fiber pull-out. Poor hole quality is the cause of nearly 60% of all scrapped parts [78]. As drilling is often one of the last machining operations, the damage that occurs at this stage results in huge economic losses when near-finished parts have to be scrapped. Understanding and detecting the type, size and location of defects that can occur during drilling operations is important for economical and sustainable process improvement [79]. This damage always occurs combined in hole-surrounding areas.

The great issue in FML drilling operations is the quality of the hole: several defects occur related to the process, mainly on the entry and exit sides of the hole, as well as dimensional and surface roughness issues of the hole wall. The detection of these defects is not trivial, especially when non-destructive methods are used. Several methods have been applied to evaluate the hole quality, such as Computed Tomography (CT)/X-ray Tomography Analysis [79–81]; C-Scan [82], Scanning Electron Microscopy (SEM), Stereoscopic Optical Microscopy (SOM), and Energy Dispersive Spectroscopy (EDS) [83]. Nguyen-Dinh et al. [84] analyzed the surface integrity of composites using x-ray tomography after trimming process. The application of the X-ray technique has enabled the measurement of the craters' volume compared with surface techniques, such as surface roughness and 3D optical topography. They observed several damaged zones with formation of craters in the material surface after trimming process, indicating material pull-out (Figure 7).

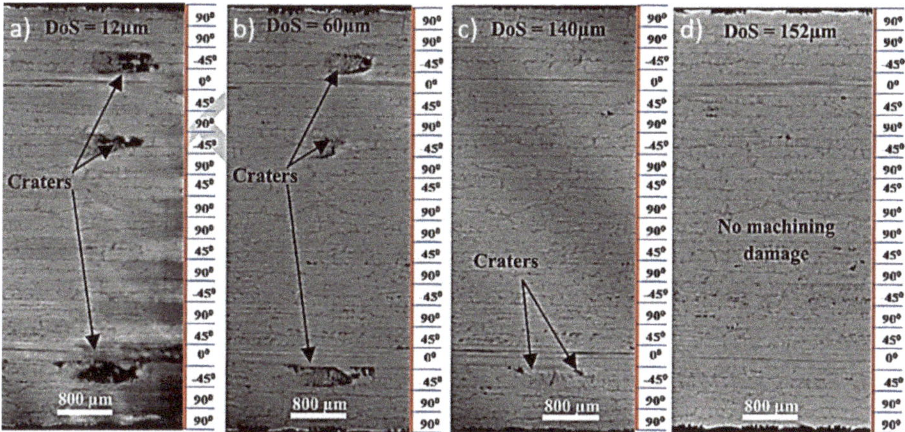

**Figure 7.** X-ray tomography images showing machining damage for cutting speed of 150 m/min and feed speed of 500 mm/min at cutting distance of 1.68 m with various depth of scanning of (**a**) 12 μm, (**b**) 60 μm, (**c**) 140 μm and (**d**) 152 μm [84].

Pejryd et al. [79] used X-ray computed tomography to detect defects caused by drilling holes in a CFRP. Surface defects and surface properties such as fiber debonding and surface roughness could be easily investigated. Figure 8a shows a typical surface image of a drilled hole based on the reconstruction of the X-ray images. Figure 8b illustrates one way of highlighting the glass fiber material. This technique allows other components to be identified by color. In this case, a red color is used to clearly distinguish it from the surrounding material.

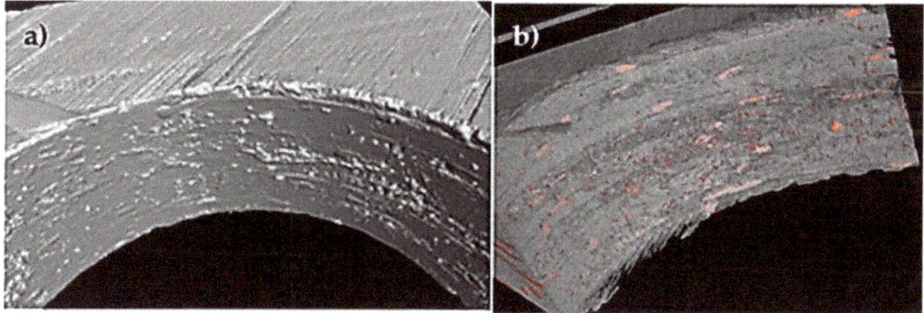

**Figure 8.** (**a**) A 3D model reconstructed from CT scan of the drilled hole, outer surface. The hole has a nominal diameter of 9.5 mm and (**b**) CT image of the inner wall of a 9.5 mm diameter hole, with glass fiber material highlighted in red [79].

Wang et al. [85] compared the holes' quality of CFRP/Al and CFRP/Ti-6Al-4V by Scanning Electron Microscopy (SEM). To reduce the damage, a two-step process of helical milling process was proposed and compared with that of conventional drilling. In the first step, milling was executed starting on the composite part, and then, in the second step, milling started on the metal part. In the conventional drilling process, they noticed that the damage is superior when the tool entry starts on the composite part, showing many uncut fibers (Figure 9a). The cutting action of the second step eliminates the damage caused by the first step, but the surface next to the hole showed fiber pull-out (Figure 9b). The images showed that the damage induced by the helical milling process in both steps is irrelevant.

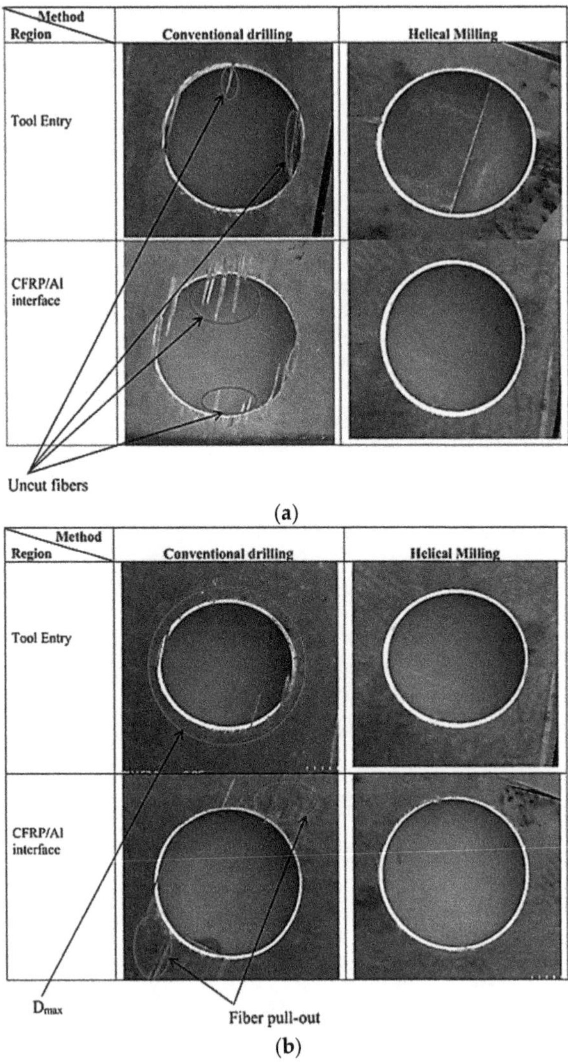



(a)



(b)

**Figure 9.** Scanning Electron Microscope (SEM) images comparing hole quality of both conventional and helical milling methods (steps 1 (**a**) and 2 (**b**)) [85].

The minimization of delamination damage is of great importance, because it is a critical parameter that determines the acceptance or rejection of composite components. To achieve this proposal, Bertolini et al. [73] analyzed the hole quality of an Al/CFRP stack obtained by different drilling processes: ultrasonic (UD) and cryogenic (CD) and compared with dry drilling/regular drilling (RD) using the SEM technique. The feed rates used were 0.05 mm/rev, 0.1 mm/rev and 0.15 mm/rev. The FML was composed of two layers: one of 5 mm thick aluminum alloy considered the entry face, and the second one composed of a 4 mm thick CFRP sheet, considered the exit face. They were evaluated solely at the exit, since the entrance appears defects free (Figure 10).

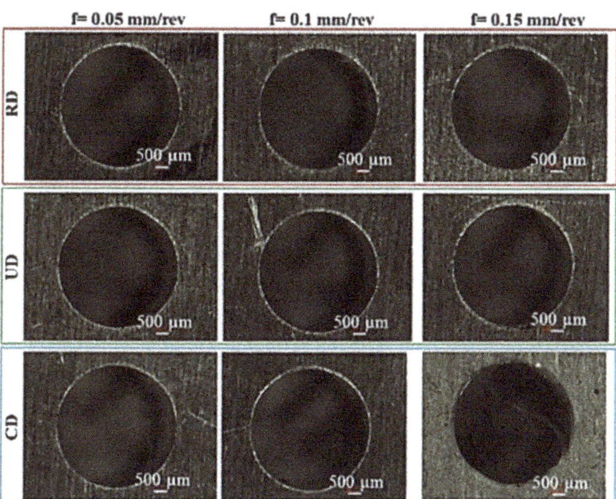

**Figure 10.** Entry delamination on the aluminum sheet at varying feed and drilling strategies [73].

Figure 11 shows the exit face where the delamination process occurred to a greater or lesser extent depending on the variable feeding and drilling strategy. Severe exit delamination occurred solely when drilling tests were conducted under CD, regardless of the adopted feed. This phenomenon can be correlated to the thrust force increase thanks to the hardening of the material as a consequence of the liquid nitrogen application. It is acknowledged that the higher the thrust force, the greater the exit delamination, because the deflection of the FML's last ply concerns a larger zone.

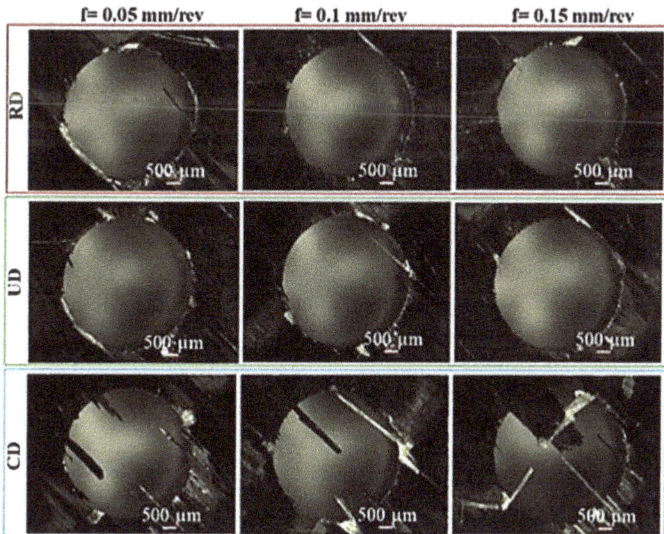

**Figure 11.** Exit delamination on the CFRP sheet at varying feed and drilling strategies [73].

These techniques already guarantee by themselves the analysis of the damage caused by the machining process, but when combined with other image analysis techniques cited here, they improve the accuracy of the analysis, complementing the information and being a determining factor for adjusting process parameters.

To reduce damage during the machining process, it is important to choose the correct tool and tool wear. Therefore, Section 4 deals with this subject in order to provide as much information as possible on the possible tools and tool wear used for multi-material machining.

## 4. Machining Tools

There are several different types of tools used in the machining operations on composites and multi-materials, depending on crucial factors, namely, the process and the material to be machined. According to these, a set of variables has to be chosen, such as the tool material, geometry or need for a coating, from which a large range can be employed. These factors will result in determined cutting forces and torques, which will impact the base material, considering its final roughness or the existence of delamination, and the tool, in particular its wear mode, in a specific way.

In order to optimize the machining process, the desirable mechanical properties for a cutting tool material are small grain size, to produce a sharp cutting edge, good toughness, to keep the sharp cutting edge without chipping or deformation under the dynamic action of the cutting forces, and high hot hardness, to provide a great resistance to abrasive wear under elevated cutting temperatures. In terms of thermal properties, since the temperature rises abruptly during the process, the tool needs to possess a good thermal conductivity to remove heat from the cutting zone, thermal stability, to maintain integrity at cutting temperatures and low chemical affinity to the workpiece material [57].

The preferable tools in the literature for drilling composite-metal stacks are carbide tools such as tungsten carbide, due to their high strength and wear resistance compared to HSS/HSS-Co tools. These also are among the ones most used, alongside Poly-Crystalline Diamond (PCD) tools [86].

### 4.1. Tool Geometry

When referring to the geometry of the tools, each machining operation has a specific type of geometry, the most fit to perform a specific kind of work. In the drilling case, there are two important angles to be considered, namely, the point angle and the helix angle. The standard helix angle for most drills is 30°; however, in spite of most drills coming with a 118° drill point angle, when it comes to drilling composites, it is recommended to use a drill bit with a point angle between 130° and 140°, values also used in aluminum drilling. A cutting tool with large helix angle, usually higher than 24°, allows a quick chip evacuation, whereas large point angles improve chip removal and reduce burr formation. The increment of these two parameters, seen in Figure 12, also minimizes surface roughness [87]. Additionally, this figure presents the various drill geometries for specific drilling operations.

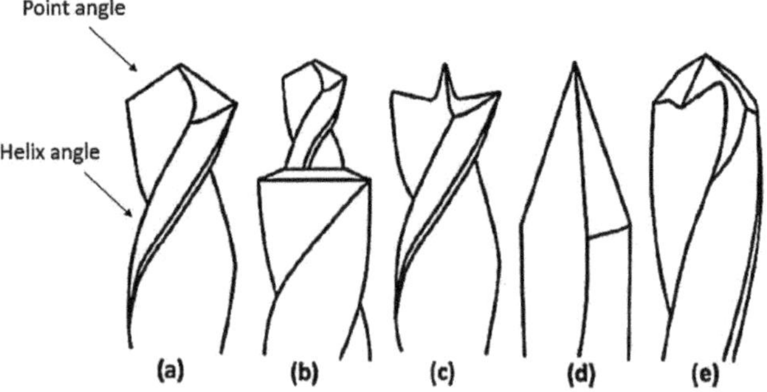



**Figure 12.** Different tool geometries used in drilling: (**a**) standard twist drill, (**b**) step drill, (**c**) Brad point drill bit, (**d**) straight flute drill bit (dagger) and (**e**) multifaceted drill bit (adapted from [54]).

The step drill is one of the drilling strategies to avoid material damage around the hole, for example delamination, as it first inflicts a primary hole, with a smaller diameter, and then it drills the intended size, so as the contact with the substrate is smoother [88].

Apart from the already exposed geometries, tools with special designs have been proposed to decrease the probability of delamination and obtain better productivity. In this matter, double helix tools minimize axial forces by utilizing two opposite helix angles [89]. Double point and multi-facet drills were also developed, producing lower thrust forces, due to an additional cutting edge [90]. In the case of milling, the multi-tooth milling cutter could significantly minimize milling defects for CFRP and decrease cutting forces [91].

When there is a need to measure the temperature reached in drilling processes, a thermocouple can be inserted into the drill bit lubricant-cooling fluid holes, as shown in Figure 13. The thermocouples do not necessarily have to be positioned in the same place for every case, meaning that one device may be in the sharp edge and other in the flank of the drill. This is a more complex procedure, but this way a more complete analysis of the temperature reached by the tool can be achieved, with information from different areas, each of which is subjected to a specific stress and, therefore, temperature [92].

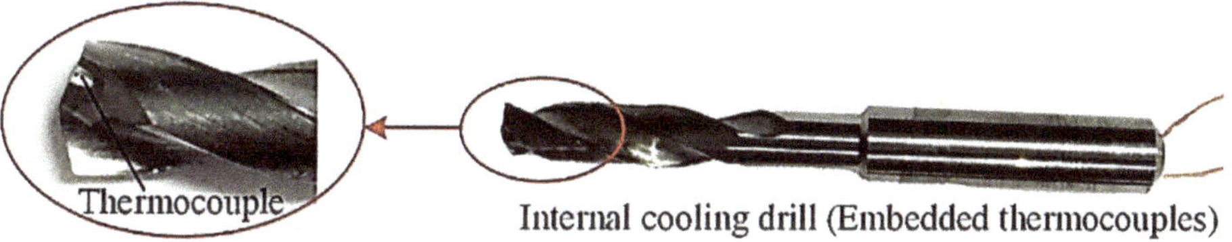


**Figure 13.** Thermocouples in the drill (adapted from [93]).

This set-up allows the measurement of the temperatures arising in the tool during the drilling. Thermocouples may also be inserted inside the multi-material: one in the composite laminate and other in the metal sheet. Although in the last years, alternative approaches to measure the temperature during the machining processes have been proposed, less invasive and relying on no-contact methods, such as infrared pyrometers or infrared cameras, the thermocouples are still the most effective method [53].

*4.2. Tool Coatings*

Two of the most prominent types of coating methods are the Physical Vapor Deposition (PVD) and Chemical Vapor Deposition (CVD). In the case of the latter, which uses higher temperatures on the process, the coating is synthesized from a mixture of different gases, depending on the type of coating being deposited [94]. On the other hand, the PVD technique intensively improves the wear resistance of different tools, effectively increasing their lifespan [95].

Studies show that the correct application of a coating is more effective than an uncoated tool in drilling multi-materials [96]; however, this is only true for higher feed rates and cutting lengths, severe conditions, as coated tools possess a much higher cost, so these are only applied when strictly necessary. Coated carbide tools such as PVD coated Boron-Aluminum-Nitride (BAM) tools have a better performance compared with uncoated tools in terms of tool wear at severe drilling conditions. In spite of this, prolonged tool life came with the price of rougher finished surfaces due to roughness of the BAM coating [86].

Figure 14 shows three different types of coatings which are common for rotary tooling such as drilling, being two monolayer (TiN and TiAlN) and one multilayer (AlTiN/TiAlN). The latter combines two layers for better heat and wear resistance [87]. From a study performed with these three coatings, the TiAlN drills produced the largest thrust and cutting forces, with the AlTiN/TiAlN resulting in the lowest values. Additionally, this coating presented a better self-lubricating effect due to its multilayer structure, which also increases its hardness, resulting in a better performance achieved during drilling [97].

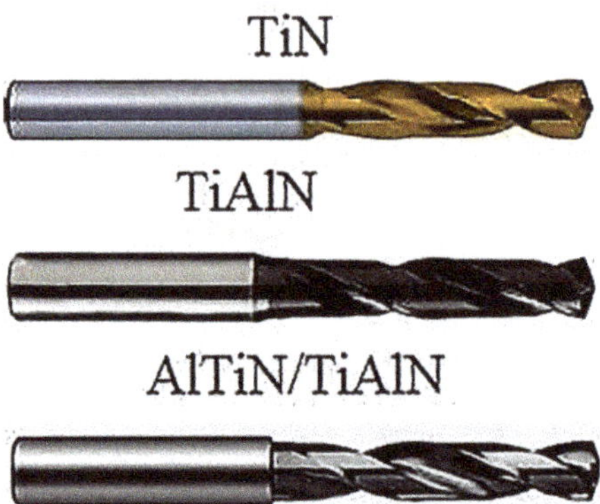


**Figure 14.** Drilling tools with different coatings (adapted from [87]).

Diamond CVD coatings are many times chosen as well, due to their enhanced characteristics when compared to the remaining options, although sometimes they experience some lack of adhesion. Nevertheless, when these problems are overcome, diamond coatings provide a longer lifetime when put side by side with other coating materials [98]. Its high thermal conductivity contributes to a temperature decrease in the cutting tool surface, and a low thermal expansion coefficient mismatch allows obtaining a good adhesion between the CVD diamond film and the ceramic substrate, as a result of a lower interface residual stress at room temperature. It also contributes to an enhanced surface hardness, which leads to a decrease in the tool wear [99]. Diamond-like carbon (DLC) coatings are also often used in the literature as an alternative to a diamond film [100].

### 4.3. Types of Tool Wear

The abrasive nature of multi-materials, with consequences for the substrate in the drilling process, such as fiber pull out, particle fracture, delamination and debonding at the fiber or particle and matrix interface, causes severe tool wear [55]. This is one of the reasons a tool coating is employed several times, so it can be possible to extend the tool life as much as possible, thus improving the process and reducing the associated costs [55].

It has been found that the tool wear increases the cutting force values [101]. This can be explained due to the loss of the tool's capacity to perform the intended job, as the contact surface with the substrate is reduced and its roughness increased, which is translated into higher cutting forces and less cutting quality. The types of wear suffered by tools during machining processes include adhesive, abrasive, fatigue and corrosive wear [57].

The adhesive wear, represented in Figure 15, can be produced by two different ways: directly, with high-speed chip particles impacting the tool, creating a micro blasting effect that reduces the cutting angle and the rigidity of the tool, and indirectly, by the incorporation of fragments of the substrate material to the tool [57]. When these fragments are removed, they can drag out tool particles, which leads to its wear by the loss of tool material, but also by the abrasion process caused by the friction of those particles with the tool rake face when they are dragged by the chip [84,102]. This abrasion action acts mainly on the flanks and outer corners of the drill bit in an irregular pattern, rounding its cutting edges and possessing various forms, such as crater wear, chisel edge wear and chipping [103–105]. Fatigue wear is the result of an excessive number of cycles performed by the tool, and the corrosive ones occur by a chemical reaction, degrading the tool material and thus keeping it from performing a proper cut.

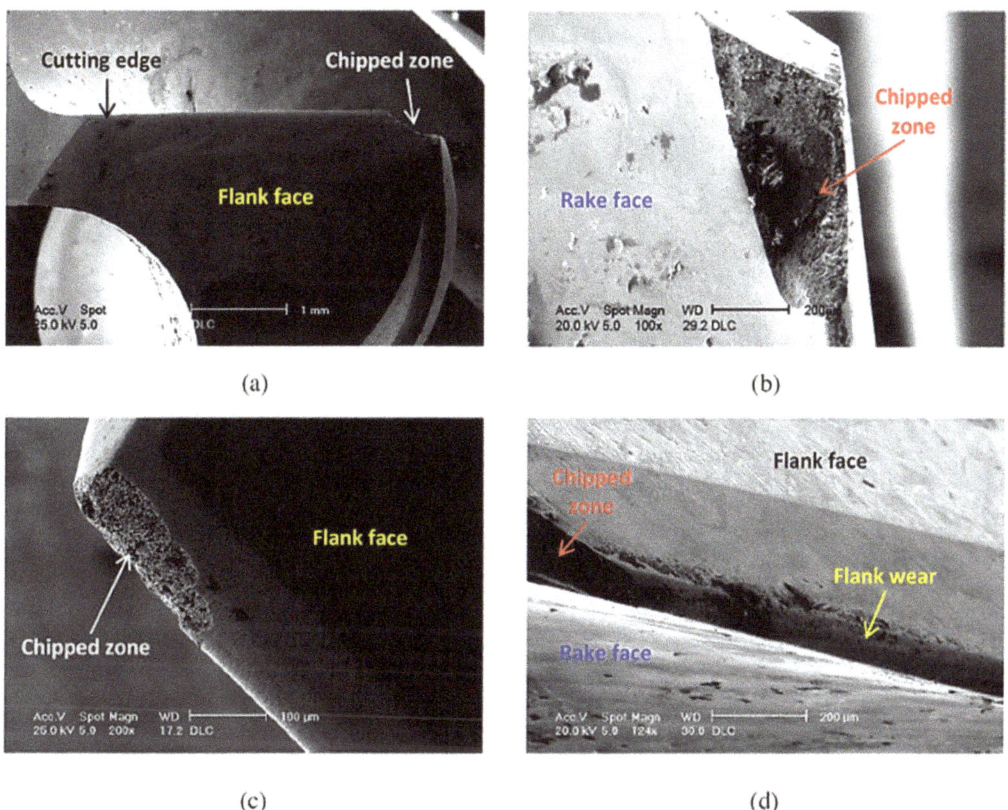



**Figure 15.** Adhesive wear in a drilling tool: (**a**) edge chipping in the flank face, (**b**) chipped zone in the rake face, (**c**) chipped zone in the flank face and (**d**) flank wear and edge rounding [105].

Sometimes, a phenomenon that can occur is the clogging of the drill flutes with chips coming from the substrate, which prevents the drill from working correctly and therefore threatens the final quality of the product [57]. In this case, as well as with certain stages of adhesive wear, cleaning the drill with sodium hydroxide is required, with the aim to eliminate these negative aggregations and accordingly return the drill to its normal shape [106]. For the milling process of multi-materials, tools with diamond inserts possess good results, due to their high hardness [101]. The same method is applied in the drilling of FRPs, with diamond-mounted points to reduce the wear at the cutting edges, due to their high abrasive wear resistance [107].

## 5. Process Sustainability

Progress towards sustainable development is increasing in all industrial sectors. Within the integrated life cycle management strategy, the manufacturing phase is one of the key performance phases due to resource consumption, emissions, and other negative impacts. To minimize environmental, ecological, and social damage, sustainable manufacturing design must take into account the three dimensions of sustainability: environmental, economic and social [108].

The machining process is an important part of the manufacturing process; therefore, it is expected to have a great impact by improving sustainability performance. Among the elements of machines, conventional coolant application has been considered as a critical limiting factor to achieving better sustainability performance [109].

The machining process of high-tech materials requires oil-based cooling/lubrication, which is one of the main non-sustainable elements, leading the R&D process in the search for alternative cooling and/or lubrication mechanisms [110]. To achieve better sustainability performance, alternative approaches to conventional flood cooling such as dry machining, minimum quantity lubrication (MQL) and cryogenic machining have been applied. Comparing MQL and dry machining, a similar result is observed; MQL does not remove the heat generated in the cutting zone sufficiently. Cryogenic machining not only reduces the disposal of large quantities of lubricating oil used in machining, but also increases tool life by 60% [109].

Some authors have studied these three cooling methods and compared their findings. Nagaraj et al. [111] compared MQL and cryogenic cooling methods to drill a CFRP composite. They concluded that cryogenic drilling gives better results than dry drilling in terms of drilled hole diameter and roundness, and sequentially they considered the MQL method as the next best alternative to dry drilling. Sharma et al. [112] presented a review to understand the influence of different types of coolants, mineral oils, vegetable oils, and nanofluid-based cutting fluids with the help of the MQL technique. They concluded that MQL reduced cutting temperature and produced metallic color chips, showed a reduction in dimensional deviation, improved tool life, produced better surface quality and showed an approximately 40% increase in material removal rate compared to aqueous flood coolant. To facilitate the analysis of the sustainability of the machining process, Hegab et al. [113] developed a detailed and general assessment model for machining processes. The four life cycle stages (pre-manufacturing, manufacturing, use and post-use) are included in the proposed algorithm. Energy consumption, machining costs, waste management, environmental impact, and personal health and safety are used to express the overall sustainability assessment index. They applied the model to three literature case studies and found that it was able to predict optimal parameters in close agreement with the experimentally measured results. Lv et al. [114] applied a model of the energy efficiency, carbon efficiency, and green degree in five types of machine tools selected for the typical machining processes (turning, milling, planning, grinding and drilling). After a careful analysis, they observed the increase in the back engagement will increase the material removal rate, thus, its impact degree is small, and increasing spindle speed and feed rate will reduce processing time, energy consumption, and emission, thus presenting a positive correlation, reaching an energy efficiency of 30%.

To achieve a sustainable machining process and estimate earnings, it is necessary to establish new models that reduce this complexity in the design and management stage, allowing the development of task manufacturing to promote sustainability in the processes involved while maintaining technical, economic and quality feasibility.

## 6. Critical Analysis and Discussion

This manuscript brings an accessible joining of most relevant improvements made in the last two decades regarding the multi-material machining operations. Through several tables containing the state-of-the-art of different themes and respective SWOT analyses comparing the different research performed around the globe, this article serves as an aggregated encyclopedia, so that less time is needed to consult the necessary information on this theme. In order to show the opportunities for each of the main areas related to the machining process on FML and high-hardness materials, SWOT analyses have been developed. The first important issue explored in this paper relates to the FML manufacturing process. To obtain an FML with excellent mechanical properties, it is important to choose the right surface treatment in order to improve the adhesion between layers of dissimilar materials. To present the SWOT analysis, four papers were compared: Roth et al. [8], Drozdziel-Jurkiewicz and Bienia [22], Dieckhoff et al. [50] and Park et al. [51]. These authors were chosen due to the new approach of surface treatments on metal sheets to produce FMLs. The SWOT analysis of the surface treatment applied to FML stacks production is presented in Figure 16.

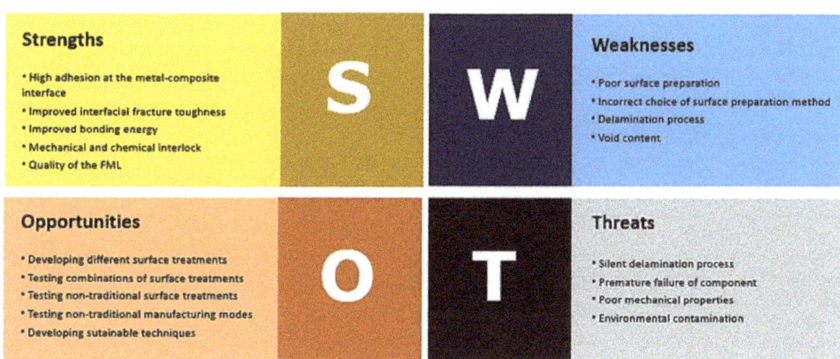



**Figure 16.** SWOT analysis showing the impact of surface treatment on the adhesion of dissimilar materials during the production of FML stacks.

The most common surface treatments are mechanical abrasion and chemical treatment or a combination of them. Generally, these treatments create a rougher surface promoting mechanical interlock, chemical interlock, or both. Thus, it is important to have a deep analysis about what parameters can influence the interfacial adhesion. Combination of surface treatments improves the bonding energy between the dissimilar materials. A mechanical surface treatment, followed by a chemical one, creates deeper valleys that improve mechanical interlock between the polymeric matrix and metal. Some chemical binders are also used to improve the chemical bonding between metal and composite. New surface treatment techniques, such as laser and plasma techniques with a sustainable bias, should be tested to replace conventional, less sustainable techniques, in order to protect the environment and reduce costs.

In a conventional manufacturing process using an autoclave, the void content can be reduced, and the mechanical properties can be improved, but the production costs are high using this manufacturing process. The thermoforming process can produce a high-quality workpiece with lower cost when compared with the autoclave process.

The second SWOT analysis is related to drilling and milling processes in high hardness materials (Figure 17). The results of two papers found in the literature, Barman et al. [63] and Bolar et al. [65], were discussed to elaborate the SWOT analysis of drilling and milling processes. The authors chose to compare the efficiency of these machining processes in the titanium alloy Ti6Al4V and in CARALL. These papers were chosen because they address and compare the drilling and milling techniques of these materials.

**Figure 17.** SWOT analysis obtained by comparing between drilling and milling processes in FML.

The helical milling process has superior response to high-hardness materials when compared to the drilling process in all parameters studied by the authors, such as machining forces, chip morphology, machining temperature, surface roughness, hole size and burr size. The drilling process is in danger of becoming obsolete compared to milling, so it is necessary to renew the drilling process, such as the development of new tools, new coatings, lubricants and coolants, as well as a more detailed study of the drilling process parameters.

The last one is a SWOT analysis on the use of lubricants/coolants in FML machining. The results of three papers found in the literature, Bertolini et al. [26], Kumar et al. [76] and Giasin et al. [78], were discussed. The authors chose to compare the efficiency of dry and cryogenic conditions during the FML drilling process of three different FML materials: GLARE, Ti/CFRP/Ti and Mg-based FML, respectively. All three papers indicated that cryogenic cooling is the best compared to MQL or dry cooling (Figure 18). Cryogenic cooling is also eco-friendly, as MQL and dry cooling are, but its advantages are better than the others. It promotes a hole quality in terms of lower average hole size, circularity error, perpendicularity error, and burr height due to a significant reduction in machining temperature and a substantial reduction in surface roughness. Although the use of cryogenic drills has shown satisfactory results, it is important to consider the type of drill selected in terms of geometry and coating, as well as the process parameters. In cryogenic machining, cutting forces and torque were more pronounced compared to dry and MQL environments. Poor choice of process parameters can lead to material failure during machining, such as delamination at the metal/composite interface, resulting in material waste and increased process costs.

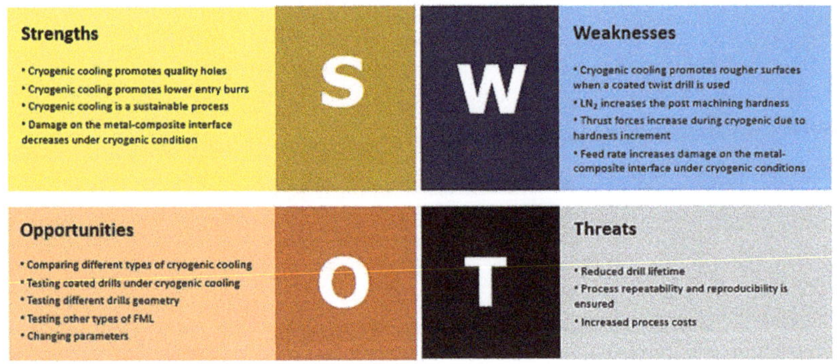


**Figure 18.** SWOT analysis of lubricants and coolants obtained by comparing between MQL, dry and cryogenic cooling.

Despite all the gathered information related to the subject in analysis, studied in detail and deepened as much as possible, this paper still possesses some limitations. Due to the considered restrictions imposed in its realization, such as the chosen databases, selected keywords, and delimited time period, there may have been eventual scientific articles that were not used for being outside these limits. Therefore, if a study was carried out before the 21st century, it was not included in this review. Nevertheless, the intention was to reassemble the data in the most up-to-date way possible, without making the mistake of aggregating information that is already outdated and surpassed by more recent studies, according to the established boundaries.

## 7. Conclusions

FMLs, or Metal Fiber Laminates, are the most sought-after materials in the aerospace and automotive industries, because they combine the mechanical properties of metallic materials, high mechanical resistance and composite materials, as well as presenting a high strength to weight ratio. Nevertheless, the continuous use of these materials in these industries requires mastery of production and finishing techniques.

To this end, it is essential to know how to select the most appropriate surface treatment process for the production of an FML. In the literature review, it was observed that there are conventional and non-conventional techniques, but the conventional techniques combined mechanical treatments by abrasion followed by electrochemical treatment of anodizing using potassium dichromate and ferric sulphate. These are still the ones that present the best results due to the production of pitting caused by the presence of sulphate ions.

By comparing the machining processes, drilling and milling, the quality of the finished part and the response of process parameters such as machining forces, chip morphology, machining temperature, surface roughness, hole size and burr size, it has been concluded that helical milling is by far the best process for machining FMLs or high-hardness materials, and may make drilling obsolete for application in these materials.

Lubrication and coolant applied during the machining process are critical factors in providing quality holes. Data from the literature have been compared and it has been found that the best lubricants/coolants are cryogenic, followed by MQL, and finally dry cooling. These three methods are designed to be sustainable, to produce no waste that pollutes the environment, to be non-toxic and to be affordable.

The tooling used also greatly influences the final quality of the hole. Literature data showed that the best drills to be used are double helix tools containing a CVD diamond-based coating in order to minimize axial forces.

Finally, the SWOT analysis discussed three main topics: surface treatment applied to the production of FML stacks, drilling and milling in FML and lubricants and coolants. The SWOT analysis confirmed what had already been analyzed and discussed in the literature.

Applications based on FMLs will grow substantially in the near future, due to the differentiating characteristics of these materials, both in terms of resistance and toughness as well as their light weight, which makes them particularly useful in mobility applications or where requirements of transport and assembly are demanding. In addition to all already referred, they are sustainable materials, as they present increased durability and due to their lower weight, they can contribute to a lower emission rate when used in mobility applications. The use of thermosetting matrices still represents a challenge in terms of recycling. Since a substantial increase in the consumption of these materials is foreseen, processing techniques need to evolve, hence, the pertinence of this work for researchers who are just starting out in this field of knowledge.

**Author Contributions:** Conceptualization, R.D.F.S.C., R.C.M.S.-C. and F.J.G.S.; methodology, R.C.M.S.-C. and F.J.G.S.; formal analysis, R.D.F.S.C., R.C.M.S.-C. and N.S.; investigation, R.C.M.S.-C.; resources, F.J.G.S.; data curation, R.D.F.S.C. and R.C.M.S.-C.; writing—original draft preparation, R.D.F.S.C. and R.C.M.S.-C.; writing—review and editing, R.C.M.S.-C., F.J.G.S. and N.S.; visualization, R.D.F.S.C., R.C.M.S.-C., F.J.G.S. and A.M.P.J., N.S.; supervision, F.J.G.S. and A.M.P.J.; project administration, F.J.G.S. and A.M.P.J.; funding acquisition, F.J.G.S. All authors have read and agreed to the published version of the manuscript.

**Funding:** This research work was developed under the "DRIVOLUTION—Transition to the factory of the future", with the reference DRIVOLUTION/BM/01/2023 research project, supported by European Structural and Investments Funds regarding the "Portugal 2020" program scope.

**Data Availability Statement:** Not applicable.

**Acknowledgments:** The authors thank the ISEP, UPorto/FEUP and Centro Paula Souza—FATEC-SJC for the infrastructure offered for the development of this project.

**Conflicts of Interest:** The authors declare no conflict of interest.

## References


1. Sinmazçelik, T.; Avcu, E.; Bora, M.Ö.; Çoban, O. A Review: Fibre Metal Laminates, Background, Bonding Types and Applied Test Methods. *Mater. Des.* **2011**, *32*, 3671–3685. [CrossRef]
2. Franz, G.; Vantomme, P.; Hassan, M.H. A Review on Drilling of Multilayer Fiber-Reinforced Polymer Composites and Aluminum Stacks: Optimization of Strategies for Improving the Drilling Performance of Aerospace Assemblies. *Fibers* **2022**, *10*, 78. [CrossRef]
3. Singh, A.P.; Sharma, M.; Singh, I. A Review of Modeling and Control during Drilling of Fiber Reinforced Plastic Composites. *Compos. B Eng.* **2013**, *47*, 118–125. [CrossRef]
4. Cortés, P.; Cantwell, W.J. The Prediction of Tensile Failure in Titanium-Based Thermoplastic Fibre-Metal Laminates. *Compos. Sci. Technol.* **2006**, *66*, 2306–2316. [CrossRef]
5. Wang, H.; Qin, X.; Li, H.; Tan, Y. A Comparative Study on Helical Milling of CFRP/Ti Stacks and Its Individual Layers. *Int. J. Adv. Manuf. Technol.* **2016**, *86*, 1973–1983. [CrossRef]
6. Baumert, E.K.; Johnson, W.S.; Cano, R.J.; Jensen, B.J.; Weiser, E.S. Fatigue Damage Development in New Fibre Metal Laminates Made by the VARTM Process. *Fatigue Fract. Eng. Mater. Struct.* **2011**, *34*, 240–249. [CrossRef]
7. Jin, K.; Wang, H.; Tao, J.; Du, D. Mechanical Analysis and Progressive Failure Prediction for Fibre Metal Laminates Using a 3D Constitutive Model. *Compos. Part A* **2019**, *124*, 105490. [CrossRef]
8. Roth, S.; Stoll, M.; Weidenmann, K.A.; Coutandin, S.; Fleischer, J. A New Process Route for the Manufacturing of Highly Formed Fiber-Metal-Laminates with Elastomer Interlayers (FMEL). *Int. J. Adv. Manuf. Technol.* **2019**, *104*, 1293–1301. [CrossRef]
9. Zhu, W.; Xiao, H.; Wang, J.; Fu, C. Characterization and Properties of AA6061-Based Fiber Metal Laminates with Different Aluminum-Surface Pretreatments. *Compos. Struct.* **2019**, *227*, 111321. [CrossRef]
10. Frankiewicz, M.; Ziółkowski, G.; Dziedzic, R.; Osiecki, T.; Scholz, P. Damage to Inverse Hybrid Laminate Structures: An Analysis of Shear Strength Test. *Mater. Sci.* **2022**, *40*, 130–144. [CrossRef]
11. Sinke, J. Manufacturing of GLARE Parts and Structures. *Appl. Compos. Mater.* **2003**, *10*, 293–305. [CrossRef]
12. Patil, N.A.; Mulik, S.S.; Wangikar, K.S.; Kulkarni, A.P. Characterization of Glass Laminate Aluminium Reinforced Epoxy—A Review. *Procedia Manuf.* **2018**, *20*, 554–562. [CrossRef]
13. Kazemi, M.E.; Shanmugam, L.; Yang, L.; Yang, J. A Review on the Hybrid Titanium Composite Laminates (HTCLs) with Focuses on Surface Treatments, Fabrications, and Mechanical Properties. *Compos. Part A Appl. Sci. Manuf.* **2020**, *128*, 105679. [CrossRef]
14. Sarasini, F.; Tirillò, J.; Ferrante, L.; Sergi, C.; Sbardella, F.; Russo, P.; Simeoli, G.; Mellier, D.; Calzolari, A. Effect of Temperature and Fiber Type on Impact Behavior of Thermoplastic Fiber Metal Laminates. *Compos. Struct.* **2019**, *223*, 110961. [CrossRef]
15. Cortés, P.; Cantwell, W.J. The Fracture Properties of a Fibre–Metal Laminate Based on Magnesium Alloy. *Compos. B Eng.* **2005**, *37*, 163–170. [CrossRef]
16. Pawar, O.A.; Gaikhe, Y.S.; Tewari, A.; Sundaram, R.; Joshi, S.S. Analysis of Hole Quality in Drilling GLARE Fiber Metal Laminates. *Compos. Struct.* **2015**, *123*, 350–365. [CrossRef]
17. Shirvanimoghaddam, K.; Hamim, S.U.; Karbalaei Akbari, M.; Fakhrhoseini, S.M.; Khayyam, H.; Pakseresht, A.H.; Ghasali, E.; Zabet, M.; Munir, K.S.; Jia, S.; et al. Carbon Fiber Reinforced Metal Matrix Composites: Fabrication Processes and Properties. *Compos. Part A Appl. Sci. Manuf.* **2017**, *92*, 70–96. [CrossRef]
18. Zhu, W.; Xiao, H.; Wang, J.; Li, X. Effect of Different Coupling Agents on Interfacial Properties of Fibre-Reinforced Aluminum Laminates. *Materials* **2021**, *14*, 1019. [CrossRef]
19. Muthu Chozha Rajan, B.; Senthil Kumar, A.; Sornakumar, T.; Senthamaraikannan, P.; Sanjay, M.R. Multi Response Optimization of Fabrication Parameters of Carbon Fiber-Reinforced Aluminium Laminates (CARAL): By Taguchi Method and Gray Relational Analysis. *Polym. Compos.* **2019**, *40*, E1041–E1048. [CrossRef]
20. Thirukumaran, M.; Jappes, J.T.W.; Siva, I.; Ramanathan, R.; Brintha, N.C. On the Interfacial Adhesion of Fiber Metal Laminates Using Surface Modified Aluminum 7475 Alloy for Aviation Industries—A Study. *J. Adhes Sci. Technol.* **2020**, *34*, 635–650. [CrossRef]
21. Kwon, D.J.; Kim, J.H.; Kim, Y.J.; Kim, J.J.; Park, S.M.; Kwon, I.J.; Shin, P.S.; DeVries, L.K.; Park, J.M. Comparison of Interfacial Adhesion of Hybrid Materials of Aluminum/Carbon Fiber Reinforced Epoxy Composites with Different Surface Roughness. *Compos. B Eng.* **2019**, *170*, 11–18. [CrossRef]
22. Droździel-Jurkiewicz, M.; Bieniaś, J. Evaluation of Surface Treatment for Enhancing Adhesion at the Metal–Composite Interface in Fibre Metal-Laminates. *Materials* **2022**, *15*, 6118. [CrossRef] [PubMed]
23. Santos, A.L.; Nakazato, R.Z.; Schmeer, S.; Botelho, E.C. Influence of Anodization of Aluminum 2024 T3 for Application in Aluminum/Cf/ Epoxy Laminate. *Compos. B Eng.* **2020**, *184*, 107718. [CrossRef]
24. Li, X.; Zhang, X.; Zhang, H.; Yang, J.; Nia, A.B.; Chai, G.B. Mechanical Behaviors of Ti/CFRP/Ti Laminates with Different Surface Treatments of Titanium Sheets. *Compos. Struct.* **2017**, *163*, 21–31. [CrossRef]
25. Süsler, S.; Bora, M.Ö.; Uçan, C.; Türkmen, H.S. The Effect of Surface Treatments on the Interlaminar Shear Failure of GLARE Laminate Included AA6061-T6 Layers by Comparing Failure Characteristics. *Compos. Interfaces* **2022**, *29*, 1–17. [CrossRef]
26. Bertolini, R.; Savio, E.; Ghiotti, A.; Bruschi, S. The Effect of Cryogenic Cooling and Drill Bit on the Hole Quality When Drilling Magnesium-Based Fiber Metal Laminates. *Procedia Manuf.* **2021**, *53*, 118–127. [CrossRef]
27. Robert, C.; Mamalis, D.; Obande, W.; Koutsos, V.; Brádaigh, C.M.Ó.; Ray, D. Interlayer Bonding between Thermoplastic Composites and Metals by In-Situ Polymerization Technique. *J. Appl. Polym. Sci.* **2021**, *138*, 51188. [CrossRef]

28. Parmar, H.; Gambardella, A.; Perna, A.S.; Viscusi, A.; della Gatta, R.; Tucci, F.; Astarita, A.; Carlone, P. Manufacturing and Metallization of Hybrid Thermoplastic-Thermoset Matrix Composites. In Proceedings of the ESAFORM 2021—24th International Conference on Material Forming, PoPuPS (University of LiFge Library), Online, 14–16 April 2021.

29. Banea, M.D.; Rosioara, M.; Carbas, R.J.C.; da Silva, L.F.M. Multi-Material Adhesive Joints for Automotive Industry. *Compos. B Eng.* **2018**, *151*, 71–77. [CrossRef]

30. Lambiase, F.; Balle, F.; Blaga, L.A.; Liu, F.; Amancio-Filho, S.T. Friction-Based Processes for Hybrid Multi-Material Joining. *Compos. Struct.* **2021**, *266*, 113828. [CrossRef]

31. Ding, Z.; Wang, H.; Luo, J.; Li, N. A Review on Forming Technologies of Fibre Metal Laminates. *Int. J. Lightweight Mater. Manuf.* **2021**, *4*, 110–126. [CrossRef]

32. Blala, H.; Lang, L.; Khan, S.; Alexandrov, S. Experimental and Numerical Investigation of Fiber Metal Laminate Forming Behavior Using a Variable Blank Holder Force. *Prod. Eng.* **2020**, *14*, 509–522. [CrossRef]

33. Heggemann, T.; Homberg, W. Deep Drawing of Fiber Metal Laminates for Automotive Lightweight Structures. *Compos. Struct.* **2019**, *216*, 53–57. [CrossRef]

34. Kalidass, K.; Raghavan, V. Numerical and Experimental Investigations on GFRP e AA 6061 Laminate Composites for Deep-Drawing Applications. *Mater. Tehnol.* **2022**, *56*, 107–114. [CrossRef]

35. Dariushi, S.; Rezadoust, A.M.; Kashizadeh, R. Effect of Processing Parameters on the Fabrication of Fiber Metal Laminates by Vacuum Infusion Process. *Polym. Compos.* **2019**, *40*, 4167–4174. [CrossRef]

36. Mamalis, D.; Obande, W.; Koutsos, V.; Blackford, J.R.; Brádaigh, C.M.Ó.; Ray, D. Novel Thermoplastic Fibre-Metal Laminates Manufactured by Vacuum Resin Infusion: The Effect of Surface Treatments on Interfacial Bonding. *Mater. Des.* **2019**, *162*, 331–344. [CrossRef]

37. Lakshmi Kala, K.; Prahlada Rao, K. Synthesis and Characterization of Fabricated Fiber Metal Laminates for Aerospace Applications. *Mater. Today Proc.* **2022**, *64*, 37–43. [CrossRef]

38. Harris, M.; Qureshi, M.A.M.; Saleem, M.Q.; Khan, S.A.; Bhutta, M.M.A. Carbon Fiber-Reinforced Polymer Composite Drilling via Aluminum Chromium Nitride-Coated Tools: Hole Quality and Tool Wear Assessment. *J. Reinf. Plast. Compos.* **2017**, *36*, 1403–1420. [CrossRef]

39. Giasin, K.; Ayvar-Soberanis, S. An Investigation of Burrs, Chip Formation, Hole Size, Circularity and Delamination during Drilling Operation of GLARE Using ANOVA. *Compos. Struct.* **2017**, *159*, 745–760. [CrossRef]

40. Giasin, K.; Ayvar-Soberanis, S.; French, T.; Phadnis, V. 3D Finite Element Modelling of Cutting Forces in Drilling Fibre Metal Laminates and Experimental Hole Quality Analysis. *Appl. Compos. Mater.* **2017**, *24*, 113–137. [CrossRef]

41. Kim, D.G.; Jung, Y.C.; Kweon, S.H.; Yang, S.H. Determination of the Optimal Milling Feed Direction for Unidirectional CFRPs Using a Predictive Cutting-Force Model. *Int. J. Adv. Manuf. Technol.* **2022**, *123*, 3571–3585. [CrossRef]

42. Patel, P.; Chaudhary, V. Damage Free Drilling of Carbon Fibre Reinforced Composites—A Review. *Aust. J. Mech. Eng.* **2021**, *370*, 1850–1870. [CrossRef]

43. El Etri, H.; Korkmaz, M.E.; Gupta, M.K.; Gunay, M.; Xu, J. A State-of-the-Art Review on Mechanical Characteristics of Different Fiber Metal Laminates for Aerospace and Structural Applications. *Int. J. Adv. Manuf. Technol.* **2022**, *123*, 2965–2991. [CrossRef]

44. Gutiérrez, J.C.H.; Campos Rubio, J.C.; de Faria, P.E.; Davim, J.P. Machining Behavior of Polymer Composites Materials for Automotive Applications. *Polimeros* **2014**, *24*, 711–719. [CrossRef]

45. Bader, B.; Türck, E.; Vietor, T. Multi material design. A current overview of the used potential in automotive industries. In *Technologies for Economical and Functional Lightweight Design*; Springer: Berlin/Heidelberg, Germany, 2019; pp. 3–13.

46. Jansson, A.; Pejryd, L. Dual-Energy Computed Tomography Investigation of Additive Manufacturing Aluminium–Carbon-Fibre Composite Joints. *Heliyon* **2019**, *5*, e01200. [CrossRef] [PubMed]

47. Min, J.; Hu, J.; Sun, C.; Wan, H.; Liao, P.; Teng, H.; Lin, J. Fabrication Processes of Metal-Fiber Reinforced Polymer Hybrid Components: A Review. *Adv. Compos. Hybrid. Mater.* **2022**, *5*, 651–678. [CrossRef]

48. Moreira, B.M.D.N.; Gouveia, R.M.; Silva, F.J.G.; Campilho, R.D.S.G. A Novel Concept of Production and Assembly Processes Integration. *Procedia Manuf.* **2017**, *11*, 1385–1395. [CrossRef]

49. Castro, T.A.M.; Silva, F.J.G.; Campilho, R.D.S.G. Optimising a Specific Tool for Electrical Terminals Crimping Process. *Procedia Manuf.* **2017**, *11*, 1438–1447. [CrossRef]

50. Dieckhoff, S.; Standfuß, J.; Pap, J.S.; Klotzbach, A.; Zimmermann, F.; Burchardt, M.; Regula, C.; Wilken, R.; Apmann, H.; Fortkamp, K.; et al. New Concepts for Cutting, Surface Treatment and Forming of Aluminium Sheets Used for Fibre-Metal Laminate Manufacturing. *CEAS Aeronaut J.* **2019**, *10*, 419–429. [CrossRef]

51. Park, S.Y.; Choi, W.J.; Choi, H.S.; Kwon, H. Effects of Surface Pre-Treatment and Void Content on GLARE Laminate Process Characteristics. *J. Mater. Process. Technol.* **2010**, *210*, 1008–1016. [CrossRef]

52. Cheng, F.; Hu, Y.; Zhang, X.; Hu, X.; Huang, Z. Adhesive Bond Strength Enhancing between Carbon Fiber Reinforced Polymer and Aluminum Substrates with Different Surface Morphologies Created by Three Sulfuric Acid Solutions. *Compos. Part A Appl. Sci. Manuf.* **2021**, *146*, 106427. [CrossRef]

53. Parodo, G.; Rubino, F.; Sorrentino, L.; Turchetta, S. Temperature Analysis in Fiber Metal Laminates Drilling: Experimental and Numerical Results. *Polym. Compos.* **2022**, *43*, 7600–7615. [CrossRef]

54. Zitoune, R.; Krishnaraj, V.; Collombet, F.; le Roux, S. Experimental and Numerical Analysis on Drilling of Carbon Fibre Reinforced Plastic and Aluminium Stacks. *Compos. Struct.* **2016**, *146*, 148–158. [CrossRef]

55. Dandekar, C.R.; Shin, Y.C. Modeling of Machining of Composite Materials: A Review. *Int. J. Mach. Tools Manuf.* **2012**, *57*, 102–121. [CrossRef]

56. Teti, R. Machining of Composite Materials. *CIRP Ann.* **2002**, *51*, 611–634. [CrossRef]

57. Krishnaraj, V.; Zitoune, R.; Collombet, F.; Davim, J.P. Challenges in Drilling of Multi-Materials. *Mater. Sci. Forum* **2013**, *763*, 145–168. [CrossRef]

58. Kumar, D.; Gururaja, S.; Jawahir, I.S. Machinability and Surface Integrity of Adhesively Bonded Ti/CFRP/Ti Hybrid Composite Laminates under Dry and Cryogenic Conditions. *J. Manuf. Process.* **2020**, *58*, 1075–1087. [CrossRef]

59. Azwan, S.; Sahira, N.I.; Abdullah, M.R.; Yahya, M.Y. Investigation on the Effect of Drilling Parameters on Tensile Loading of Fibre-Metal Laminates. In *Proceedings of the IOP Conference Series: Materials Science and Engineering*; IOP Publishing Ltd.: Bristol, UK, 2019; Volume 670.

60. Zitoune, R.; Krishnaraj, V.; Collombet, F. Study of Drilling of Composite Material and Aluminium Stack. *Compos. Struct.* **2010**, *92*, 1246–1255. [CrossRef]

61. Denkena, B.; Boehnke, D.; Dege, J.H. Helical Milling of CFRP-Titanium Layer Compounds. *CIRP J. Manuf. Sci. Technol.* **2008**, *1*, 64–69. [CrossRef]

62. Pereira, R.B.D.; Brandão, L.C.; de Paiva, A.P.; Ferreira, J.R.; Davim, J.P. A Review of Helical Milling Process. *Int. J. Mach. Tools Manuf.* **2017**, *120*, 27–48. [CrossRef]

63. Barman, A.; Adhikari, R.; Bolar, G. Evaluation of Conventional Drilling and Helical Milling for Processing of Holes in Titanium Alloy Ti6Al4V. *Mater. Today Proc.* **2020**, *28*, 2295–2300. [CrossRef]

64. Hemant, K.; Kona, A.; Karthik, S.A.; Bolar, G. Experimental Investigation into Helical Hole Milling of Fiber Metal Laminates. *Mater. Today Proc.* **2020**, *27*, 208–216. [CrossRef]

65. Bolar, G.; Sridhar, A.K.; Ranjan, A. Drilling and Helical Milling for Hole Making in Multi-Material Carbon Reinforced Aluminum Laminates. *Int. J. Lightweight Mater. Manuf.* **2022**, *5*, 113–125. [CrossRef]

66. Aamir, M.; Giasin, K.; Tolouei-Rad, M.; Vafadar, A. A Review: Drilling Performance and Hole Quality of Aluminium Alloys for Aerospace Applications. *J. Mater. Res. Technol.* **2020**, *9*, 12484–12500. [CrossRef]

67. Iyer, R.; Koshy, P.; Ng, E. Helical Milling: An Enabling Technology for Hard Machining Precision Holes in AISI D2 Tool Steel. *Int. J. Mach. Tools Manuf.* **2007**, *47*, 205–210. [CrossRef]

68. Wang, C.Y.; Chen, Y.H.; An, Q.L.; Cai, X.J.; Ming, W.W.; Chen, M. Drilling Temperature and Hole Quality in Drilling of CFRP/Aluminum Stacks Using Diamond Coated Drill. *Int. J. Precis. Eng. Manuf.* **2015**, *16*, 1689–1697. [CrossRef]

69. Giasin, K. The Effect of Drilling Parameters, Cooling Technology, and Fiber Orientation on Hole Perpendicularity Error in Fiber Metal Laminates. *Int. J. Adv. Manuf. Technol.* **2018**, *97*, 4081–4099. [CrossRef]

70. Shyha, I.S.; Soo, S.L.; Aspinwall, D.K.; Bradley, S.; Perry, R.; Harden, P.; Dawson, S. Hole Quality Assessment Following Drilling of Metallic-Composite Stacks. *Int. J. Mach. Tools Manuf.* **2011**, *51*, 569–578. [CrossRef]

71. Boubekri, N.; Shaikh, V. Minimum Quantity Lubrication (MQL) in Machining: Benefits and Drawbacks. *J. Ind. Intell. Inf.* **2014**, *3*, 205–209. [CrossRef]

72. Biermann, D.; Hartmann, H. Reduction of Burr Formation in Drilling Using Cryogenic Process Cooling. *Procedia CIRP* **2012**, *3*, 85–90. [CrossRef]

73. Bertolini, R.; Alagan, N.T.; Gustafsson, A.; Savio, E.; Ghiotti, A.; Bruschi, S. Ultrasonic Vibration and Cryogenic Assisted Drilling of Aluminum-CFRP Composite Stack-An Innovative Approach. *Procedia CIRP* **2022**, *108*, 94–99. [CrossRef]

74. Liu, H.; Birembaux, H.; Ayed, Y.; Rossi, F.; Poulachon, G. Recent Advances on Cryogenic Assistance in Drilling Operation: A Critical Review. *J. Manuf. Sci. Eng.* **2022**, *144*, 100801. [CrossRef]

75. Pal, A.; Chatha, S.S.; Sidhu, H.S. Performance Evaluation of the Minimum Quantity Lubrication with $Al_2O_3$- Mixed Vegetable-Oil-Based Cutting Fluid in Drilling of AISI 321 Stainless Steel. *J. Manuf. Process.* **2021**, *66*, 238–249. [CrossRef]

76. Kumar, D.; Gururaja, S. Investigation of Hole Quality in Drilled Ti/CFRP/Ti Laminates Using $CO_2$ Laser. *Opt. Laser Technol.* **2020**, *126*, 106130. [CrossRef]

77. Giasin, K.; Ayvar-Soberanis, S.; Hodzic, A. The Effects of Minimum Quantity Lubrication and Cryogenic Liquid Nitrogen Cooling on Drilled Hole Quality in GLARE Fibre Metal Laminates. *Mater. Des.* **2016**, *89*, 996–1006. [CrossRef]

78. Bonhin, E.P.; David-Müzel, S.; Guidi, E.S.; Botelho, E.C.; Ribeiro, M.V. Influence of Drilling Parameters on Thrust Force and Burr on Fiber Metal Laminate (Al 2024-T3/Glass Fiber Reinforced Epoxy). *Procedia CIRP* **2021**, *101*, 338–341. [CrossRef]

79. Pejryd, L.; Beno, T.; Carmignato, S. Computed Tomography as a Tool for Examining Surface Integrity in Drilled Holes in CFRP Composites. *Procedia CIRP* **2014**, *13*, 43–48. [CrossRef]

80. Saoudi, J.; Zitoune, R.; Mezlini, S.; Gururaja, S.; Seitier, P. Critical Thrust Force Predictions during Drilling: Analytical Modeling and X-ray Tomography Quantification. *Compos. Struct.* **2016**, *153*, 886–894. [CrossRef]

81. Saoudi, J.; Zitoune, R.; Gururaja, S.; Salem, M.; Mezleni, S. Analytical and Experimental Investigation of the Delamination during Drilling of Composite Structures with Core Drill Made of Diamond Grits: X-ray Tomography Analysis. *J. Compos. Mater.* **2018**, *52*, 1281–1294. [CrossRef]

82. Hocheng, H.; Tsao, C.C. Computerized Tomography and C-Scan for Measuring Drilling-Induced Delamination in Composite Material Using Twist Drill and Core Drill. *Key Eng. Mater.* **2007**, *339*, 16–20. [CrossRef]

83. Álvarez, M.; Salguero, J.; Sánchez, J.A.; Huerta, M.; Marcos, M. SEM and EDS Characterisation of Layering TiOx Growth onto the Cutting Tool Surface in Hard Drilling Processes of Ti-Al-V Alloys. *Adv. Mater. Sci. Eng.* **2011**, *2011*, 414868. [CrossRef]

84. Nguyen-Dinh, N.; Zitoune, R.; Bouvet, C.; Leroux, S. Surface Integrity While Trimming of Composite Structures: X-ray Tomography Analysis. *Compos. Struct.* **2019**, *210*, 735–746. [CrossRef]

85. Wang, G.D.; Melly, S.K.; Li, N. Experimental Studies on a Two-Step Technique to Reduce Delamination Damage during Milling of Large Diameter Holes in CFRP/Al Stack. *Compos. Struct.* **2018**, *188*, 330–339. [CrossRef]

86. Ashrafi, S.A.; Sharif, S.; Farid, A.A.; Yahya, M.Y. Performance Evaluation of Carbide Tools in Drilling CFRP-Al Stacks. *J. Compos. Mater.* **2014**, *48*, 2071–2084. [CrossRef]

87. Giasin, K.; Gorey, G.; Byrne, C.; Sinke, J.; Brousseau, E. Effect of Machining Parameters and Cutting Tool Coating on Hole Quality in Dry Drilling of Fibre Metal Laminates. *Compos. Struct.* **2019**, *212*, 159–174. [CrossRef]

88. John, K.M.; Kumaran, S.T.; Kurniawan, R.; Moon Park, K.; Byeon, J.H. Review on the Methodologies Adopted to Minimize the Material Damages in Drilling of Carbon Fiber Reinforced Plastic Composites. *J. Reinf. Plast. Compos.* **2019**, *38*, 351–368. [CrossRef]

89. Karpat, Y.; Polat, N. Mechanistic Force Modeling for Milling of Carbon Fiber Reinforced Polymers with Double Helix Tools. *CIRP Ann. Manuf. Technol.* **2013**, *62*, 95–98. [CrossRef]

90. Kuo, C.; Wang, C.; Ko, S. Wear Behaviour of CVD Diamond-Coated Tools in the Drilling of Woven CFRP Composites. *Wear* **2018**, *398–399*, 1–12. [CrossRef]

91. Bi, G.; Wang, F.; Fu, R.; Chen, P. Wear Characteristics of Multi-Tooth Milling Cutter in Milling CFRP and Its Impact on Machining Performance. *J. Manuf. Process.* **2022**, *81*, 580–593. [CrossRef]

92. Yashiro, T.; Ogawa, T.; Sasahara, H. Temperature Measurement of Cutting Tool and Machined Surface Layer in Milling of CFRP. *Int. J. Mach. Tools Manuf.* **2013**, *70*, 63–69. [CrossRef]

93. An, Q.; Dang, J.; Li, J.; Wang, C.; Chen, M. Investigation on the Cutting Responses of CFRP/Ti Stacks: With Special Emphasis on the Effects of Drilling Sequences. *Compos. Struct.* **2020**, *253*, 112794. [CrossRef]

94. Casais, R.; Baptista, A.M.; Silva, F.J.; Andrade, F.; Sousa, V.; Marques, M.J. Experimental Study on the Wear Behavior of B4C and TiB$_2$ Monolayered PVD Coatings under High Contact Loads. *Int. J. Adv. Manuf. Technol.* **2022**, *120*, 6585–6604. [CrossRef]

95. Silva, F.; Martinho, R.; Andrade, M.; Baptista, A.; Alexandre, R. Improving the Wear Resistance of Moulds for the Injection of Glass Fibre-Reinforced Plastics Using PVD Coatings: A Comparative Study. *Coatings* **2017**, *7*, 28. [CrossRef]

96. Montoya, M.; Calamaz, M.; Gehin, D.; Girot, F. Evaluation of the Performance of Coated and Uncoated Carbide Tools in Drilling Thick CFRP/Aluminium Alloy Stacks. *Int. J. Adv. Manuf. Technol.* **2013**, *68*, 2111–2120. [CrossRef]

97. Giasin, K.; Hawxwell, J.; Sinke, J.; Dhakal, H.; Köklü, U.; Brousseau, E. The Effect of Cutting Tool Coating on the Form and Dimensional Errors of Machined Holes in GLARE®Fibre Metal Laminates. *Int. J. Adv. Manuf. Technol.* **2020**, *107*, 2817–2832. [CrossRef]

98. Silva, F.J.G.; Fernandes, A.J.S.; Costa, F.M.; Baptista, A.P.M.; Pereira, E. Unstressed PACVD Diamond Films on Steel Pre-Coated with a Composite Multilayer. *Surf. Coat. Technol.* **2005**, *191*, 102–107. [CrossRef]

99. Martinho, R.P.; Silva, F.J.G.; Baptista, A.P.M. Cutting Forces and Wear Analysis of Si$_3$N$_4$ Diamond Coated Tools in High Speed Machining. *Vacuum* **2008**, *82*, 1415–1420. [CrossRef]

100. Silva, F.J.G.; Casais, R.C.B.; Baptista, A.P.M.; Marques, M.J.; Sousa, V.M.C.; Alexandre, R. Comparative Study of the Wear Behavior of B4C Monolayered and CrN/CrCN/DLC Multilayered Physical Vapor Deposition Coatings Under High Contact Loads: An Experimental Analysis. *J. Tribol.* **2022**, *144*, 031701. [CrossRef]

101. Ciecieląg, K.; Zaleski, K. Comparative Study In The Passive Force And Cutting Torque In The Milling Process Of Polymer Matrix Composites And Aluminum Alloys. *Adv. Sci. Technol.—Res. J.* **2013**, *7*, 6–12. [CrossRef]

102. Barik, T.; Pal, K. Prediction of TiAlN- and TiN-Coated Carbide Tool Wear in Drilling of Bidirectional CFRP Laminates Using Wavelet Packets of Thrust–Torque Signatures. *J. Braz. Soc. Mech. Sci. Eng.* **2022**, *44*, 364. [CrossRef]

103. Romoli, L.; Lutey, A.H.A. Quality Monitoring and Control for Drilling of CFRP Laminates. *J. Manuf. Process.* **2019**, *40*, 16–26. [CrossRef]

104. Kumar, M.S.; Prabukarthi, A.; Krishnaraj, V. Study on Tool Wear and Chip Formation during Drilling Carbon Fiber Reinforced Polymer (CFRP)/Titanium Alloy (Ti6Al4V) Stacks. *Procedia Eng.* **2013**, *64*, 582–592.

105. D'Orazio, A.; El Mehtedi, M.; Forcellese, A.; Nardinocchi, A.; Simoncini, M. Tool Wear and Hole Quality in Drilling of CFRP/AA7075 Stacks with DLC and Nanocomposite TiAlN Coated Tools. *J. Manuf. Process.* **2017**, *30*, 582–592. [CrossRef]

106. Benezech, L.; Landon, Y.; Rubio, W. Study of Manufacturing Defects and Tool Geometry Optimisation for Multi-Material Stack Drilling. *Adv. Mat. Res.* **2012**, *423*, 1–11. [CrossRef]

107. Biermann, D.; Bathe, T.; Rautert, C. Core Drilling of Fiber Reinforced Materials Using Abrasive Tools. *Procedia CIRP* **2017**, *66*, 175–180. [CrossRef]

108. Peralta, M.E.; Marcos, M.; Aguayo, F.; Lama, J.R.; Córdoba, A. Sustainable Fractal Manufacturing: A New Approach to Sustainability in Machining Processes. *Procedia Eng.* **2015**, *132*, 926–933. [CrossRef]

109. Lu, T.; Kudaravalli, R.; Georgiou, G. Cryogenic Machining through the Spindle and Tool for Improved Machining Process Performance and Sustainability: Pt. I, System Design. *Procedia Manuf.* **2018**, *21*, 266–272. [CrossRef]

110. Pušavec, F.; Stoić, A.; Kopač, J. Sustainable Machining Process—Myth or Reality. *Strojarstvo* **2010**, *52*, 197–204.

111. Nagaraj, A.; Uysal, A.; Jawahir, I.S. An Investigation of Process Performance When Drilling Carbon Fiber Reinforced Polymer (CFRP) Composite under Dry, Cryogenic and MQL Environments. *Procedia Manuf.* **2020**, *43*, 551–558. [CrossRef]

112. Sharma, A.K.; Tiwari, A.K.; Dixit, A.R. Effects of Minimum Quantity Lubrication (MQL) in Machining Processes Using Conventional and Nanofluid Based Cutting Fluids: A Comprehensive Review. *J. Clean. Prod.* **2016**, *127*, 1–18. [CrossRef]

113. Hegab, H.A.; Darras, B.; Kishawy, H.A. Towards Sustainability Assessment of Machining Processes. *J. Clean. Prod.* **2018**, *170*, 694–703. [CrossRef]

114. Lv, L.; Deng, Z.; Liu, T.; Wan, L.; Huang, W.; Yin, H.; Zhao, T. A Composite Evaluation Model of Sustainable Manufacturing in Machining Process for Typical Machine Tools. *Processes* **2019**, *7*, 110. [CrossRef]

*Review*

# A Comprehensive Review on the Conventional and Non-Conventional Machining and Tool-Wear Mechanisms of INCONEL®

A. F. V. Pedroso [1], V. F. C. Sousa [1], N. P. V. Sebbe [1], F. J. G. Silva [1,2,*], R. D. S. G. Campilho [1,2], R. C. M. Sales-Contini [1,3] and A. M. P. Jesus [2,4]

[1] ISEP, Polytechnic Institute of Porto, R. Dr. António Bernardino de Almeida, 4249-015 Porto, Portugal

[2] Associate Laboratory for Energy, Transports and Aerospace (LAETA-INEGI), Rua Dr Roberto Frias, 400, 4200-465 Porto, Portugal

[3] Aeronautical Structures Laboratory, Faculdade de Tecnologia de São José dos Campos Prof. Jessen Vidal, Centro Paula Souza, São José dos Campos, 1350 Distrito Eugênio de Melo, São José dos Campos 12247-014, SP, Brazil

[4] Faculty of Engineering, Department of Mechanical Engineering, University of Porto, Rua Dr. Roberto Frias, 400, 4200-465 Porto, Portugal

* Correspondence: fgs@isep.ipp.pt; Tel.: +351-228340500

**Citation:** Pedroso, A.F.V.; Sousa, V.F.C.; Sebbe, N.P.V.; Silva, F.J.G.; Campilho, R.D.S.G.; Sales-Contini, R.C.M.; Jesus, A.M.P. A Comprehensive Review on the Conventional and Non-Conventional Machining and Tool-Wear Mechanisms of INCONEL®. *Metals* **2023**, *13*, 585. https://doi.org/10.3390/met13030585

Academic Editor: Yadir Torres Hernández

Received: 14 February 2023
Revised: 2 March 2023
Accepted: 9 March 2023
Published: 13 March 2023

**Abstract:** Nickel-based superalloys, namely INCONEL® variants, have had an increase in applications throughout various industries like aeronautics, automotive and energy power plants. These superalloys can withstand high-temperature applications without suffering from creep, making them extremely appealing and suitable for manufactured goods such as jet engines or steam turbines. Nevertheless, INCONEL® alloys are considered difficult-to-cut materials, not only due to their superior material properties but also because of their poor thermal conductivity ($k$) and severe work hardening, which may lead to premature tool wear (TW) and poor final product finishing. In this regard, it is of paramount importance to optimise the machining parameters, to strengthen the process performance outcomes concerning the quality and cost of the product. The present review aims to systematically summarize and analyse the progress taken within the field of INCONEL® machining sensitively over the past five years, with some exceptions, and present the most recent solutions found in the industry, as well as the prospects from researchers. To accomplish this article, ScienceDirect, Springer, Taylor & Francis, Wiley and ASME have been used as sources of information as a result of great fidelity knowledge. Books from Woodhead Publishing Series, CRC Press and Academic Press have been also used. The main keywords used in searching information were: "Nickel-based superalloys", "INCONEL® 718", "INCONEL® 625" "INCONEL® Machining processes" and "Tool-wear mechanisms". The combined use of these keywords was crucial to filter the huge information currently available about the evolution of INCONEL® machining technologies. As a main contribution to this work, three SWOT analyses are provided on information that is dispersed in several articles. It was found that significant progress in the traditional cutting tool technologies has been made, nonetheless, the machining of INCONEL® 718 and 625 is still considered a great challenge due to the intrinsic characteristics of those Ni-based-superalloys, whose machining promotes high-wear to the tools and coatings used.

**Keywords:** nickel-based superalloys; INCONEL® 718; INCONEL® 625; INCONEL® machining processes and tool-wear mechanisms

## 1. Introduction

With the increasing requirement to achieve the best thermal efficiency in the field of aeronautics [1,2] and energy power plants steam turbines [3], applications in which aluminium and steel would succumb to creep [4] as a result of thermally induced crystal vacancies [5], nickel-based (Table 1) alloys became a very attractive solution for high-temperature operation [6–9]. Ni-Cr-Fe superalloys (Figure 1, blue zone), better known as INCONEL® (trademark registered by the International Nickel Company of Delaware and New York [10]), are materials resistant to oxidation, caustic and high-purity water corrosion, and stress-corrosion cracking (SCC) [11], optimal for service in extreme environments subjected to high mechanical loads [12], within numerous applications and characteristics (Table 2).

**Table 1.** Some physical properties of nickel (adapted from [7]).

| Characteristic | | Value | Units |
|---|---|---|---|
| Z | | 58.71 | AMU |
| Crystal structure | | FCC | [-] |
| Lattice constant | @ 25 °C | 0.35238 | nm |
| $\rho$ | | 8908 | kg/m$^3$ |
| $T_m$ | | 1453 | °C |
| $T_C$ | | 353 | |
| $c_p$ | | 0.44 | kJ/kg K |
| $\alpha$ | | $13.3 \times 10^{-6}$ | K$^{-1}$ |
| | @ 100 °C | 82.8 | |
| k | @ 300 °C | 63.6 | W/m K |
| | @ 500 °C | 61.9 | |
| $\rho_R$ | @ 20 °C | $6.97 \times 10^{-8}$ | $\Omega$m |
| $B_{max}$ | | 0.617 | |
| $B_r$ | | 0.300 | T |
| $H_C$ | | 239 | A/m |
| E | | 206.0 | |
| G | | 73.6 | GPa |
| $\nu$ | | 0.30 | [-] |

Caption: $B_{max}$—saturation magnetization; $B_r$—residual magnetization; $c_p$—specific heat at constant pressure; E—Young's Modulus; FCC—face-centred cubic; G—shear modulus; $H_C$—coercive force; $T_C$—Curie temperature; $T_m$—melting temperature; Z—atomic mass; $\alpha$—thermal expansion coefficient; $\rho$—volumetric mass density; $\rho_R$—electrical resistivity, $\nu$—Poisson's coefficient.

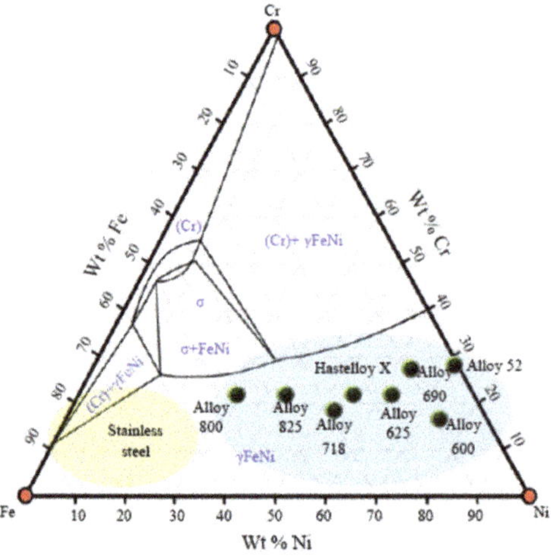


**Figure 1.** Fe-Ni-Cr ternary phase diagram [13] (Caption: *wt%*—element weight percentage).

A brief insight is provided with the most known alloys, including the INCONEL® 600, a Ni-Cr alloy that offers high levels of resistance to several corrosive elements. In high-temperature situations, INCONEL® 600 will not succumb to Cl-ion SCC or general oxidation, but it can still undergo corrosion by sulphuration deterioration in the high-temperature flue gas. This was a research topic by Wei, et al. [14]. INCONEL® 600 also suffers from severe hydrogen embrittlement at 250 °C [15]. Nonetheless, this alloy is recommended for use in furnace components and chemical processing equipment [16]. Moreover, INCONEL® 600 is also effectively used in the food industry and nuclear engineering due to its constant crystalline structure in applications that would cause permanent distortion to other alloys [17].

**Table 2.** Summary of applications and characteristics of some nickel-based superalloys (adapted from [18,19]).

| Superalloy | Sub-Grouped Material | Industry Applications | Characteristics |
|---|---|---|---|
| Ni-based alloys | INCONEL® (587, 597, 600, 601, 617, 625, 706, 690, 718, X750, 901) | Aircraft motors, nuclear reactors, gas turbines, spacecraft, pumps, furnaces, heat-treating equipment, petrochemical processing equipment, chemical processing, submarine, and manufacturing industry. | 600: Solid solution strengthened; 625: Acid resistant, good weldability; 690: Low cobalt content for nuclear applications, and low resistivity; 718: Gamma phase ($\gamma'$) double prime solution strengthened with good weldability. |
| | INCONEL® 722 | Acids in the chemical industry | - |
| | INCONEL® 751 | - | Increased Al content for improved failure strength in the 870 °C range. |
| | INCONEL® 792 | - | Increased Al content for improved high-temperature corrosion properties, especially used in gas turbines. |
| | INCONEL® 903 | Petrochemical tubing. | - |
| | INCONEL® 939 | - | $\gamma'$ prime strengthened with good weldability. |

The INCONEL® 601 alloy, similarly to INCONEL® 600, resists various forms of high-temperature corrosion and oxidization [20]. Nevertheless, this Ni-Cr alloy has an addition of aluminium which results in higher mechanical properties, even in extremely hot environments. INCONEL® 601 can prevent the significant strains ($\varepsilon$) that would appear under operating loads when exposed to high-temperature environments. The applications go from the use in furnaces to heat-treating equipment like retorts, baskets and gas-turbine components [21] to petrochemical processing equipment. The INCONEL® 625, which will have a special focus during this study, is a rare alloy that gains strength without having to undergo an extensive strengthening heat treatment [22]. It is a Ni-Cr-Mo alloy with an addition of Nb. The Nb reacts with Mo, causing the alloy's matrix to stiffen and increase its strength [23]. Like most INCONEL® alloys, the INCONEL® 625 has high resistance to several corrosive elements [24], withstanding harsh environments that would severely affect the performance of other alloys. This alloy is particularly effective when it comes to staving-off crevice corrosion and pitting. The INCONEL® 625 is a versatile alloy that is effectively used in the marine engineering, aerospace, chemical, and energy industries, among other applications [25,26]. The INCONEL® 690 alloy, unlike others in the group, is a high Ni and Cr alloy (Cr gives it particularly strong resistance to corrosion [15,27] that occurs in aqueous atmospheres). Along with its ability to resist the corrosion caused by oxidizing acids and salts, INCONEL® 690 can also withstand the sulfidation that takes place at extremely high temperatures ($T$). One of the most known INCONEL® alloys is the 718 alloy. This alloy, along with the formerly mentioned 625 alloy, will also have a special focus. INCONEL® 718 differs from other INCONEL® variants in structure and response, since it is designed for operation at $T \leq 650\,°C$ [28]. The 718 alloy is obtained by precipitation hardening [29,30]. It contains substantial levels of Fe, Mo, and Nb, as well as trace amounts of Ti and Al [31]. It has good weldability, which is not matched by most INCONEL® alloys [32], and combines anti-corrosive elements with a high level of strength and flexibility. It is particularly resistant to post-weld cracking, maintaining its structure in both high-temperature and aqueous environments as well, being most widely used in different industries, such as petrochemical, aeronautics, energy, and aerospace [33]. INCONEL® alloys tend to form a thick and stable passivating oxide layer to protect the surface from further attack, retaining strength over a wide $T$ range, making INCONEL® an attractive material for high-temperature applications [34]. The strength at high-temperatures of INCONEL® alloys may be developed by solid solution strengthening or precipitation strengthening, depending on the alloy [35]. In those processes, small amounts of niobium combine with nickel to form the intermetallic compound $Ni_3Nb$ or $\gamma$-prime, which consists of small cubic crystals that inhibit slip and creep effectively at elevated $T$ [36].

Concerning the roles of major phases or composition elements that contribute to the INCONEL® alloys, an overview is presented in Table 3 on the effects of Ni-alloying to better comprehend the compositions presented in Table 4.

**Table 3.** Major roles of solutes in different types of INCONEL® alloys [26].

| Element | Fe-Base | Co-Base | Ni-Base |
|---|---|---|---|
| Cr | Improves hot corrosion and oxidation resistance. Solid-solution hardening. | $M_{23}C_6$ and $M_7C_3$ carbide precipitation ($M_xC_y$ metallic carbide). Improves hot corrosion and oxidation resistance. Promotes Topologically Close-Packed (TCP) phases, also called Frank–Kasper phases. | $M_{23}C_6$ and $M_7C_3$ carbide precipitation. Improves hot corrosion and oxidation resistance. Moderate solid-solution hardening. A moderate increase in $\gamma'$ volume fraction ($vt\%$). Tends to stabilize the $Ni_2Cr$ phase in alloys containing more than 20% Cr. Promotes TCP phases. |
| Al | Induces $\gamma'$ precipitation. Retards formation of hexagonal $\eta$-$Ni_3Ti$ phase. | Improves oxidation resistance. Forms intermetallic $\beta$-CoAl. | Moderate solid-solution hardening. Induces $\gamma'$ precipitation. Improves oxidation resistance. |
| Ti | $\gamma'$ precipitation. TiC carbide precipitation. | TiC carbide precipitation. Formation of $Co_3Ti$ intermetallic. Formation of $Ni_3Ti$ with sufficient Ni. Reduces surface stability. | Moderate solid-solution hardening. $\gamma'$ precipitation. TiC carbide precipitation. Retards the precipitation of $Ni_2(Cr, Mo)$ phase particles. |
| Mo | Solid solution hardening. Forms $M_6C$ carbide precipitates. | Solid solution hardening. Forms $Co_3Mo$ intermetallic precipitates. Promotes TCP phases. | High solid-solution hardening. A moderate increase in $\gamma'$ $vt\%$. $M_6C$ and MC carbide formation. Promotes formation of $Ni_2(Cr, Mo)$ phase particles. Promotes $\sigma$ and $\mu$-TCP phases. |
| W | Solid solution hardening. $M_6C$ carbide precipitation. | Solid solution hardening. Formation of $Co_3W$ intermetallic. Promotes TCP phases. | High-solid solution hardening. A moderate increase in $\gamma'$ $vt\%$. $M_6C$ carbide formation. Increases $\rho$. Promotes the formation of $Ni_2(Cr, Mo, W)$ particles. Promotes $\sigma$ and $\mu$-TCP phases. |
| Ta | $\gamma''$ precipitation. Forms TaC carbide precipitates. | $M_6C$ and MC carbide precipitation. Formation of $Co_2Ta$ intermetallic. Reduces surface stability. | High-solid solution hardening. TaC carbide precipitation. A large increase in $\gamma'$ $vt\%$. Improves oxidation resistance. |

**Table 3.** *Cont.*

| Element | Fe-Base | Co-Base | Ni-Base |
|---|---|---|---|
| Nb | $\gamma''$ precipitation.<br>NbC carbide precipitation.<br>$\delta$-Ni$_3$Nb precipitation. | M$_6$C and MC carbide precipitation.<br>Formation of Co$_2$Nb intermetallic.<br>Reduces surface stability. | High-solid solution hardening.<br>A large increase in $\gamma'$ $vt\%$.<br>NbC carbide formation.<br>$\gamma''$ precipitation.<br>$\delta$-Ni$_3$Nb precipitation. |
| Re | - | - | Moderate solid-solution hardening.<br>Increases $\gamma/\gamma'$ lattice mismatch.<br>Retards coarsening.<br>Decreases oxidation resistance. |
| Fe | Not applicable. | Improves workability. | Promotes $\sigma$ and Laves TCP phases.<br>Improves workability.<br>Raises $\gamma$ solidus $T$. |
| Co | - | Not applicable. | A moderate increase in $\gamma'$ $vt\%$ in some alloys.<br>Raises $\gamma'$ solvus $T$. |
| Ni | FCC matrix stabilizer.<br>Inhibits TCP phase precipitation. | FCC stabilizer.<br>Decreases hot corrosion resistance. | Not applicable. |
| C | Carbide formation.<br>Stabilizes FCC matrix. | Carbide formation.<br>Decreases ductility. | Carbide formation.<br>Moderate solid-solution Hardening. |
| B | Improves creep strength and ductility.<br>Retards formation of grain-boundary $\eta$-Ni$_3$Ti | Improves creep strength.<br>and ductility | Moderate solid-solution hardening.<br>Inhibits carbide coarsening.<br>Improves grain-boundary strength.<br>Improves creep strength and ductility |
| Zr | Improves creep strength and ductility.<br>Retards formation of grain-boundary $\eta$-Ni$_3$Ti. | ZrC carbide formation.<br>Improves creep strength and ductility.<br>Reduces surface stability. | Moderate solid-solution hardening.<br>Inhibits carbide coarsening.<br>Improves grain-boundary strength.<br>Improves creep strength and ductility. |
| Hf | - | - | Improves creep strength and ductility.<br>Improves grain-boundary strength.<br>HfC formation.<br>Promotes eutectic $\gamma/\gamma'$ formation. |
| V | Improves notch ductility at elevated $T$.<br>Improves hot workability. | | Imparts extra passivation to some alloys in certain liquid media. |

**Table 4.** Chemical composition of relevant INCONEL® alloys.

| INCONEL® | | 182 [37] | 600 [38] | 601 [39] | 625 [40] | 690 [38] | 713C [41] | 718 [42] | X750 [43] | 800 [44] | 825 [45] |
|---|---|---|---|---|---|---|---|---|---|---|---|
| | Ni | 62 | Bal. | 62.6 | 60.76 | Bal. | 71.76 | 53.98 | 71.32 | 35 | 38.25 |
| | Cr | 15 | 15.2 | 23.05 | 21.69 | 29.9 | 12.7 | 18.11 | 16.22 | 23 | 22.70 |
| | Fe | Bal. | 11.0 | - | 4.21 | 11.6 | 1.6 | Bal. | 8.04 | 39.5 | 31.08 |
| | Mo | - | - | - | 8.62 | - | 4.6 | 3.00 | - | - | 2.77 |
| | Nb & Ta | 2.0 | - | - | 3.38 | - | 2.2 | 5.44 | 0.9 | - | - |
| | Co | - | - | - | - | - | 0.06 | - | 0.01 | - | 0.04 |
| | Mn | 5.5 | 0.23 | 0.1 | 0.31 | 0.25 | 0.04 | - | 0.21 | - | 0.12 |
| | Cu | 0.5 | - | 0.1 | - | - | - | - | 0.03 | - | 2.78 |
| | Al | - | - | 1.4 | 0.53 | - | 5.9 | 0.53 | 0.68 | 0.15–0.6 | 0.05 |
| *wt%* | Ti | 1.0 | 0.3 | - | 0.21 | 0.3 | 0.71 | 1.01 | 2.47 | 0.15–0.6 | 0.65 |
| | Si | 1.0 | 0.29 | 0.37 | 0.40 | 0.33 | 0.08 | - | 0.08 | - | 0.36 |
| | C | 0.06 | 0.022 | 0.025 | 0.04 | 0.025 | 0.188 | - | 0.04 | $\leq$0.1 | 0.04 |
| | S | 0.02 | 0.025 | - | - | 0.025 | 0.006 | - | - | - | 0.02 |
| | P | 0.02 | 0.086 | - | - | 0.01 | 0.005 | - | - | - | 0.01 |
| | B | - | - | - | - | - | 0.014 | - | - | - | - |
| | W | - | - | - | - | - | 0.06 | - | - | - | 0.36 |
| | Zr | - | - | - | - | - | 0.14 | - | - | - | - |
| | V | - | - | - | - | - | - | - | - | - | 0.06 |
| | N | - | 0.024 | - | - | 0.02 | - | - | - | - | - |

Caption: Bal.—Balance.

As it was possible to find out, some of the consequences of alloying Ni with certain chemical elements make INCONEL® alloys [46] a difficult-to-machine material [47] (Figure 2) and difficult to metal shape [48], identically to stainless steel [49,50]. As opposed to other alloys, like Al-alloys [36,51], Magnesium (Mg) alloys [51], steel alloys [52] or Ti-alloys [51], INCONEL® alloys do not benefit from better-established wear mechanisms between the pair tool-workpiece.

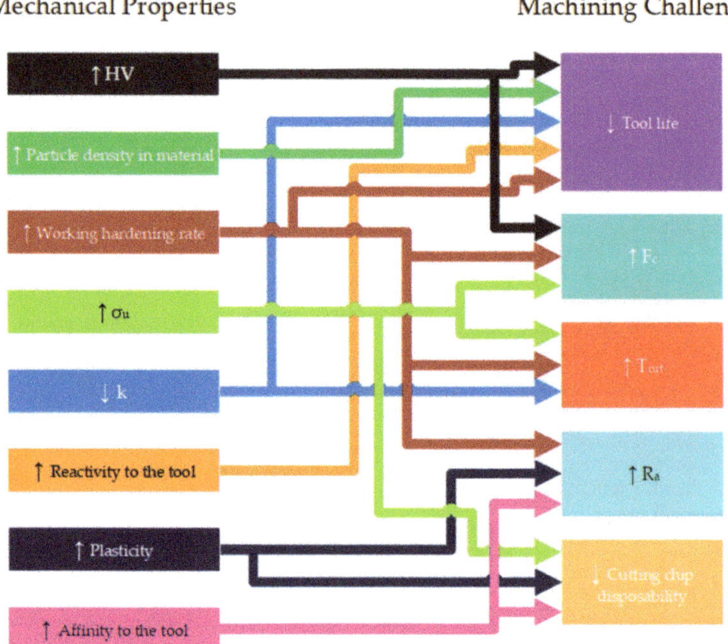


**Figure 2.** Relationship between mechanical properties and machining challenges with INCONEL® (adapted from [19]).

Table 5 presents the most relevant and used models of the better-established wear mechanisms referred to, based on physics and experiments for heat partition coefficient $R_{\mathrm{chip}}$, for common materials like aluminium and low carbon mild steels.

**Table 5.** Predictive models based on physics and experiment for heat partition coefficient $R_{\mathrm{chip}}$ [53].

| A Predictive Model for Heat Partition Coefficient $R_{\mathrm{chip}}$ | Equation | Establishment Basis |
|---|---|---|
| Loewen—Shaw [54] | $R_{L-SH} = \dfrac{q_F \cdot \frac{l_c}{\lambda_T} \cdot A - \Delta\theta_{p\,\max} + \theta_0}{q_F \cdot \frac{l_c}{\lambda_T} \cdot A + q_F \cdot \dfrac{0.377 \cdot l_c}{\lambda_W \cdot \sqrt{\frac{v_{ch} \cdot l_c}{4 \cdot \alpha_W}}}}$ | Dry-cutting process of AISI 1113 steel with $K_2S$ cemented carbide tool (cutting speed, $v_c = 30\text{–}182$ m/min). |
| Shaw [55] | $R_{SH} = \dfrac{1}{1 + \left(0.754 \cdot \frac{\lambda_T}{\lambda_W}\right) / A \cdot \sqrt{\frac{v_{ch} \cdot l_c}{2 \cdot \alpha_W}}}$ | Dry-cutting process of AISI 1113 steel with high-speed steel (HSS) tool/$K_2S$ cemented carbide tool ($v_c = 30\text{–}182$ m/min). |
| Kato—Fujii [56] | $R_{KF} = \dfrac{1}{1 + \frac{\lambda_T}{\lambda_W} \cdot \sqrt{\frac{\alpha_W}{\alpha_T}}}$ | Surface grinding process of stainless steel/carbon steel with Al-oxide wheel. |

**Table 5.** *Cont.*

| A Predictive Model for Heat Partition Coefficient $R_{\text{chip}}$ | Equation | Establishment Basis |
| --- | --- | --- |
| List—Sutter [57] | $$R_{L-SU} = \cfrac{1}{1+0.754\cdot\cfrac{\frac{\lambda_T\cdot\sqrt{v_{ch}\cdot l_c}}{\lambda_W\cdot\sqrt{\alpha_W}}}{\frac{2}{\pi}\cdot\left[\ln\left(\frac{2w}{l_c}\right)+\frac{1}{3}\cdot\frac{l_c}{w}+\frac{1}{2}\right]}}$$ | Dry-cutting process of AISI 1018 mild steel with uncoated carbide tool ($v_c$ = 23–60 m/min; undeformed chip thickness ($h_{ch}$) range 0.26–0.38 mm). |
| Gecim—Winer [58] | $$R_{G-W} = \cfrac{0.807\cdot\lambda_W\cdot\sqrt{\frac{v_{ch}\cdot l_c}{\alpha_W}}}{\lambda_T+0.807\cdot\lambda_W\cdot\sqrt{\frac{v_{ch}\cdot l_c}{\alpha_W}}}$$ | Based on the thermal behaviour of the two-dimensional, transient $T$ distribution in the vicinity of a small, stationary, circular heat source equation of the average $T$ of the moving and stationary heat sources between a frictional contact. |
| Reznikov [59] | $$R_R = \cfrac{1}{1+1.5\cdot\frac{\lambda_T}{\lambda_W}\cdot\sqrt{\frac{\alpha_W}{\alpha_T}}}$$ | Based on the Green function to analyse the chip deformation and friction work along the tool rake face. |
| Berliner—Krajnov [60] | $$R_{B-K} = \cfrac{1}{1+0.45\cdot\frac{\lambda_T}{\lambda_W}\cdot\sqrt{\frac{\pi\cdot\alpha_W}{v_{ch}\cdot l_c}}}$$ | |
| Tian—Kennedy [60] | $$R_{T-K} = \cfrac{1}{1+\frac{\lambda_T}{\lambda_W}\cdot\sqrt{\frac{1+\frac{v_{ch}\cdot l_c}{\alpha_T}}{1+\frac{v_{ch}\cdot l_c}{\alpha_W}}}}$$ | Consideration of Peclet numbers for the tool and workpiece materials in sliding tribological contact. |

Caption: $A$—area shape factor; $l_c$—tool-chip contact length; $q_F$—frictional heat flux generated in SDZ; $R_{\text{chip}}$—heat partition coefficient into the moving chip from the secondary deformation zone (SDZ); $v_{ch}$—chip moving speed; $\alpha_T$—tool thermal diffusivity; $w$—tool-chip contact area; $\alpha_W$—workpiece thermal diffusivity; $\Delta\theta_{p\,\text{max}}$—maximum tool-chip interface temperature rise due to heat generation in PDZ; $\theta^0$—environment temperature; $\lambda_T$—tool thermal conductivity; $\lambda_W$—workpiece thermal conductivity.

It is suggested to consult the work of Zhao, et al. [53] to better understand the additional variables described in Table 5.

Figure 3 explains how superficial hardness is affected in INCONEL® alloys when machined after cold work processes, compared to some more stable materials like Cu, Al and mild steel. Machining (or surface) cold working may result from mechanical machining (milling, lathing, grinding) [61] or surface treatment (sandblasting, shot-peening), and may introduce residual tensile or compressive stresses into the surface of materials. Compressive stresses generated by shot-peening processes prevent the occurrence of stress corrosion cracks. In the case of plastic strain, tensile stresses appear instead, and the resulting stress levels may be extremely high [62]. A curious detail patent in Figure 3 is the similarity behaviour between INCONEL® 718 and 625 alloys after the 20% cold reduction.

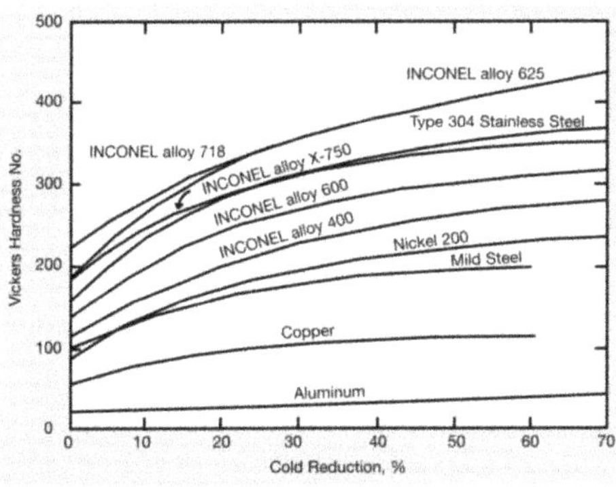


**Figure 3.** Effect of cold work on hardness for different INCONEL® alloys and comparison with other materials [1].

As a consequence of low *k* [36] of nickel-based alloys, which significantly influences heat distribution during the machining process, the surface integrity is affected when applying traditional cold forming techniques, due to the rapid work hardening (Figure 3) in the chip formation region [63]. This phenomenon leads to plastic deformation of either the INCONEL® workpiece or the tool, on subsequent machining passes [64], eventually resulting in built-up-edge (BUE) formation [31] (Figure 4) and consequentially in premature tool failure [65]. For this reason, age-hardened INCONEL® alloys, such as the 718 alloy, are typically machined using an aggressive but slow cut with a hard tool, minimizing the number of passes required [66].

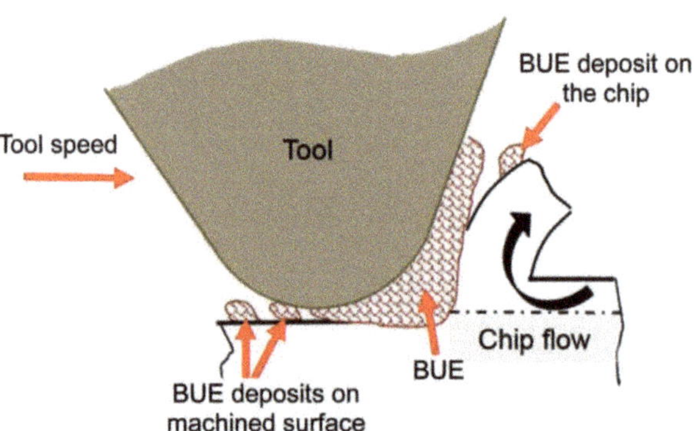


**Figure 4.** Schematic diagram of BUE formation in micromachining processes [67].

The BUE phenomenon occurs because of an accumulation of hot debris generated by the chip-start cutting process and deposited on the tool surface during machining, leading afterwards to adhesion and abrasion TW. From an experimental point of view, some authors noted that the BUE is significantly affected by the state of stress around the tool cutting edge and happens under extreme contact conditions at the tool–chip interface as high friction, high pressure, and high sliding velocity [68]. INCONEL® alloys are well known to abrade tools and develop BUE [69], especially the 718 alloy. Also during the machining of INCONEL® 625, heat concentration is likely to occur at the cutting edges, resulting in early tool failure and consequent BUE [70].

## 2. Method of Research

The research and information compiling method are illustrated in the flowchart of Figure 5, which is simple to visually interpret and track down all the inherent steps in the making of this specific paper. In the flowchart, all the consulted databases and most used keywords are found (in the topic of this document), to find information about conventional and non-conventional machining and tool-wear mechanisms of INCONEL® alloys.

Additionally, will be provided three attachments containing abbreviations, symbols and units used within the article.

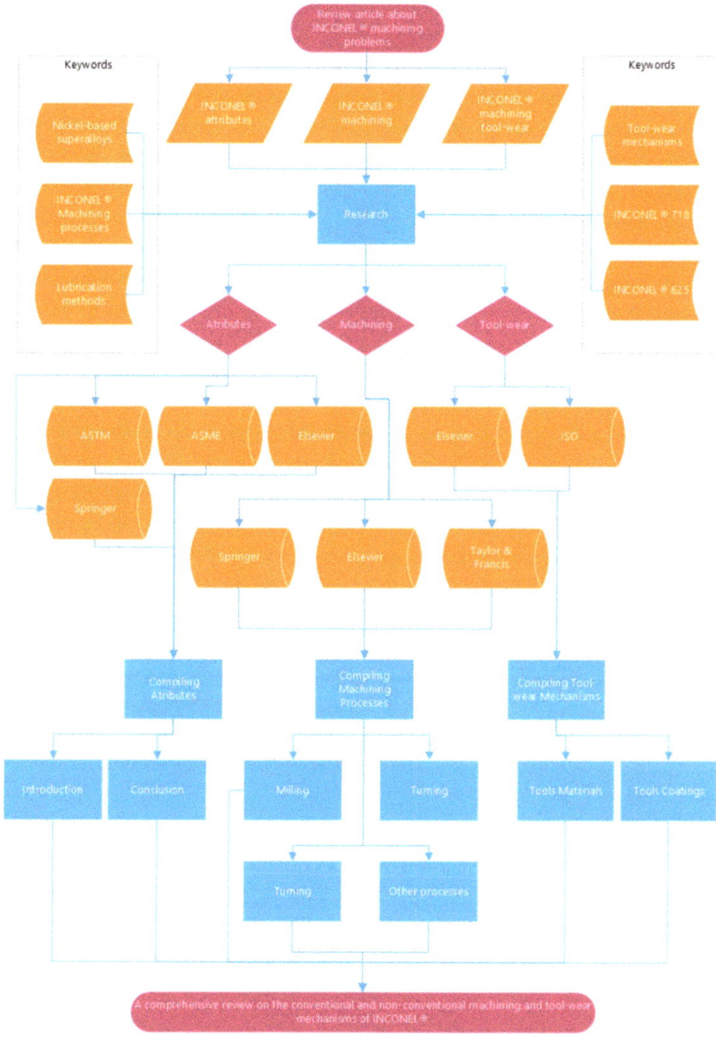

**Figure 5.** Research method accomplished to achieve a better redacting result to the review paper.

## 3. Literature Review

### 3.1. Conventional Manufacturing Processes

The machining process of chip-start cutting is a technological process able to transform a wrought stock into a component, using a cutting tool. The surplus material from the wrought stock, or just stock, is removed in the form of chips; a consequence of the mechanical action of a cutting wedge with higher hardness than the material of the component that is meant to manufacture. In the following literature review, Milling, Turning, Drilling and Boring will be the discussed processes, in which chip-start cutting is a key and common factor to all these traditional processes.

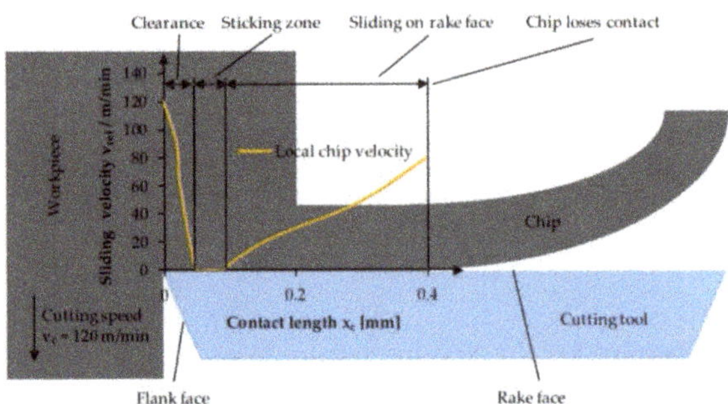



**Figure 6.** Evolution of the sliding velocity along the tool-material interface [71].

Making use of Figure 6, and taking into account that the chip-start cutting process is very much equal to the machining of INCONEL® 718 and 625, Bonnet et al. [72] described the different friction parts on the rake face in the machining of steel. Directly behind the cutting edge, the chip velocity rapidly shrinks to zero. For a certain contact length, the chip material has a sliding velocity of zero, which starts to increase for the rest of the tool–chip interface, before the chip loses contact with the tool [72].

Due to the friction created around the chip creation process, three distinct heat zones are created within the vicinities of the cutting wedge. In Figure 7, the three different thermal affected regions between the tool-workpiece are visible. In Figure 7a there is a thermo-mechanical deformation of the primary shear region (or primary deformation zone, PDZ) where the majority of the energy is converted into heat due to the internal friction of the material to be cut. In Figure 7b there is a tool-material interface region, or SDZ, of the tool rake surface and the chip rear face where heat is generated by the rubbing between the chip and the tool and finally. In Figure 7c the contact between the flank of the tool and the already machined surface takes place, called tertiary deformation zone (TDZ).

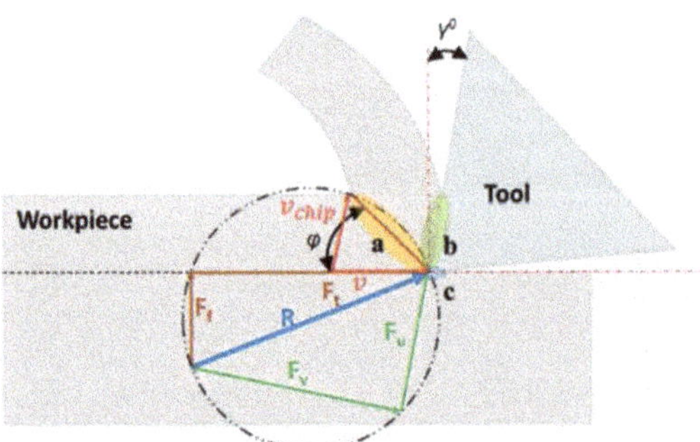



**Figure 7.** Regions of heat generation during metal orthogonal machining (adapted from [73]): (Caption: $\varphi$—shear plane angle; $\gamma^0$—Rake angle, $F_f$—Feed force, $F_t$—tangential force).

A novel approach to improve the efficiency of the traditional chip-start cutting process is laser-assisted machining (LAM), illustrated in Figure 8, which consists of preheating the material to cut and lowering the superficial hardness to facilitate tool cutting. This solution is common to turning, milling and grinding. Kim and Lee [74] also worked on a machining preheat approach for the INCONEL® 718 alloy, which includes a magnetic induction coil instead of a laser.

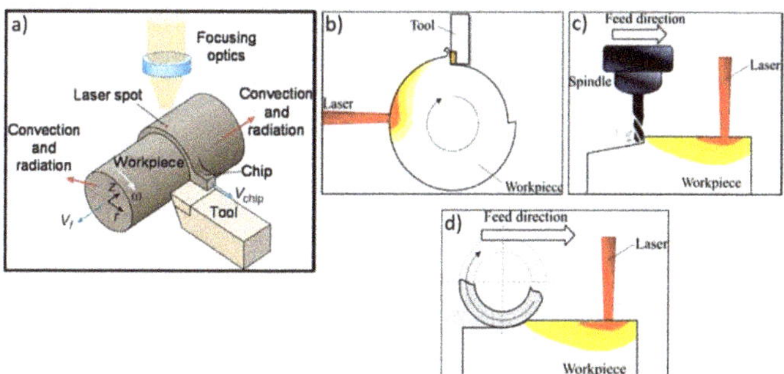

**Figure 8.** (**a**) Schematic of LAM indicating heat-losses by convection and radiation, (**b**) Schematic of LAM turning, (**c**) LAM milling, and (**d**) LAM grinding (adapted from [75]) (Caption: $V_f$—feed velocity).

### 3.1.1. Milling

Milling is the nomenclature given to the machining process that uses rotary cutting tools to remove excess material from the wrought stock. Nowadays, with the use of CNCs, milling can be done at a maximum of six degrees of freedom (DOF).

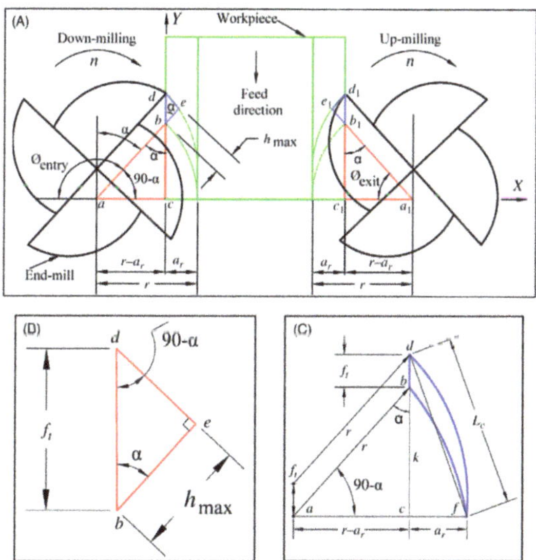

**Figure 9.** Chip formation showing (**A**) chip formation showing cutter tooth entry angle in down-milling and cutter tooth exit angle in up-milling, (**B**) maximum chip thickness, $h_{max}$, and (**C**) chip length, $L_c$ [76].

Traditional milling tends to have lower $a_p$ values and higher $a_e$ values compared to more advanced milling techniques. However, this would cause a concentration of all heat generated in a small portion of the cutting edge, which in this case is the tip of the tool. It would require more axial passes too. This problem can be well managed in aluminium and steel alloys, but not with refractory materials like INCONEL® alloys. Many milling approaches can be tackled to enhance INCONEL® machining, such as up and down-milling, studied by Hadi et al. [77] in INCONEL® 718 machining, illustrated by Figure 9. Another interesting and efficient technique [78] that enriches milling INCONEL® 718 and 625 is trochoidal milling, illustrated by Figure 10, which consists of making the centre of the cutting tool walk a "helical horizontal" path. This procedure not only prevents tool jamming due to workpiece heat-dilation, but it also enables cutting bigger $a_e$ dimensions, with lower $a_p$, improving heat-spread over the entire tool with more radial passes.

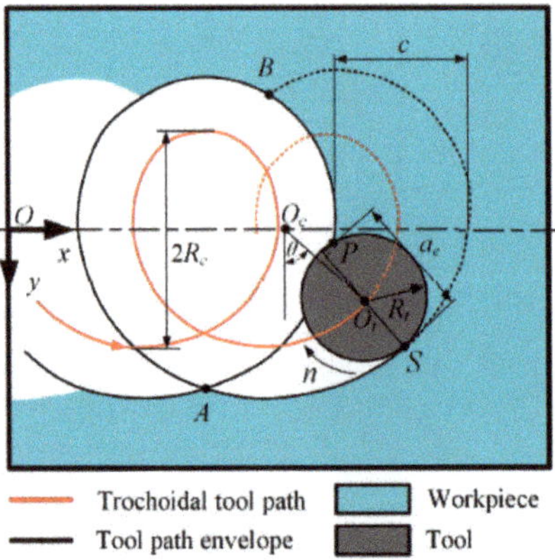


Trochoidal tool path ——— Workpiece

——— Tool path envelope ——— Tool

**Figure 10.** Example of a geometric model of trochoidal milling [79].

Table 6 presents the latest experimental challenges and developments in the machining of INCONEL® with the milling process.

**Table 6.** Critical challenges and developments in the milling process of INCONEL®.

| Author | Challenges | Remarks |
|---|---|---|
| Guimaraes, et al. [64] | Evaluation of how the machining processes deteriorate the surface integrity to extend the service life of the INCONEL® 625 components as long as possible. The influence of tool geometry, feed rate ($f$), and tool rotational speed ($s$) on surface integrity were evaluated for the milling process. | The results indicate that $s$ has the greatest influence on specific cutting pressure ($\beta$). The parameters $s$ and $f$ were the main factors that affected the thickness of the cold worked zone. The results suggest that $R_a$ after machining is driven by mechanical-thermal loadings and causes beneficial results related to corrosion resistance and compressive residual stress. |
| Pleta, et al. [80] | Assessment of the INCONEL® 718 trochoidal milling process and optimization for manufacturing scenarios. To accomplish this goal, the modelling of the cutting forces ($F_c$) must be investigated with semi-mechanistic methods. Furthermore, machining parameters are investigated as to how they relate to the improvement of tool life and $F_c$ utilizing the Taguchi method. | It is found that TW increases the depth of the machining affected zone as does increasing $h_{ch}$. Nutational rate ($\dot{\varphi}$) and rotational rate ($\dot{\theta}$) have the largest interactions with both $F_c$ and tool $VB$. It was found that $h_{ch}$ and TW increased the depth of the plastically deformed and elongated grains in both the radial and axial orientations. |
| Shankar, et al. [81] | This investigation has designed a tool condition monitoring system (TCM) while milling INCONEL® 625 based on sound and vibration signatures. The experiments were carried out based on response surface methodology (RSM). | The process parameters such as $s$, $f$, $a_p$ and vegetable-based cutting fluids were optimized based on $R_a$ and flank wear ($VB$). It was determined that the sound pressure and vibration signatures have a direct relation with $VB$. The statistical features values were extracted from the experimental data and the cutting tool $VB$ was predicted with a mean square error (MSE) of 8.4212%. The $R_a$ of the machined surface varies from 0.081–0.273 μm. The $VB$ of the cutting tool varies from 0.0187–0.0254 mm. Based on the desirability function, $s$ = 221 rpm, $f$ = 0.02 mm/rev and $a_p$ = 0.17 mm were identified as optimal process parameters. The average values of sound pressure for a brand-new, normal life (or useful life), and dull tools are 0.01955, 0.2513 and 0.4858 Pa, respectively. Similarly, the vibration signal range for a brand new, normal life (or useful life), and dull tools is 0.029–0.4394 g, 0.0780–1.32 g and 0.120–5 g. |

**Table 6.** *Cont.*

| Author | Challenges | Remarks |
| --- | --- | --- |
| Alonso, et al. [82] | Slot milling operations were performed to investigate the influence of $s$ and machining direction in INCONEL® 718 alloy. | It was observed at higher $s$, lower values of $R_a$ and lower torque ($M$) values were obtained. Moreover, the main novelty of this work is the influence of the anisotropy of Wire and Arc Additive Manufacturing (WAAM) INCONEL® 718 alloy on its machinability. Milling along the extruder travel direction offers better dimensional tolerance values with lower cutting $M$, being the more efficient way. |
| Boozarpoor, et al. [83] | Turn-milling technology was utilized to machine cylindrical samples of INCONEL® 718 alloy. The effect of process factors such as tool rotational speed (or spindle, $s$), workpiece rotational speed ($v_c$), $f$ and eccentricity ($e$) on surface roughness ($R_a$) and tensile residual stress was analysed. | The results showed that $f$ is the most influential parameter that determines the value of $R_a$ and residual stress. Furthermore, by considering production rate as a constraint, it was logically discussed that a setting of 1000 rpm cutter-speed, 300 rpm work rotational speed, 0.12 mm/rev feed rate and 0.2 mm eccentricity can guarantee maximum production rate as well minimum $R_a$ and tensile residual stress. |
| Anburaj and Pradeep Kumar [70] | Face milling was carried out on INCONEL® 625 and twenty-seven iterations ($L_{27}$) were conducted using Design of Experiments (DOE), including three levels of $v_c$ and $f_z$ with constant $a_p$. The study included three lubrication conditions such as dry-machining, normal coolant (wet) and cryogenic $CO_2$ ($l$) coolant. The output responses such as cutting temperature ($T_{cut}$), cutting feed force ($F_x$), cutting normal force ($F_y$), cutting axial force ($F_z$) and $R_a$ were evaluated. | The results were optimized using the TOPSIS technique with ANOVA tests, as the results of the highest closeness coefficient ($C_i$) value indicated the cryogenic $CO_2$ ($l$) coolant environment, and the input optimized parameters were $v_c = 80$ m/min and $f_z = 0.05$ mm/tooth. Other parameters such as $T_{cut} = 57.38$ °C, $F_x = 201.5$ N, $F_y = 251.1$ N, $F_z = 335.9$ N, and $R_a = 0.159$ μm were found optimal parameters, in the 19th iteration having obtained $C_i = 0.928835$. |

### 3.1.2. Turning

Opposed to milling, the turning process takes place in a lathe for components with a revolution axis, i.e., turbine shafts. The workpiece spins in a lathe while the fixed tools, with or without inserts, remove the surplus material. It is patent in Figure 11 the normal movement of the tool while turning a sort of shaft and some intrinsic characteristics of the inserts used. Some of the main problems in turning INCONEL® 718 and 625 are the specific cutting energy (SCE) and rapid augment of surface hardening upon cutting material. Moreover, since shafts must comply with certain geometric specs for the better functionality of the component, $R_a$ is a key factor to be studied, varying $v_c$, $f$ and $a_p$.

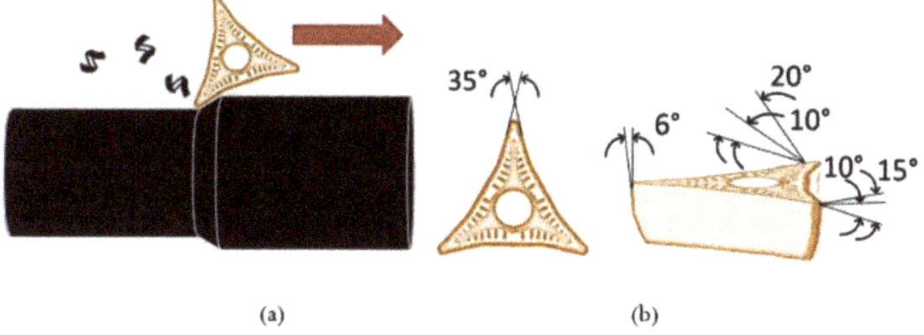

(a)      (b)

**Figure 11.** (a) Turning example; (b). Insert A-type (view of basic side cutting edge angle, rake angle, and secondary angles for chip breakage) [84].

Table 7 presents the latest experimental challenges and developments in the machining of INCONEL® with the turning process.

**Table 7.** Critical challenges and developments in the turning process of INCONEL®.

| Author | Challenges | Remarks |
|---|---|---|
| Waghmode and Dabade [23] | Examination and observation of the response parameters like $F_c$ and $R_a$ on input parameters such as $v_c$, $f$ and $a_p$. Experimentation was conducted as per Taguchi's L9 orthogonal array during INCONEL® 625 alloy turning operation. The results are analysed using the ANOVA method. | ANOVA suggests that the $a_p$ has a 55.05% contribution to $F_c$. In thrust force ($F_z$), the contribution of $a_p$ is 55.83% which is the parameter with the most influence. As $a_p$ increases, $F_z$ increase. The main contribution for the feed force is provided by the $a_p$, with 83.39%. Parameters $f$ and $a_p$ highly affect $R_a$, having 20.61 and 51.68% contribution, respectively. As $f$ increases, $R_a$ increases too. With parameter $v_c$, it shows an inverse trend. |
| Kosaraju, et al. [25] | A multi-objective optimization based on the Taguchi-based Grey Relation Analysis (TGRA) method was employed to find the optimal levels of turning INCONEL® 625 parameters for the objective of lower $F_c$ and better $R_a$ under dry-cutting conditions. | From the statistical analysis, the results show that $f$ is identified as the most significant parameter for the turning operation according to the weighted sum of $F_c$ and $R_a$. The optimal combination of control factors and their levels are $v_c$ = 75 m/min, $f$ = 0.103 mm/rev and $a_p$ = 0.2 mm. |
| Vignesh and Ramanujam [75] | Evaluate the influence of laser-assisted high-speed machining (LAHSM) on the $F_c$ ($F_z$ in Figure 8), $R_a$, TW and the chip morphology during the INCONEL® 718 turning process. | LAHSM optimal parameters were $v_c$ = 80 m/min, $f$ = 0.08 mm/rev and laser power, $P_{Laser}$ = 1300 W. $F_c$, $R_a$ and TW values were better over the conventional turning process ones, leading to a reduction of $F_c$ by 24.5%, $R_a$ by 56% and TW by 29%. |
| Raykar, et al. [85] | High-pressure cooling (HPC) of cutting tools can be very effective when machining difficult-to-cut materials like INCONEL® 718. An analysis of microhardness and degree of work hardening (DWH) is carried out to evaluate the machinability of the INCONEL® 718 alloy while turning in a high-pressure coolant environment. | Microhardness has a sudden change between 30–120 μm bellow the machined surface. After this region, the microhardness value is similar to the bulk, which is found at 270–300 μm below the machined surface for all samples. The microhardness near the machined surface is found to be 1.11× the bulk microhardness. Microhardness variation is unaffected by the changes in the process parameters since none is statistically significant at a 95% confidence level. A significant microhardness value is noticed at a depth of 90 μm. |
| Infante-García, et al. [86] | The increase of the undeformed chip cross-section and the high SCE of INCONEL® 718 give rise to load peaks during machining. Different tests involving multipass finishing turning have been carried out to study the magnitude of the load peaks for different cutting conditions. | Initial results have shown a significant peak in the machining loads, predicted by Altintas force law [87]. This peak is related to the tool tip radius and the cutting parameters, after the second and successive passes. The main factor that contributes to that is the increase of undeformed chip cross section during a short interval. Thus, the progression of TW is significantly influenced. The machining load at the end of a turning pass can significantly increase during a short interval. Consequently, this effect may influence TW progression leading to premature tool failure. |
| Makhesana, et al. [88] | The turning tests to INCONEL® 625 are conducted under dry, MQL, and nanofluid-MQL (nMQL) environments and the machining results are compared considering $R_a$, chip morphology, TW, $T_{cut}$, $P_{in}$ and microhardness | Sunflower oil blended with $MoS_2$ resulted in 56, 42, and 22% improved $R_a$ compared to dry, MQL, and nMQL (Graphite) conditions, respectively. Also, the efficiency of nMQL with graphite and $MoS_2$ is evaluated by slower TW progression. Also, MQL, nMQL with $MoS_2$, and nMQL (graphite) resulted in lower $T_{cut}$ by 18, 35, and 25%, respectively, compared to dry turning. The effective performance of nMQL is credited to the better penetration ability of the applied lubricant. Furthermore, the MQL application with compressed air facilitated chip removal and heat dissipation during machining. |

**Table 7.** *Cont.*

| Author | Challenges | Remarks |
|---|---|---|
| Airao, et al. [89] | Examination of the machinability of the INCONEL® 718 alloy in conventional and ultrasonic assisted turning (UAT) under dry, wet, minimal quantity lubrication (MQL), and $CO_2$ ($l$) strategies. The experiments are performed in an in-house developed ultrasonic-assisted turning (UAT) setup, keeping all the machining parameters constant. | The $CO_2$ ($l$) reduces edge chipping, nose wear, adhesion, and abrasion wear. The conventional turning under $CO_2$ ($l$) reduces the $VB$ by 32–60% and power consumption ($P_{in}$) by 5–31% compared to dry, wet, and MQL strategies. Similarly, the UAT under $CO_2$ ($l$) reduces the $VB$ by 32–53% and $P_{in}$ by 11–40% compared to dry, wet, and MQL strategies. The UAT reduces $R_a$ compared to conventional turning when used under MQL and $CO_2$ ($l$). The $CO_2$ ($l$), in conjunction with ultrasonic vibration, significantly reduces SCE and TW without compromising $R_a$. Moreover, this combination also helps in enhancing the chip breakability and reducing $\varepsilon$ concentration. |

### 3.1.3. Drilling

Drilling is a cutting process where a drill bit is spun to cut a circular hole in a component. In INCONEL® applications, drilling is important to create micro holes that will permit the cooling of gas turbines, as illustrated by Figure 12 and studied by Venkatesan et al. [90] on the hole quality assessment in INCONEL® 625 alloy parts.

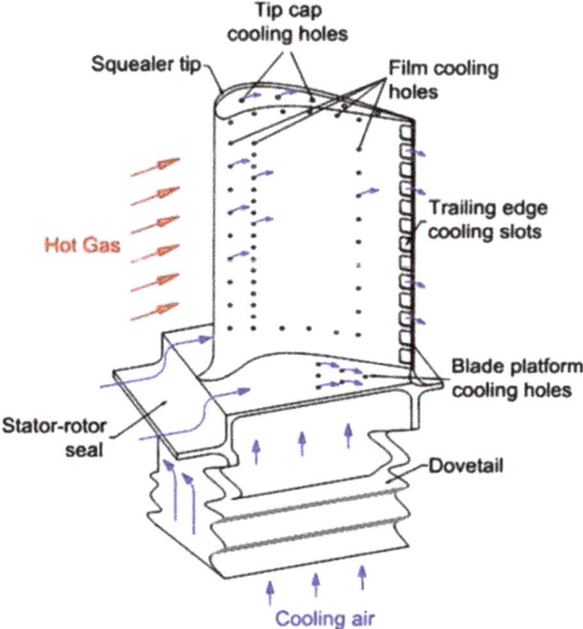


**Figure 12.** Gas turbine blade cooling schematic [91].

The INCONEL® 718 alloy has many challenges in deep-hole-drilling as well since the process is prone to drill jamming due to material expansion inside the holes. Table 8 presents the latest experimental challenges and developments in the machining of INCONEL® alloys with the drilling process.

**Table 8.** Critical challenges and developments in the drilling process of INCONEL®.

| Author | Challenges | Remarks |
| --- | --- | --- |
| Neo, et al. [92] | The traditional carbide-tipped gun drills often get worn at an accelerated rate and require repetitive re-sharpening or replacement when drilling INCONEL® 718 alloys. This occurrence lowers productivity and increases costs. Furthermore, it is also a challenging task to meet the stringent hole-straightness requirement of 1/1000 mm for deep-hole drilling (DHD). | In contrast to traditional carbide-tipped gun drills, the developed PCBN-tipped gun drill can operate at higher $v_c$ and reduce drilling forces, drilling $M$, and TW. Furthermore, the surface quality of drilled holes is much better than those drilled by traditional carbide gun drills. For a stable drilling condition, it is also recommended to perform deep-hole drilling with the developed PcBN gun drill on INCONEL® 718 at $v_c \geq 50$ m/min. |
| Venkatesan, et al. [90] | Drilling micro-cooling holes in turbine blades on INCONEL® 625 is one of the noteworthy applications of micro-drilling, and few are the investigations on the hole quality assessment and drill bit tool life. The multi-response optimization of test parameters in micro-drilling conditions is presented, using the approach on the Taguchi $L_{27}$ design. Micro-drilling is performed in a 2 mm plate thickness of INCONEL® 625 with an uncoated micro-drill. | After each drilling test, the hole diameter, circularity error, overcut, taper ratio, cylindricity and hole damage factor are measured, and the results are examined. The deviation in hole diameter, cylindricity, circularity error, roundness, overcut, taper ratio, and hole damage factor obtained is increased to 5.5, 87.2, 50.5, 5.7, 77.4, 20.0 and 5.4%, respectively, from 1st to 25th hole. Better hole quality features with the least deformed layer thickness and low burr formation at entry, and exit, are consistently achieved with $s$ = 21,000 rpm and $f$ = 6 mm/min for a given tool diameter. The optimized parameters are a tool diameter of 0.8 mm, (the most suitable tool diameter for hole quality and productivity in micro-drilling), $s$ = 21,000 rpm and $f$ = 10 mm/min. Chip clogging, entrance burr, and exit burr are obtained before the tool failure. These new findings have brought out a highly beneficial database for aerospace industries without losing the quality of the hole in production. |
| Sahoo, et al. [93] | Use of cryogenically treated drill bits in INCONEL® 718 alloy machining. Drill bits are conditioned under two different environments i.e., cryogenically treated with single-tempered drill bits, and cryogenically treated with double-tempered drill bits. The Taguchi method was used for trial design and optimization of factors along with the Whale optimization algorithm. | Results show that $f$ is the most contributing parameter to maximize $F_z$, while $s$ is the most contributing parameter to maximising $M$. Also, tool condition is the most contributing factor to minimising $R_a$. By maximizing the $F_z$, up to 184 N, and $M$, up to 0.72 Nm, during drilling operation on INCONEL® 718, using cryogenically treated and single tempered drill bit, it was found that the optimal parameter settings were $s$ = 215 rpm and $f$ = 0.106 mm/rev. $R_a$ was minimized to 3.77 μm. It was also seen that cryogenically treated and the single-tempered drill bit was more influential in attaining maximum $F_z$ and $M$, while cryogenically treated and the double-tempered drill bit was more influential in attaining minimum $R_a$. |

### 3.1.4. Boring

Boring is the manufacturing process in which previously drilled holes are enlarged by a single-point cutting tool. Not much information is available about the boring process on INCONEL® alloys whereby it is only presented in the study carried out by Ratnam et al. [94], whose challenges involved the investigation of the machining parameters' effect on $R_a$, TW, $F_c$ on the cutting tool and workpiece vibration during dry boring of INCONEL® 718 with TiCN-Al$_2$O$_3$-TiN coated inserts using response surface methodology (RSM). It was found that the use of accelerometers, radioactive sensors and piezoelectric actuators does not make it possible to measure rotating objects' vibrations. On the other hand, the LDVs are capable to measure rotating objects' vibrations with a simple experimental arrangement. Parameters $s$ and $f$ were found to have a significant influence on $R_a$. The parameter

$a_p$ was found to be significant on $VB$, $F_c$, and workpiece vibration amplitude ($VA$). The optimal machining parametric combination was obtained using the desirability function. Cutting condition parameters such as $s = 360$ rpm, $f = 0.14$ mm/rev and $a_p = 0.4949$ mm was obtained for $VA = 38.7$ µm, minimum $F = 117.8$ N, $VB = 0.3$ mm and minimum $R_a = 2.55$ µm. The proposed RSM approach was an easy method to obtain maximum information with a smaller number of experiments. and successfully used by different authors in the improvement of process parameters.

*3.2. Non-Conventional Manufacturing Processes—Electrical Discharge Machining (EDM)*

This review also presents some new insights into a non-conventional process, which is Electrical Discharge Machining (EDM). The process is a non-conventional machining method that allows the production of pieces with complex shapes, and it can be used in materials such as INCONEL® 718 and 625. This particular manufacturing technique removes material from the wrought-stock thanks to melting and vaporising cavities using electrical discharges that come from a scrolling wire [95], as illustrated in Figure 13.

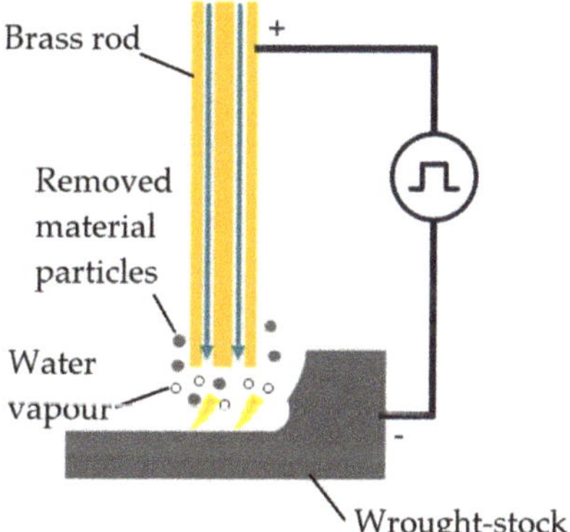


**Figure 13.** Schematic representation of the EDM experiment setup (adapted from [96]).

Table 9 presents the latest experimental challenges and developments in the machining of INCONEL® alloys with EDM.

**Table 9.** Critical challenges and developments in the EDM process of INCONEL®.

| Author | Challenges | Remarks |
| --- | --- | --- |
| Mohapatra, et al. [39] | Comparative study of the chemical and mechanical properties of INCONEL® 718, 625, 825 and 601. Evaluation of the machinability of different grades of alloys using EDM. | An increase in the peak current resulted in improved material removal rate (MRR) by keeping the gap voltage ($V_g$), pulse-on time ($T_{on}$), duty factor ($\tau$) and dielectric circulation flushing pressure ($F_p$) fixed at a constant value, for the different INCONEL® grades. An increase in $R_a$ is observed by incrementing the peak discharge current ($I_p$). In the case of surface crack density (SCD), with an increase in $I_p$, SCD ends to decrease, again followed by an increasing trend; and then finally remained constant. |
| Manikandan, et al. [97] | INCONEL® 625 is one of the hard-to-machine materials extensively used in high-temperature applications. It has better strength and lower $k$, compared with more common materials, which leads to poor machinability by traditional processes. To overcome such kind of disadvantages, unconventional manufacturing methods have been developed and proposed to be suitable substitutes, such as WEDM. | This work details a single-aspect optimization problem of WEDM of INCONEL® 625 with the help of Taguchi analysis. In this investigation, the MRR and Overcut (OC) were deemed as the performance characteristics, and the Taguchi response showed that the best parameter combination to maximize the performance of $R_a$ is $T_{on} = 30$ µs, $T_{off} = 15$ µs and $I_p = 3$ A. The best set of process parameters for attaining better OC is $T_{on} = 30$ µs, $T_{off} = 5$ µs and $I_p = 1$ A. Contour plots were also explored to reveal the combined influence of process parameters on the preferred performance measures. |

**Table 9.** *Cont.*

| Author | Challenges | Remarks |
| --- | --- | --- |
| Hussain, et al. [98] | Cost reduction in machining through increased MRR using optimum machining process parameters. An experimental study is presented for optimizing process parameters $T_{on}$, $I_p$ and $T_{off}$ to maximize MRR for subtracting material from INCONEL® 625 work part using Taguchi DOE and its analysis. | A Taguchi orthogonal array $L_9$ is applied for experimental design and analysis. Optimized values of performance factors obtained by analysis are $I_p$ = 12 A, $T_{on}$ = 160 µs, and $T_{off}$ = 35 µs. The maximum MRR, of 24.48 mm³/mm, was obtained with optimum values of performance parameters. Under these conditions, it was found that, $R^2$ = 98% which reflects a high confidence level in the experiment. It is also found that both $T_{off}$ and $I_p$ have a considerable impact on MRR. This is because higher values of $T_{on}$ and $I_p$ enhance the amount of energy discharge on the workpiece, which leads to increased material melting and vaporisation. |
| Liu, et al. [95] | Enhancing the machining process of INCONEL® 718 using zinc-diffused coating brass wire electrode and Taguchi-Data Envelopment Analysis-based Ranking (DEAR). | MRR, kerf width ($K_w$), and $R_a$ were considered as the quality measures. In the WEDM process, the brass wire electrode worked as expected and the optimal arrangement of input factors was found as $T_{on}$ = 140 µs, $T_{off}$ = 50 µs, $SV$ = 60 V, and $WT$ = 5 kg, which are the most relevant factors with a $C_i$ = 0.989. |
| Rahul, et al. [99] | INCONEL® 718, 625, 825 and 601 machinability was experimentally analysed during the execution of EDM. The experimental design was planned according to a 5-factor/4-level $L_{16}$ orthogonal array. The following process variables were considered: $V_g$, $I_p$, pulse-on time ($T_{on}$), duty factor ($\tau$) and $F_p$. Machinability was assessed in consideration of MRR, electrode wear rate, $R_a$ and surface crack density (SCD) at the already worked surface. The satisfaction function approach integrated with the Taguchi method was attempted to determine optimal parameter settings. | The MRR efficiency was found to vary from: 1.3389–29.3128 mm³/min for INCONEL® 625; 1.1844–31.5995 mm³/min for INCONEL® 718. $R_a$ was found to vary from: 4.7–11.5333 µm for INCONEL® 625; 6–12.3667 µm for INCONEL® 718. Optimal parameters setting determined as: [$V_g$ = 80 V, $I_p$ = 7 A, $T_{on}$ = 200 µs, $\tau$ = 75%, $F_p$ = 0.6 bar] for INCONEL® 625 [$V_g$ = 70 V, $I_p$ = 7 A, $T_{on}$ = 500 µs, $\tau$ = 80%, $F_p$ = 0.6 bar] for INCONEL® 718. A significant carbon enrichment due to the thermo-mechanical effect of EDM was noticed on the machined surface of INCONEL® 718 and 625, during EDM operation, attributed to the formation of carbides and grain refinement, which increased micro-strain as well as dislocation density. The increase in $I_p$ resulted in improved MRR (while keeping $V_g$, $T_{on}$, $\tau$ and $F_p$ constant) for different INCONEL® alloys. $R_a$ was observed to increase as $I_p$ increased. |
| Farooq, et al. [100] | Several developments in WEDM have been reported, but the influence of $F_p$ attributes has not been thoroughly investigated. The influence of four process variables, namely servo voltage, $F_p$, nozzle diameter ($\emptyset_N$), and nozzle–workpiece distance ($W_D$), were analysed on INCONEL® 718 concerning geometrical errors (angular and radial deviations), spark gap ($S_G$) formation, and $R_a$. In this regard, detailed statistical and microscopical analyses are employed with mono and multi-objective process optimization by employing the TGRA method. | Process parameters are more influential on radius deviation as compared to angular error. At $F_p$ = 4 kg/cm², the radius deviation is around 4%, which increased to 5.4% at $F_p$ = 12 kg/cm². A low $F_p$ value agglomerates debris, which resulted in a coarser surface due to the re-solidification, with a $R_a$ = 2.12 µm. However, with the increase of $F_p$ to 12 kg/cm², the surface quality improved and resulted in $R_a$ = 1.93 µm. An increasing trend is observed where $S_G$ increased with the magnitude of $F_p$. Using $F_p$ = 4 kg/cm² resulted in $S_G$ = 107 µm. With the increase up to $F_p$ = 12 kg/cm², values of $S_G$ = 111.5 µm were attained. Similarly, an increase in $\emptyset_N$ from 4 mm to 8 mm not only increased $S_G$ from 108.5 µm to 110 µm but also increased angular error from 0.255% to 0.6%. The high $\emptyset_N$ increased the entry of dielectric flow, which hindered the stability of the thermo-electric erosion process near corners. The $F_p$ and $\emptyset_N$ parameters showed a significant effect on the $R_a$. The optimized parametric settings are $SV$ = 50 V, $F_p$ = 4 kg/cm², $\emptyset_N$ = 8 mm, and $W_D$ = 10 mm. The confirmatory experiment reduced the process's limitations to an $S_G$ = 109 µm spark gap, 0.956% angular error, 3.49% cylindricity error, and $R_a$ = 2.2 µm. |

### 3.3. Tool Wear

With the tool operation in machining, wear starts to be a key factor in quality and productivity, namely with INCONEL® alloys, such as 718 and 625. To identify a worn milling tool, the ISO 8688-2 [101] standard predicts that a tool presenting either $VB = 300$ μm or $VB_{max} = 500$ μm on the flank is considered a worn tool [69]. For turning tools or inserts, the ISO 3685 standard is the one to consult [102]. Taking into account a novel lubrication method, Bartolomeis et al. [69] observed that the tool wear behaviour mechanisms during EL conditions were abrasion and microchipping on the cutting edge, due to the tendency of INCONEL® 718 to develop BUE.

Figure 14 packs the initial causes of wear, the various wear mechanisms that lead to different types of wear and the final consequences from the tool-wear due to INCONEL® machining. As a complement to Figure 14, Table 10 presents the typical TW mathematical models, used to characterize the numerous TW mechanisms. It is suggested to consult the work of Wang et al. [37] to better understand the additional variables described in Table 10.

**Table 10.** Typical TW mathematical models (adapted from [37]).

| Authors | TW Model | Comments |
|---|---|---|
| Taylor [103] | $C = v_c \cdot T_{tool}^n$ or $T_{tool} = \dfrac{C}{v_c^p \cdot f^q \cdot a_p^r}$ | Taylor's empirical tool life model. |
| Archard [104] | $V = k \cdot \dfrac{P \cdot L}{3 \cdot \sigma_S} = k \cdot \dfrac{P \cdot L}{H}$ | Abrasive wear model. |
| Usui [105,106] | $\dfrac{dw}{dt} = A_1 \cdot \sigma_n \cdot v_s \cdot e^{-\frac{B_1}{T}}$ | Diffusive wear model. |
| Takeyama [107] | $\dfrac{dw}{dt} = G(v, f) + D \cdot e^{-\frac{Q}{RT}}$ | Abrasive and adhesive wear model. |
| Childs [108] | $\dfrac{dw}{dt} = \dfrac{A}{H} \cdot \sigma_n \cdot v_s$ | Abrasive and adhesive wear model. |
| Schmidt [109] | $\dfrac{dw}{dt} = B \cdot e^{-\frac{Q}{RT}}$ | Diffusive wear model. |
| Luo [110] | $\dfrac{dw}{dt} = \dfrac{A}{H} \cdot \dfrac{F_n}{v_c \cdot f} \cdot v_s + B \cdot e^{-\frac{Q}{RT}}$ | Abrasive, adhesive, and diffusive wear model. |
| Astakov [111,112] | $hs = \dfrac{dh_r}{dS} = \dfrac{100 \cdot (h_r - h_{r-i})}{(1 - l_i) \cdot f}$ | Surface wear model. |
| Attanasio [113,114] | $\begin{cases} \dfrac{dw}{dt} = D(T) \cdot e^{-\frac{Q}{RT}} \\ D(T) = D_1 \cdot T^3 + D_2 \cdot T^2 + D_3 \cdot T + D_4 \end{cases}$ | Diffusive wear model, presenting the $T$-dependent diffusive coefficient. |
| Pálmai [115] | $\dfrac{dW}{dt} = \dfrac{v_c}{W} \cdot \left[ A_\alpha + A_{th} \cdot e^{-\frac{B}{v_c^x + K \cdot W}} \right]$ | TW model, considering the effects of wear-induced cutting, force, and $T$ rise on TW. |
| Halila [116,117] | $W = N \cdot \sum_{i \geq i \min}^{I} \sum_{j=1} P_r^R(R_i) \cdot P_r^\phi(\phi_j) \cdot \dfrac{R_i^2 \cdot P}{2 \cdot H_t \cdot \tan(\phi_j)} \cdot v_c$ | TW model is dependent on the material removal rate. |

### 3.4. Tool Materials

As previously mentioned, due to poor $k$ from INCONEL® alloys, which lead to a substantial increase of $T$ in the three heat-zones when machining, the used tools are more prone to premature wear since the heat generated will end up creating BUE, which will rapidly degrade coatings and the tool material itself. The TW mechanisms, which include abrasive wear, adhesive wear, and plastic deformation, are following described. Severe TW is one of the key reasons for machining inefficiency [118].

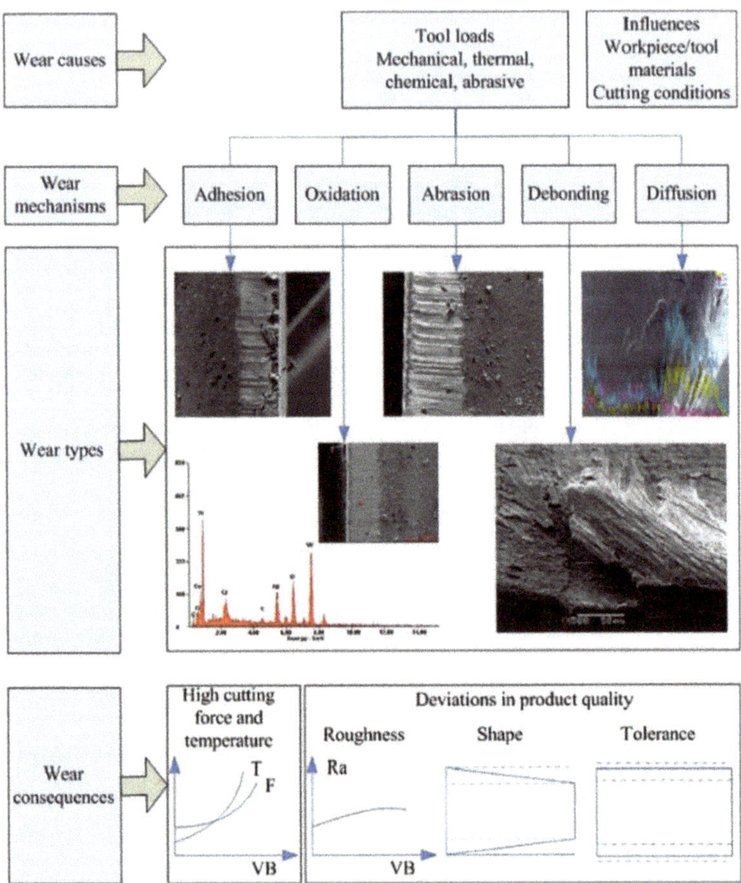

**Figure 14.** The wear causes, wear mechanisms, wear types, and wear consequences in the cutting of Ni based superalloys [118].

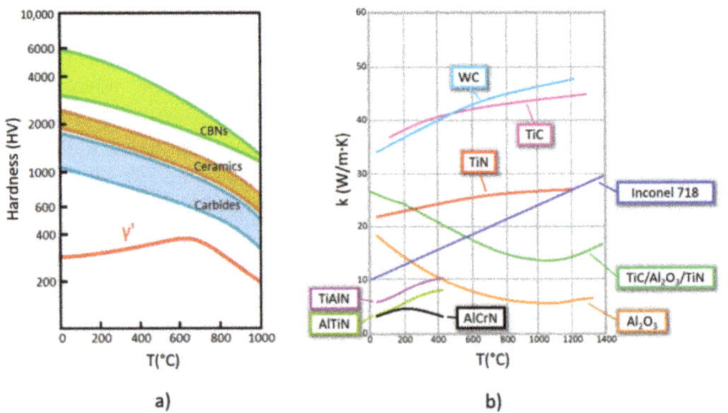

**Figure 15.** (**a**) Hot hardness characteristic curve of CBN, Ceramic and Carbide tool materials compared with the γ′ structure of INCONEL® 718, (**b**) Thermal conductivity of tungsten carbide (WC), INCONEL® 718 and different coatings for carbide tools against *T* (adapted from [119]).

Tool materials, depending on their application, may vary as hard metal, high-speed steel (HSS, and its variant HSS-Co), cermets, ceramics (where carbides are inserted), PcBN and many more. Table 11 presents the latest experimental observations and performance of non-coated tools in the machining of INCONEL® alloys.

**Table 11.** Machinability performance during non-coated tools assisted machining of INCONEL®.

| Author | Machining Type Material | Tool Material and Cutting Conditions | Remarks |
|---|---|---|---|
| Infante-García, et al. [86] | Turning INCONEL® 718 | PcBN inserts<br><br>$v_c$ = 300 m/min<br>$f$ = 0.07, 0.1 and 0.15 mm/rev<br>$a_p$ = 0.15, 0.25 and 0.5 mm | A significant peak in the machining forces, predicted by the Altintas force law [87], is related to the tool tip radius and the cutting parameters, after the second and successive passes. The rapid development and magnitude of the peak loads influence the tool wear progression, as a premature notch developed along the cutting edge, observed after the test with $a_p$ = 0.5 mm and $f$ = 0.15 mm/rev. The low machinability of INCONEL® 718, along with the brittleness of PCBN tools, leads to a premature failure of the cutting tool when this phenomenon is not considered in conventional turning. |
| Breidenstein, et al. [120] | Turning INCONEL® 718 | PcBN inserts.<br><br>$V_c$ = 200 m/min<br>$f$ = 0.1 mm/rev<br>$a_p$ = 0.2 mm | The ISO 3685:1993 [102] standard was used as a reference for the $VB$ measurement. Pulsed laser ablation (PLA) using ns-laser leads to a transformation of cBN to hBN in all considered laser parameter combinations. The cutting tool hardness is significantly decreased by the hBN formation, from over 3400 HV down to 1700 HV, However, not all laser parameters reduce hardness by the same amount. Tools with laser-prepared cutting edges achieve comparable tool wear to reference tools when applying appropriate laser parameters. There is an indication that the transformed hBN acts as a solid lubricant, which leads to a decrease in the cutting forces. |
| Rakesh and Chakradhar [121] | Turning INCONEL® 625 | Uncoated inserts ISO ref: CNMP120408<br><br>Level 1<br>$v_c$ = 40 m/min<br>$f$ = 0.05 mm/rev<br>$a_p$ = 0.2 mm<br>Dry machining<br><br>Level 2<br>$v_c$ = 50 m/min<br>$f$ = 0.1 mm/rev<br>$a_p$ = 0.4 mm<br>MQL<br><br>Level 3<br>$v_c$ = 60 m/min<br>$f$ = 0.16 mm/rev<br>$a_p$ = 0.6 mm<br>nMQL<br><br>Level 4<br>$v_c$ = 70 m/min<br>$f$ = 0.2 mm/rev<br>$a_p$ = 0.8 mm<br>Cryo N$_2$ ($l$)<br><br>Level 5<br>$v_c$ = 80 m/min<br>$f$ = 0.25 mm/rev<br>$a_p$ = 1 mm | The ISO 3685:1993 [102] standard was used as a reference for the $VB$ measurement. Among all four cooling conditions, the minimum $R_a$, $VB$, and main $F_c$ were obtained for cryogenic machining using N$_2$ ($l$)-air mixture as a coolant. This result indicates that cryogenic coolants are well-suitable for the machining of INCONEL® 625 alloys. The lowest $R_a$ = 0.481 μm was obtained under cryogenic machining with the parameter setting of $v_c$ = 60 m/min, $f$ = 0.15 mm/rev, and $a_p$ = 0.6 mm. Reductions of 38.49%, 34.56% and 30.08% in $R_a$ were achieved with cryogenic machining when compared to dry, MQL and nMQL, respectively, under the same parameters. The minimum value of $VB$ was observed at the lowest levels of cutting parameters, irrespective of the cooling conditions. Also, the lowest $VB$ = 85.52 μm was noticed under cryogenic machining, when machining with parameter settings of $v_c$ = 60 m/min, $f$ = 0.05 mm/rev and $a_p$ = 0.6 mm. The $VB$ reductions by cryogenic machining, when compared to dry, MQL and nMQL, are 20.32%, 11.42%, and 8.81%, respectively. |

**Table 11.** *Cont.*

| Author | Machining Type Material | Tool Material and Cutting Conditions | Remarks |
|---|---|---|---|
| Zhang, et al. [103] | Turning INCONEL® 718 | WC insert<br><br>$v_c$ = 80, 160 and 240 m/min<br>$f$ = 0.005, 0.01 and 0.015 mm/rev<br>$a_p$ = 0.15, 0.2 and 0.25 mm | A high-speed ultrasonic vibration cutting (HUVC) method has been proposed for the precision machining of INCONEL® 718. The TGRA $L_9$ array method was used for analysis. Owing to the limitations of the cooling pressure and duty cycle, the useful highspeed stable region for INCONEL® 718 was $v_c$ = 80–300 m/min. In this range, compared to the conventional effective cutting region, the cutting efficiency was significantly improved. The HUVC-Taylor's equation developed in this study aimed to provide a comprehensive understanding of the most recently proposed high-speed ultrasonic precision machining methods and provided guidance for appropriate practical applications in the future. The impact effect due to the tool-workpiece separation was a factor that needs to be suppressed. This effect was the core reason for the failed region in ultra-high-speed machining. In this regard, developing the impact-resistant tool could be seen as the next meaningful work for further cutting speed enhancement. |
| Hoier, et al. [122] | Turning INCONEL® 718 | WC-Co<br><br>$v_c$ = 30 m/min<br>$f$ = 0.075 mm/rev<br>$a_p$ = 1 mm<br>$t$ = 16.2 min | $VB$ was measured according to the instructions of the ISO 3685:1999 standard [102] It was characterized using white-light interferometry (WLI), scanning electron microscopy (SEM), and electron backscatter diffraction (EBSD). Wear topography and surface-induced plastic deformation were evaluated. Abrasion marks and grooves on different length scales indicate that $VB$ was caused by abrasion during sliding contact of the tool with the workpiece. Examination of worn WC grains employing EBSD proved to be a suitable method to assess the contribution of plastic deformation to TW in metal cutting. This method is of particular interest to developing a deeper understanding of relative wear rates associated with the machining of different workpiece classes and alloys. The parameter $VB$ = 185 μm is below the maximum limit imposed by the standard. |

### 3.5. Tool Coatings

Some tool materials have enough hardness to cut through INCONEL®, as it is shown in Figure 15, although others require a coating to protect the core material from abrasion when machining. The binomial substrate/coating is selected, as a function of specific requisites, from each application which often demands from the cutting tool antagonistic characteristics like tenacity and hardness. The usage of coated tools is highly advantageous from a production point of view, not because it is only possible to escalate $v_c$ or $s$ values, but also to promote better quality to the fabricated components, and some of them can be multilayer, in which each layer has its unique function. The preferred manufacturing processes to make coated tools are metallurgical powder processes, chemical vapour deposition (CVD) or even physical vapour deposition (PVD).

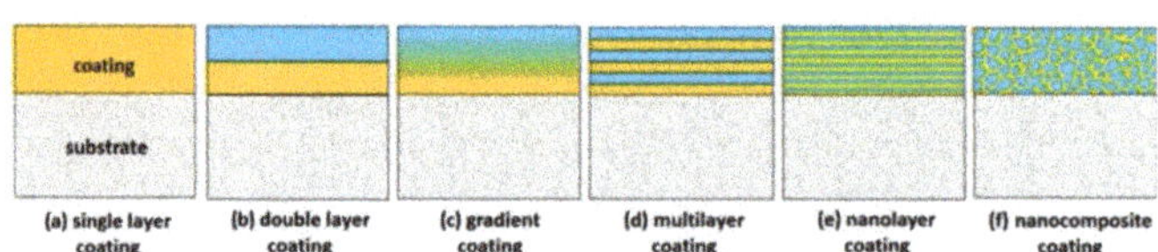


**Figure 16.** Scheme of how different types of coatings look when applied on the substrate [123].

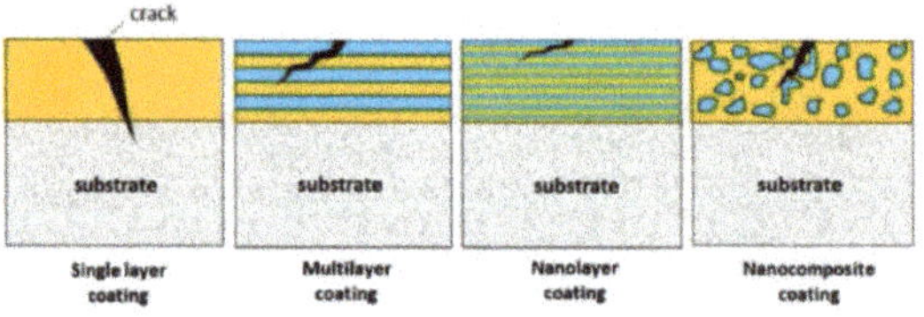


**Figure 17.** Crack propagation behaviour for each of the common coating structures [124].

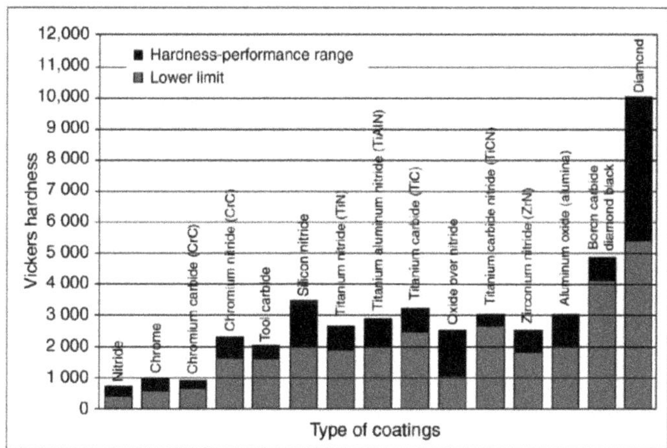

**Figure 18.** The hardness of different coating materials with a lower limit and suitable performance range [125].

A schematic of how different coatings can appear in a tool is presented in Figure 16, whereas Figure 17 demonstrates how crack propagation occurs inside the coatings due to TW. Figure 18 shows the lower limits and the hardness performance range of some of the most used tool coatings. Examples of preferred coatings to machine INCONEL® 718 include TiAlN, TiAlCrN, TiCN, TiN/AlTiN, and TiAlCrSiYN/TiAlCrN [126].

Table 12 presents the latest experimental observations and performance of coated tools in the machining of INCONEL®.

**Table 12.** Machinability performance during coated tools assisted machining of INCONEL®.

| Author | Machining Type Material | Tool Coatings and Cutting Conditions | Remarks |
|---|---|---|---|
| Zhao, et al. [53] | Turning INCONEL® 718 | TiAlN coated carbide tool. WC uncoated tool $v_c$ = 40, 80 and 120 m/min $f$ = 0.1 mm/rev | Irrespectively of the tool type, the tool-chip contact length was decreased with the increase of $v_c$. The tool-chip contact length and crater wear for the PVD AlTiN coated carbide tool was decreased compared with that of the uncoated carbide tool in dry orthogonal cutting of INCONEL® 718. A PVD AlTiN coated carbide tool decreased by 5.94 μm or 5.26% the $h_{ch}$, compared with an uncoated tool at $v_c$ = 120 m/min. The maximum $T_{cut}$ for the PVD AlTiN coated carbide tool decreased by 14.00 °C or 2.11%, 25.00 °C or 3.64%, and 39.00 °C or 5.47%, compared with that for an uncoated WC tool in dry orthogonal cutting of INCONEL® 718 at $v_c$ = 40, 80 and 120 m/min, respectively. For uncoated carbide tools, the relative error ($\Sigma\|\delta\|$) from the predictive model-$R_{G-W}$, compared to the measured $T$ at $v_c$ = 40, 80 and 120 m/min was 31.83%. $\Sigma\|\delta\|$ from other predictive models were 56.16% (model-$R_{L-SH}$), 53.54% (model-$R_{SH}$), 96.06% (model-$R_{K-F}$), 157.62% (model-$R_{L-SU}$), 119.32% (model-$R_{G-W}$), 41.52% (model-$R_R$) and 96.87% (model-$R_{T-K}$). For PVD AlTiN coated carbide tools, $\Sigma\|\delta\|$ from the predictive model-$R_R$, compared with the measured $T$ at $v_c$ = 40, 80 and 120 m/min was 52.61%. $\Sigma\|\delta\|$ from other predictive models were 147.44% (model-$R_{L-SH}$), 159.99% (model-$R_{SH}$), 76.48% (model-$R_{K-F}$), 63.00% (model-$R_{L-SU}$), 128.72% (model-$R_{G-W}$), 136.25% (model-$R_{B-K}$) and 76.22% (model-$R_{T-K}$). |

**Table 12.** *Cont.*

| Author | Machining Type Material | Tool Coatings and Cutting Conditions | Remarks |
|---|---|---|---|
| Montazeri, et al. [126] | Turning INCONEL® 718 | Uncoated carbide tool, TiAlN-coated carbide tool, Al-Si coated carbide tool. <br><br> $v_c$ = 50 m/min <br> $f$ = 0.1 mm/rev <br> $a_p$ = 0.15 mm | Results showed that the proposed soft Al-Si coating can provide a solution to the outlined machining challenges of INCONEL® 718. The tool life of the Al-Si coating was more than 6× higher than that of the uncoated tool and around 3× higher than the TiAlN coating, and the $F_c$ of the soft Al-Si coating was around ½ that of the uncoated tool. These improvements can be attributed to better lubricity and frictional behaviour in the tool-chip interface due to the superior lubricity of the Al-Si coating, which resulted in lower adhesion and BUE formation, less contact pressure at the cutting zone and lower friction. |
| Agarwal, et al. [127] | Milling INCONEL® 718 | TiAlN-coated carbide insert <br><br> $s$ = 501, 902 and 1203 rpm <br> $f$ = 0.25 mm/tooth <br> $a_p$ = 1 mm <br> Emulsion Flood-Cooling (EFC) | The proposed method was implemented in the form of an automated computational program, and a series of experiments were performed to analyse the progression of the $VB$ area of the tool over the volume of material removed. Based on the outcomes of the present study, it has been realized that the image processing method presented in this study can accurately and efficiently evaluate the $VB$ width and wear area. Also, the proposed methodology was able to replicate the well-known curve of the $VB$ area versus the volume of material removed. |
| Liu, et al. [128] | Turning INCONEL® 625 | PVD-TiAlN coated carbide tools <br><br> $v_c$ = 25–175 m/min <br> $f$ = 0.02–0.3 mm/rev <br> $a_p$ = 0.5 mm <br> expected $t \geq 10$ min | The turning experiment of INCONEL® 625 with a PVD-TiAlN-coated carbide tool exceeded $t$ = 10 min. It was found that the main TW morphologies of the tool were the BUE, crater wear, chipping, tipping, and fracturing. The main TW mechanisms were abrasion, adhesion, and oxidation. Adhesion wear under low $v_c$ yielded a BUE wear morphology, whereas adhesion wear and oxidation wear under high $v_c$ resulted in crater wear. As $v_c$ and $f$ further increased, the tool began to peel, tip, or even fracture. A two-dimensional TW map based on $v_c$ and $f$ was drawn. In the wear map, three tool failure limits, i.e., BUE limit; crater wear limit; and chipping, tipping, and fracture limit were determined. A safety zone was planned to determine the optimum cutting parameter range. A method of optimizing the cutting parameters by combining the wear map, tool wear, and $R_a$ of the machined workpiece was proposed. The optimum cutting parameters were $v_c$ = 60 m/min, $f$ = 0.1 mm/rev, and $a_p$ = 0.5 mm. A complete tool life experiment was performed with the optimized cutting parameters and a tool life distribution model obeying a normal distribution was established. When reliability was 0.9, the recommended tool-life was 70 min. |
| Criado, et al. [129] | Turning INCONEL® 718 | TiAlN + TiN coated carbide insert. <br><br> $v_c$ = 250–300 m/min <br> $f$ = 0.1–0.15 mm/rev <br> $a_p$ = 0.15 mm <br><br> TiN-coated PcBN insert <br><br> $v_c$ = 50–70 m/min <br> $f$ = 0.1–0.15 mm/rev <br> $a_p$ = 0.25 mm | Carbide tools have a longer tool life than PcBN tools, but PcBN tools higher allow speeds between 4x and 6x. In terms of machined surface per edge, it has been proven that, at $f$ = 0.15 mm/ver, more machined volume is obtained, while for $f$ = 0.1 mm/rev, the machined surface quality is maintained in the PcBN tools. For this reason, the viability of using PcBN tools in finishing operations in INCONEL® 718 is demonstrated. The best combination found for the PcBN tool was at $v_c$ = 250 m/min and $f$ = 0.15 mm/rev. It was found that the tool life increases at low cutting speeds, although $f$ does not affect it significantly. The finished surface machined with the PcBN tool obtains a more constant behaviour and excellent $R_a$ during most of the machining. However, no significant changes were observed depending on the cutting conditions. |
| Saleem and Mumtaz [130] | Milling INCONEL® 625 | PVD TiAlN$_2$ coated carbide inserts. <br><br> Level 1 <br> $v_c$ = 35 m/min <br> $f$ = 0.08 mm/rev <br> $a_p$ = 0.25 mm <br><br> Level 2 <br> $v_c$ = 45 m/min <br> $f$ = 0.2 mm/rev <br> $a_p$ = 0.5 mm | The Taguchi L$_8$ array method was used in the analysis. $a_p$ is found to be the statistically significant parameter for tool life with a 95% confidence level. Tool life is most affected by $a_p$ followed by $f_z$ with percentile contributions (PCR) of 45.43% and 18.425%, respectively, with lower values of the parameters resulting in better performance in general. A maximum tool life of 42.8 min was achieved when machining was done employing $f_z$ = 0.08 mm/tooth and $a_p$ = 0.25 mm with $v_c$ = 45 m/min. A SEM analysis indicates adhesion, BUE, attrition and chipping to be the main wear mechanism in general. |

## 4. Discussion

After all that has been presented throughout this paper, a SWOT analysis was performed to discuss present perceptions of the INCONEL® machinability (Table 13), tool-wear (Table 14) and coatings utility to the tools (Table 15).

**Table 13.** SWOT analysis about INCONEL® machinability.

| | | Positive Factors | Negative Factors |
|---|---|---|---|
| Internal factors | | **Strengths** Components with creep resistance in high-temperature operation. | **Weakness** Residual stress, Microhardness, Poor $R_a$, High $F_c$. |
| External factors | | **Opportunities** Making use of new non-conventional processes that improve machinability, like EDM and additive manufacturing (AM). Automobile industry and turbine blade manufacturing. | **Threats** Machining cost-effectiveness due to high $t$ and TW. |

**Table 14.** SWOT analysis of TW resultant from INCONEL® machining.

| | | Positive Factors | Negative Factors |
|---|---|---|---|
| Internal factors | | **Strengths** PcBN, carbide tools, ceramics and cermets can withstand high-temperature machining | **Weakness** Hardness drops with $T$ rising. |
| External factors | | **Opportunities** With laser assistance, hBN has better lubricity, lower hardness, and higher tenacity than cBN. | **Threats** BUE, Abrasion, Adhesion, Debonding, Diffusion, Oxidation. |

**Table 15.** SWOT analysis about cutting tools coatings.

| | | Positive Factors | Negative Factors |
|---|---|---|---|
| Internal factors | | **Strengths** Coatings prolong the effective tool's life. | **Weakness** Hardness drops with $T$ rising. After a while, the coating starts to debond from the substrate |
| External factors | | **Opportunities** New coatings with better lubricity are being researched and developed (R&D'd). | **Threats** Some high-temperature applications with INCONEL® are not justifiable. Preference for uncoated PcBN tools. BUE, Abrasion, Adhesion, Debonding, Diffusion, Oxidation. |

## 5. Conclusions

Despite significant progress in the traditional cutting tool technologies, the machining of INCONEL® 718 and 625 is still considered a great challenge because of the intrinsic characteristics of those Ni-superalloys. It is notable, nevertheless, that there has been a pursuit to bring ease to conventional processes, resulting from the evolution of techniques and tool materials, to get better machinability with the Ni-based superalloys. From another point of view, by introducing non-conventional processes and assists like EDM, and hybrid techniques such as LAM and UAT, the evolution differential has a great potential to bring down manufacturing costs. Likewise, the conventional processes have had several improvements in the last five years, as reviewed along all states of the art, calling upon Taguchi DOE methods for improving tool-wear and for improving $R_a$, either with a lubrication environment or not. This is important for the own component's wear resistance. A constructive criticism is made of the usage of TGRA and DOE methods, which were several times noticed to be used by different research in the states of the art of this paper. It is efficient to take advantage of such powerful methods to evaluate a series of parameters in an $L_x$ array, and through the combination between them, to achieve the best result of $R_a$. Nonetheless, it is known that one of the main challenges in tackling INCONEL® machining is the high costs of the manufacturing processes, due to the elapsed time milling, and turning, and this key factor has been neglected. With this review paper, it is suggested to the forthcoming authors to take advantage of TGRA and ANOVA analyses, concerning the achievement of low-cost solutions when machining INCONEL®, at the same time quality is preserved by taking $R_a$ parameter into account as it has been done so far. The present work highlighted a large amount of information regarding INCONEL® 718 traditional machining and applications, within the academic and scientific community, compared to its counterpart INCONEL® 625. On the other hand, the INCONEL® 625 showed advancements in non-conventional processes due to difficulties at the onset of chip cutting. Henceforward, research work is planned with the prospect of delivering a review paper regarding the evolution of lubrication environments, allied to the traditional machining of INCONEL®.

**Author Contributions:** Conceptualization: A.F.V.P., F.J.G.S. and R.D.S.G.C.; methodology: A.F.V.P., F.J.G.S. and R.D.S.G.C.; validation: V.F.C.S., N.P.V.S. and A.M.P.J.; formal analysis: F.J.G.S., V.F.C.S. and R.C.M.S.-C.; investigation: A.F.V.P.; data curation: F.J.G.S., R.C.M.S.-C. and A.M.P.J.; writing—original draft preparation: A.F.V.P.; writing—review and editing: V.F.C.S., F.J.G.S., R.D.S.G.C and R.C.M.S.-C.; visualization: R.C.M.S.-C. and A.M.P.J.; supervision: F.J.G.S. and R.D.S.G.C.; project administration: F.J.G.S.; funding acquisition: F.J.G.S. All authors have read and agreed to the published version of the manuscript.

**Funding:** The work is developed under the "DRIVOLUTION—Transition to the factory of the future", with the reference DRIVOLUTION/BL/01/2023 research project, supported by European Structural and Investments Funds with the "Portugal2020" program scope.

**Data Availability Statement:** No new data was created.

**Acknowledgments:** The authors thank ISEP and INEGI for their support.

**Conflicts of Interest:** The authors declare no conflict of interest.

## References

1. Ghiban, B.; Elefterie, C.F.; Guragata, C.; Bran, D. Requirements of Inconel 718 alloy for aeronautical applications. *AIP Conf. Proc.* **2018**, *1932*, 030016. [CrossRef]
2. Qadri, S.; Harmain, G.; Wani, M. Influence of Tool Tip Temperature on Crater Wear of Ceramic Inserts During Turning Process of Inconel-718 at Varying Hardness. *Tribol. Ind.* **2020**, *42*, 310–326. [CrossRef]
3. Nomoto, H. Development in materials for ultra-supercritical and advanced ultra-supercritical steam turbines. In *Advances in Steam Turbines for Modern Power Plants*, 2nd ed.; Chapter 13; Tanuma, T., Ed.; Woodhead Publishing: Sawston, UK, 2022; pp. 309–327.
4. Ghosh, R.N. Creep Life Predictions of Engineering Components: Problems & Prospects. *Procedia Eng.* **2013**, *55*, 599–606. [CrossRef]
5. Kassner, M.E. Chapter 10—Creep Fracture. In *Fundamentals of Creep in Metals and Alloys*, 3rd ed.; Kassner, M.E., Ed.; Butterworth-Heinemann: Boston, MA, USA, 2015; pp. 233–260.
6. Kassner, M.E. Chapter 11—$\gamma/\gamma'$ Nickel-Based Superalloys. In *Fundamentals of Creep in Metals and Alloys*, 3rd ed.; Kassner, M.E., Ed.; Butterworth-Heinemann: Boston, MA, USA, 2015; pp. 261–273.
7. Weber, J.H.; Banerjee, M.K. Nickel and Nickel Alloys: An Overview. In *Reference Module in Materials Science and Materials Engineering*; Elsevier: Amsterdam, The Netherlands, 2019.
8. Liu, L.; Zhang, J.; Ai, C. Nickel-Based Superalloys. In *Encyclopedia of Materials: Metals and Alloys*; Caballero, F.G., Ed.; Elsevier: Oxford, UK, 2022; pp. 294–304.

9.  Ashby, M.F. *Materials Selection in Mechanical Design*; Elsevier Science: Amsterdão, Paises Baixos, 2016.
10. International Nickel Company. *Monel, Inconel, Nickel, and Nickel Alloys*; International Nickel Company: Toronto, ON, Canada, 1947.
11. Dai, H.; Shi, S.; Yang, L.; Hu, J.; Liu, C.; Guo, C.; Chen, X. Effects of elemental composition and microstructure inhomogeneity on the corrosion behavior of nickel-based alloys in hydrofluoric acid solution. *Corros. Sci.* **2020**, *176*, 108917. [CrossRef]
12. Deng, D. Additively Manufactured Inconel 718: Microstructures and Mechanical Properties. Licentiate Thesis, Comprehensive Summary. Linköping University Electronic Press, Linköping, Sweden, 2018.
13. Joshi, G.R.; Badheka, V.J.; Darji, R.S.; Oza, A.D.; Pathak, V.J.; Burduhos-Nergis, D.D.; Burduhos-Nergis, D.P.; Narwade, G.; Thirunavukarasu, G. The Joining of Copper to Stainless Steel by Solid-State Welding Processes: A Review. *Materials* **2022**, *15*, 7234. [CrossRef]
14. Wei, C.; Wang, Z.; Chen, J. Sulfuration corrosion failure analysis of Inconel 600 alloy heater sleeve in high-temperature flue gas. *Eng. Fail. Anal.* **2022**, *135*, 106111. [CrossRef]
15. Xu, D.; Guo, S. Chapter 2—Corrosion Types and Elemental Effects of Ni-Based and FeCrAl Alloys. In *Corrosion Characteristics, Mechanisms and Control Methods of Candidate Alloys in Sub- and Supercritical Water*; Xu, D., Guo, S., Eds.; Springer: Singapore, 2022; pp. 23–49.
16. Vinod, K.; Udaya Ravi, M.; Yuvaraja, N. A Study of Surface Morphology and Wear Rate Prediction of Coated Inconel 600, 625 and 718 Specimens. *Int. J. Sci. Acad. Res. (IJSAR)* **2023**, *3*, 1–9. [CrossRef]
17. Dhananchezian, M. Influence of variation in cutting velocity on temperature, surface finish, chip form and insert after dry turning Inconel 600 with TiAlN carbide insert. *Mater. Today Proc.* **2021**, *46*, 8271–8274. [CrossRef]
18. Mahesh, K.; Philip, J.T.; Joshi, S.N.; Kuriachen, B. Machinability of Inconel 718: A critical review on the impact of cutting temperatures. *Mater. Manuf. Process.* **2021**, *36*, 753–791. [CrossRef]
19. Yin, Q.; Liu, Z.; Wang, B.; Song, Q.; Cai, Y. Recent progress of machinability and surface integrity for mechanical machining Inconel 718: A review. *Int. J. Adv. Manuf. Technol.* **2020**, *109*, 215–245. [CrossRef]
20. Singh, N.; Routara, B.C.; Nayak, R.K. Study of machining characteristics of Inconel 601 with cryogenic cooled electrode in EDM using RSM. *Mater. Today Proc.* **2018**, *5*, 24277–24286. [CrossRef]
21. Harish, U.; Mruthunjaya, M.; Yogesha, K.; Siddappa, P.; Anil, K.K. Enhancing the performance of inconel 601 alloy by erosion resistant WC-CR-CO coated material. *J. Eng. Sci. Technol.* **2022**, *17*, 0379–0390.
22. Singh, J.B. Physical Metallurgy of Alloy 625. In *Alloy 625: Microstructure, Properties and Performance*; Chapter 3; Singh, J.B., Ed.; Springer Nature: Singapore, 2022; pp. 67–110.
23. Waghmode, S.P.; Dabade, U.A. Optimization of process parameters during turning of Inconel 625. *Mater. Today Proc.* **2019**, *19*, 823–826. [CrossRef]
24. Singh, J.B. Chapter 7—Corrosion Behavior of Alloy 625. In *Alloy 625: Microstructure, Properties and Performance*; Singh, J.B., Ed.; Springer Nature: Singapore, 2022; pp. 241–291.
25. Kosaraju, S.; Vijay Kumar, M.; Sateesh, N. Optimization of Machining Parameter in Turning Inconel 625. *Mater. Today Proc.* **2018**, *5*, 5343–5348. [CrossRef]
26. Singh, J.D. Chapter 1 Introduction. In *Alloy 625: Microstructure, Properties and Performance*; Singh, J.B., Ed.; Springer Nature Singapore: Singapore, 2022; pp. 1–27.
27. Yin, M.; Shao, Y.; Kang, X.; Long, J.; Zhang, X. Fretting corrosion behavior of WC-10Co-4Cr coating on Inconel 690 alloy by HVOF thermal spraying. *Tribol. Int.* **2023**, *177*, 107975. [CrossRef]
28. Zhang, Q.; Zhang, J.; Zhuang, Y.; Lu, J.; Yao, J. Hot Corrosion and Mechanical Performance of Repaired Inconel 718 Components via Laser Additive Manufacturing. *Materials* **2020**, *13*, 2128. [CrossRef]
29. Ding, J.; Xue, S.; Shang, Z.; Li, J.; Zhang, Y.; Su, R.; Niu, T.; Wang, H.; Zhang, X. Characterization of precipitation in gradient Inconel 718 superalloy. *Mater. Sci. Eng. A* **2021**, *804*, 140718. [CrossRef]
30. Teixeira, Ó.; Silva, F.J.G.; Atzeni, E. Residual stresses and heat treatments of Inconel 718 parts manufactured via metal laser beam powder bed fusion: An overview. *Int. J. Adv. Manuf. Technol.* **2021**, *113*, 3139–3162. [CrossRef]
31. Montazeri, S.; Aramesh, M.; Veldhuis, S.C. Novel application of ultra-soft and lubricious materials for cutting tool protection and enhancement of machining induced surface integrity of Inconel 718. *J. Manuf. Process.* **2020**, *57*, 431–443. [CrossRef]
32. Manikandan, S.G.K.; Sivakumar, D.; Kamaraj, M. 1—Physical metallurgy of alloy 718. In *Welding the Inconel 718 Superalloy*; Manikandan, S.G.K., Sivakumar, D., Kamaraj, M., Eds.; Elsevier: Amsterdam, The Netherlands, 2019; pp. 1–19.
33. Fayed, E.M.; Saadati, M.; Shahriari, D.; Brailovski, V.; Jahazi, M.; Medraj, M. Optimization of the Post-Process Heat Treatment of Inconel 718 Superalloy Fabricated by Laser Powder Bed Fusion Process. *Metals* **2021**, *11*, 144. [CrossRef]
34. Jeyapandiarajan, P.; Anthony, X.M. Evaluating the Machinability of Inconel 718 under Different Machining Conditions. *Procedia Manuf.* **2019**, *30*, 253–260. [CrossRef]
35. Pröbstle, M.; Neumeier, S.; Hopfenmüller, J.; Freund, L.P.; Niendorf, T.; Schwarze, D.; Göken, M. Superior creep strength of a nickel-based superalloy produced by selective laser melting. *Mater. Sci. Eng. A* **2016**, *674*, 299–307. [CrossRef]
36. Evans, R. 2—Selection and testing of metalworking fluids. In *Metalworking Fluids (MWFs) for Cutting and Grinding*; Astakhov, V.P., Joksch, S., Eds.; Woodhead Publishing: Sawston, UK, 2012; pp. 23–78.
37. Wang, C.; Ming, W.; Chen, M. Milling tool's flank wear prediction by temperature dependent wear mechanism determination when machining Inconel 182 overlays. *Tribol. Int.* **2016**, *104*, 140–156. [CrossRef]

38. Zhang, P.; Li, J.; Yu, H.L.; Tu, X.H.; Li, W. An experimental study on the fretting wear behavior of Inconel 600 and 690 in pure water. *Wear* **2021**, *486–487*, 203995. [CrossRef]
39. Mohapatra, S.; Rahul; Kumar Sahoo, A. Comparative study of Inconel 601, 625, 718, 825 super-alloys during Electro-Discharge Machining. *Mater. Today Proc.* **2022**, *56*, 226–230. [CrossRef]
40. Liu, X.; Fan, J.; Cao, K.; Chen, F.; Yuan, R.; Liu, D.; Tang, B.; Kou, H.; Li, J. Creep anisotropy behavior, deformation mechanism, and its efficient suppression method in Inconel 625 superalloy. *J. Mater. Sci. Technol.* **2023**, *133*, 58–76. [CrossRef]
41. Kurniawan, R.; Park, G.C.; Park, K.M.; Zhen, Y.; Kwak, Y.I.; Kim, M.C.; Lee, J.M.; Ko, T.J.; Park, C.-S. Machinability of modified Inconel 713C using a WC TiAlN-coated tool. *J. Manuf. Process.* **2020**, *57*, 409–430. [CrossRef]
42. Tanaka, H.; Sugihara, T.; Enomoto, T. High Speed Machining of Inconel 718 Focusing on Wear Behaviors of PCBN Cutting Tool. *Procedia CIRP* **2016**, *46*, 545–548. [CrossRef]
43. Şirin, Ş.; Kıvak, T. Effects of hybrid nanofluids on machining performance in MQL-milling of Inconel X-750 superalloy. *J. Manuf. Process.* **2021**, *70*, 163–176. [CrossRef]
44. Gupta, M.K.; Song, Q.; Liu, Z.; Sarikaya, M.; Jamil, M.; Mia, M.; Singla, A.K.; Khan, A.M.; Khanna, N.; Pimenov, D.Y. Environment and economic burden of sustainable cooling/lubrication methods in machining of Inconel-800. *J. Clean. Prod.* **2021**, *287*, 125074. [CrossRef]
45. Yadav, R.K.; Abhishek, K.; Mahapatra, S.S.; Nandi, G. A study on machinability aspects and parametric optimization of Inconel 825 using Rao1, Rao2, Rao3 approach. *Mater. Today Proc.* **2021**, *47*, 2784–2789. [CrossRef]
46. *ASTM Standard B637-16*; Standard Specification for Precipitation-Hardening and Cold Worked Nickel Alloy Bars, Forgings, and Forging Stock for Moderate or High Temperature Service. ASTM International: West Conshohocken, PA, USA, 2016; Volume 7. [CrossRef]
47. Astakhov, V.P.; Godlevskiy, V. 3—Delivery of metalworking fluids in the machining zone. In *Metalworking Fluids (MWFs) for Cutting and Grinding*; Astakhov, V.P., Joksch, S., Eds.; Woodhead Publishing: Sawston, UK, 2012; pp. 79–134.
48. Soffel, F.; Eisenbarth, D.; Hosseini, E.; Wegener, K. Interface strength and mechanical properties of Inconel 718 processed sequentially by casting, milling, and direct metal deposition. *J. Mater. Process. Technol.* **2021**, *291*, 117021. [CrossRef]
49. Sierra-Soraluce, A.; Li, G.; Santofimia, M.J.; Molina-Aldareguia, J.M.; Smith, A.; Muratori, M.; Sabirov, I. Effect of microstructure on tensile properties of quenched and partitioned martensitic stainless steels. *Mater. Sci. Eng. A* **2023**, *864*, 144540. [CrossRef]
50. Ding, H.; Zou, B.; Wang, X.; Liu, J.; Li, L. Microstructure, mechanical properties and machinability of 316L stainless steel fabricated by direct energy deposition. *Int. J. Mech. Sci.* **2023**, *243*, 108046. [CrossRef]
51. Danish, M.; Ginta, T.L.; Yasir, M.; Rani AM, A. Chapter 1—Light alloys and their machinability. In *Machining of Light Alloys: Aluminum, Titanium, and Magnesium*; Carou, D., Davim, J., Eds.; Academic Press: Cambridge, MA, USA, 2018.
52. Kumar Wagri, N.; Petare, A.; Agrawal, A.; Rai, R.; Malviya, R.; Dohare, S.; Kishore, K. An overview of the machinability of alloy steel. *Mater. Today Proc.* **2022**, *62*, 3771–3781. [CrossRef]
53. Zhao, J.; Liu, Z.; Wang, B.; Hu, J. PVD AlTiN coating effects on tool-chip heat partition coefficient and cutting temperature rise in orthogonal cutting Inconel 718. *Int. J. Heat Mass Transf.* **2020**, *163*, 120449. [CrossRef]
54. Loewen, E.G.; Shaw, M.C. On the Analysis of Cutting-Tool Temperatures. *Trans. Am. Soc. Mech. Eng.* **2022**, *76*, 217–225. [CrossRef]
55. Shaw, M.C.; Cookson, J. *Metal Cutting Principles*; Oxford University Press: New York, NY, USA, 2005; Volume 2.
56. Kato, T.; Fujii, H. Energy Partition in Conventional Surface Grinding. *J. Manuf. Sci. Eng.* **1999**, *121*, 393–398. [CrossRef]
57. List, G.; Sutter, G.; Bouthiche, A. Cutting temperature prediction in high speed machining by numerical modelling of chip formation and its dependence with crater wear. *Int. J. Mach. Tools Manuf.* **2012**, *54–55*, 1–9. [CrossRef]
58. Gecim, B.; Winer, W.O. Transient Temperatures in the Vicinity of an Asperity Contact. *J. Tribol.* **1985**, *107*, 333–341. [CrossRef]
59. Reznikov, A.; Reznikov, A. Thermophysical aspects of metal cutting processes. *Mashinostroenie Mosc.* **1981**, *212*.
60. Grzesik, W. *Advanced Machining Processes of Metallic Materials: Theory, Modelling and Applications*; Elsevier: Amsterdam, The Netherlands, 2008.
61. Andresen, P.L. 5—Understanding and predicting stress corrosion cracking (SCC) in hot water. In *Stress Corrosion Cracking of Nickel Based Alloys in Water-Cooled Nuclear Reactors*; Féron, D., Staehle, R.W., Eds.; Woodhead Publishing: Sawston, UK, 2016; pp. 169–238.
62. Féron, D.; Guerre, C.; Herms, E.; Laghoutaris, P. 9—Stress corrosion cracking of Alloy 600: Overviews and experimental techniques. In *Stress Corrosion Cracking of Nickel Based Alloys in Water-Cooled Nuclear Reactors*; Féron, D., Staehle, R.W., Eds.; Woodhead Publishing: Sawston, UK, 2016; pp. 325–353.
63. Dai, X.; Zhuang, K.; Pu, D.; Zhang, W.; Ding, H. An Investigation of the Work Hardening Behavior in Interrupted Cutting Inconel 718 under Cryogenic Conditions. *Materials* **2020**, *13*, 2202. [CrossRef] [PubMed]
64. Guimaraes, M.C.R.; Fogagnolo, J.B.; Paiva, J.M.; Veldhuis, S.C.; Diniz, A.E. Evaluation of milling parameters on the surface integrity of welded inconel 625. *J. Mater. Res. Technol.* **2022**, *20*, 2611–2628. [CrossRef]
65. Li, X.; Liu, X.; Yue, C.; Liang, S.Y.; Wang, L. Systematic review on tool breakage monitoring techniques in machining operations. *Int. J. Mach. Tools Manuf.* **2022**, *176*, 103882. [CrossRef]
66. Zheng, H.; Liu, K. Machinability of Engineering Materials. In *Handbook of Manufacturing Engineering and Technology*; Nee, A., Ed.; Springer London: London, UK, 2013; pp. 1–34.
67. Wang, Z.; Kovvuri, V.; Araujo, A.; Bacci, M.; Hung, W.N.P.; Bukkapatnam, S.T.S. Built-up-edge effects on surface deterioration in micromilling processes. *J. Manuf. Process.* **2016**, *24*, 321–327. [CrossRef]

68. Nouari, M.; Haddag, B.; Moufki, A.; Atlati, S. Chapter 2—Investigation on the built-up edge process when dry machining aeronautical aluminum alloys. In *Machining of Light Alloys: Aluminum, Titanium, and Magnesium*; Carou, D., Davim, J., Eds.; Academic Press: Cambridge, MA, USA, 2018.

69. Bartolomeis, A.D.; Newman, S.T.; Shokrani, A. Initial investigation on Surface Integrity when Machining Inconel 718 with Conventional and Electrostatic Lubrication. *Procedia CIRP* **2020**, *87*, 65–70. [CrossRef]

70. Anburaj, R.; Pradeep Kumar, M. Experimental studies on cryogenic $CO_2$ face milling of Inconel 625 superalloy. *Mater. Manuf. Process.* **2021**, *36*, 814–826. [CrossRef]

71. Lakner, T.; Hardt, M. A Novel Experimental Test Bench to Investigate the Effects of Cutting Fluids on the Frictional Conditions in Metal Cutting. *J. Manuf. Mater. Process.* **2020**, *4*, 45. [CrossRef]

72. Bonnet, C.; Valiorgue, F.; Rech, J.; Claudin, C.; Hamdi, H.; Bergheau, J.M.; Gilles, P. Identification of a friction model—Application to the context of dry cutting of an AISI 316L austenitic stainless steel with a TiN coated carbide tool. *Int. J. Mach. Tools Manuf.* **2008**, *48*, 1211–1223. [CrossRef]

73. Kuzu, A.T.; Berenji, K.R.; Ekim, B.C.; Bakkal, M. The thermal modeling of deep-hole drilling process under MQL condition. *J. Manuf. Process.* **2017**, *29*, 194–203. [CrossRef]

74. Kim, E.J.; Lee, C.M. A Study on the Optimal Machining Parameters of the Induction Assisted Milling with Inconel 718. *Materials* **2019**, *12*, 233. [CrossRef] [PubMed]

75. Vignesh, M.; Ramanujam, R. Chapter 9—Laser-assisted high speed machining of Inconel 718 alloy. In *High Speed Machining*; Gupta, K., Paulo Davim, J., Eds.; Academic Press: Cambridge, MA, USA, 2020; pp. 243–262.

76. Okafor, A.C. Chapter 5—Cooling and machining strategies for high speed milling of titanium and nickel super alloys. In *High Speed Machining*; Gupta, K., Paulo Davim, J., Eds.; Academic Press: Cambridge, MA, USA, 2020; pp. 127–161.

77. Hadi, M.A.; Ghani, J.A.; Haron, C.H.C.; Kasim, M.S. Comparison between Up-milling and Down-milling Operations on Tool Wear in Milling Inconel 718. *Procedia Eng.* **2013**, *68*, 647–653. [CrossRef]

78. Bo, P.; Fan, H.; Bartoň, M. Efficient 5-axis CNC trochoidal flank milling of 3D cavities using custom-shaped cutting tools. *Comput.-Aided Des.* **2022**, *151*, 103334. [CrossRef]

79. Liu, D.; Zhang, Y.; Luo, M.; Zhang, D. Investigation of Tool Wear and Chip Morphology in Dry Trochoidal Milling of Titanium Alloy Ti–6Al–4V. *Materials* **2019**, *12*, 1937. [CrossRef]

80. Pleta, A.; Nithyanand, G.; Niaki, F.A.; Mears, L. Identification of optimal machining parameters in trochoidal milling of Inconel 718 for minimal force and tool wear and investigation of corresponding effects on machining affected zone depth. *J. Manuf. Process.* **2019**, *43*, 54–62. [CrossRef]

81. Shankar, S.; Mohanraj, T.; Pramanik, A. Tool Condition Monitoring While Using Vegetable Based Cutting Fluids during Milling of Inconel 625. *J. Adv. Manuf. Syst.* **2019**, *18*, 563–581. [CrossRef]

82. Alonso, U.; Veiga, F.; Suárez, A.; Gil Del Val, A. Characterization of Inconel 718®superalloy fabricated by wire Arc Additive Manufacturing: Effect on mechanical properties and machinability. *J. Mater. Res. Technol.* **2021**, *14*, 2665–2676. [CrossRef]

83. Boozarpoor, M.; Teimouri, R.; Yazdani, K. Comprehensive study on effect of orthogonal turn-milling parameters on surface integrity of Inconel 718 considering production rate as constrain. *Int. J. Lightweight Mater. Manuf.* **2021**, *4*, 145–155. [CrossRef]

84. Amigo, F.J.; Urbikain, G.; Pereira, O.; Fernández-Lucio, P.; Fernández-Valdivielso, A.; de Lacalle, L.N.L. Combination of high feed turning with cryogenic cooling on Haynes 263 and Inconel 718 superalloys. *J. Manuf. Process.* **2020**, *58*, 208–222. [CrossRef]

85. Raykar, S.J.; Chaugule, Y.G.; Pasare, V.I.; Sawant, D.A.; Patil, U.N. Analysis of microhardness and degree of work hardening (DWH) while turning Inconel 718 with high pressure coolant environment. *Mater. Today Proc.* **2022**, *59*, 1088–1093. [CrossRef]

86. Infante-García, D.; Diaz-Álvarez, J.; Cantero, J.-L.; Muñoz-Sánchez, A.; Miguélez, M.-H. Influence of the undeformed chip cross section in finishing turning of Inconel 718 with PCBN tools. *Procedia CIRP* **2018**, *77*, 122–125. [CrossRef]

87. Meyer, R.; Köhler, J.; Denkena, B. Influence of the tool corner radius on the tool wear and process forces during hard turning. *Int. J. Adv. Manuf. Technol.* **2012**, *58*, 933–940. [CrossRef]

88. Makhesana, M.A.; Patel, K.M.; Krolczyk, G.M.; Danish, M.; Singla, A.K.; Khanna, N. Influence of $MoS_2$ and graphite-reinforced nanofluid-MQL on surface roughness, tool wear, cutting temperature and microhardness in machining of Inconel 625. *CIRP J. Manuf. Sci. Technol.* **2023**, *41*, 225–238. [CrossRef]

89. Airao, J.; Nirala, C.K.; Khanna, N. Novel use of ultrasonic-assisted turning in conjunction with cryogenic and lubrication techniques to analyze the machinability of Inconel 718. *J. Manuf. Process.* **2022**, *81*, 962–975. [CrossRef]

90. Venkatesan, K.; Nagendra, K.U.; Anudeep, C.M.; Cotton, A.E. Experimental Investigation and Parametric Optimization on Hole Quality Assessment During Micro-drilling of Inconel 625 Superalloy. *Arab. J. Sci. Eng.* **2021**, *46*, 2283–2309. [CrossRef]

91. Cherrared, D. Numerical simulation of film cooling a turbine blade through a row of holes. *J. Therm. Eng.* **2017**, *3*, 1110. [CrossRef]

92. Neo, D.W.K.; Liu, K.; Kumar, A.S. High throughput deep-hole drilling of Inconel 718 using PCBN gun drill. *J. Manuf. Process.* **2020**, *57*, 302–311. [CrossRef]

93. Sahoo, A.K.; Jeet, S.; Bagal, D.K.; Barua, A.; Pattanaik, A.K.; Behera, N. Parametric optimization of CNC-drilling of Inconel 718 with cryogenically treated drill-bit using Taguchi-Whale optimization algorithm. *Mater. Today Proc.* **2022**, *50*, 1591–1598. [CrossRef]

94. Ratnam, C.; Adarsha Kumar, K.; Murthy, B.S.N.; Venkata Rao, K. An experimental study on boring of Inconel 718 and multi response optimization of machining parameters using Response Surface Methodology. *Mater. Today Proc.* **2018**, *5*, 27123–27129. [CrossRef]

95. Liu, L.; Thangaraj, M.; Karmiris-Obratański, P.; Zhou, Y.; Annamalai, R.; Machnik, R.; Elsheikh, A.; Markopoulos, A.P. Optimization of Wire EDM Process Parameters on Cutting Inconel 718 Alloy with Zinc-Diffused Coating Brass Wire Electrode Using Taguchi-DEAR Technique. *Coatings* **2022**, *12*, 1612. [CrossRef]

96. Kliuev, M.; Kutin, A.; Wegener, K. Electrode wear pattern during EDM milling of Inconel 718. *Int. J. Adv. Manuf. Technol.* **2021**, *117*, 2369–2375. [CrossRef]

97. Manikandan, N.; Binoj, J.S.; Thejasree, P.; Sasikala, P.; Anusha, P. Application of Taguchi method on Wire Electrical Discharge Machining of Inconel 625. *Mater. Today Proc.* **2021**, *39*, 121–125. [CrossRef]

98. Hussain, A.; Kumar Sharma, A.; Preet Singh, J. Maximizing MRR of Inconel 625 machining through process parameter optimization of EDM. *Mater. Today Proc.* **2022**; *In Press.* [CrossRef]

99. Rahul; Datta, S.; Biswal, B.B.; Mahapatra, S.S. Machinability analysis of Inconel 601, 625, 718 and 825 during electro-discharge machining: On evaluation of optimal parameters setting. *Measurement* **2019**, *137*, 382–400. [CrossRef]

100. Farooq, M.U.; Anwar, S.; Kumar, M.S.; AlFaify, A.; Ali, M.A.; Kumar, R.; Haber, R. A Novel Flushing Mechanism to Minimize Roughness and Dimensional Errors during Wire Electric Discharge Machining of Complex Profiles on Inconel 718. *Materials* **2022**, *15*, 7330. [CrossRef]

101. *ISO 8688-2:1989(E)*; Tool Life Testing in Milling—Part 2: End milling. ISO: London, UK, 1989; 26p.

102. *ISO 3685:1993(E)*; Tool-Life Testing with Single-Point Turning Tools. ISO: London, UK, 1993; 53p.

103. Zhang, X.; Peng, Z.; Liu, L.; Zhang, X. A Tool Life Prediction Model Based on Taylor's Equation for High-Speed Ultrasonic Vibration Cutting Ti and Ni Alloys. *Coatings* **2022**, *12*, 1553. [CrossRef]

104. Archard, J.F. Contact and Rubbing of Flat Surfaces. *J. Appl. Phys.* **1953**, *24*, 981–988. [CrossRef]

105. Usui, E.; Shirakashi, T.; Kitagawa, T. Analytical Prediction of Three Dimensional Cutting Process—Part 3: Cutting Temperature and Crater Wear of Carbide Tool. *J. Eng. Ind.* **1978**, *100*, 236–243. [CrossRef]

106. Usui, E.; Shirakashi, T.; Kitagawa, T. Analytical prediction of cutting tool wear. *Wear* **1984**, *100*, 129–151. [CrossRef]

107. Takeyama, H.; Murata, R. Basic Investigation of Tool Wear. *J. Eng. Ind.* **1963**, *85*, 33–37. [CrossRef]

108. Childs, T.H.C.; Maekawa, K.; Obikawa, T.; Yamane, Y. *Metal Machining: Theory and Applications*; Arnold: London, UK, 2000.

109. Schmidt, C.; Frank, P.; Weule, H.; Schmidt, J.; Yen, Y.; Altan, T. Tool wear prediction and verification in orthogonal cutting. In Proceedings of the 6th CIRP Workshop on Modeling of Machining, Hamilton, ON, Canada, 19–20 May 2003; pp. 93–100.

110. Luo, X.; Cheng, K.; Holt, R.; Liu, X. Modeling flank wear of carbide tool insert in metal cutting. *Wear* **2005**, *259*, 1235–1240. [CrossRef]

111. Astakhov, V.P. Effects of the cutting feed, depth of cut, and workpiece (bore) diameter on the tool wear rate. *Int. J. Adv. Manuf. Technol.* **2007**, *34*, 631–640. [CrossRef]

112. Astakhov, V.P. Chapter 4 Cutting tool wear, tool life and cutting tool physical resource. In *Tribology and Interface Engineering Series*; Briscoe, B.J., Ed.; Elsevier: Amsterdam, The Netherlands, 2006; Volume 52, pp. 220–275.

113. Attanasio, A.; Ceretti, E.; Fiorentino, A.; Cappellini, C.; Giardini, C. Investigation and FEM-based simulation of tool wear in turning operations with uncoated carbide tools. *Wear* **2010**, *269*, 344–350. [CrossRef]

114. Attanasio, A.; Ceretti, E.; Rizzuti, S.; Umbrello, D.; Micari, F. 3D finite element analysis of tool wear in machining. *CIRP Ann.* **2008**, *57*, 61–64. [CrossRef]

115. Pálmai, Z. Proposal for a new theoretical model of the cutting tool's flank wear. *Wear* **2013**, *303*, 437–445. [CrossRef]

116. Halila, F.; Czarnota, C.; Nouari, M. A new abrasive wear law for the sticking and sliding contacts when machining metallic alloys. *Wear* **2014**, *315*, 125–135. [CrossRef]

117. Halila, F.; Czarnota, C.; Nouari, M. New stochastic wear law to predict the abrasive flank wear and tool life in machining process. *Proc. Inst. Mech. Eng. Part J. J. Eng. Tribol.* **2014**, *228*, 1243–1251. [CrossRef]

118. Wang, R.; Yang, D.; Wang, W.; Wei, F.; Lu, Y.; Li, Y. Tool Wear in Nickel-Based Superalloy Machining: An Overview. *Processes* **2022**, *10*, 2380. [CrossRef]

119. De Bartolomeis, A.; Newman, S.T.; Jawahir, I.S.; Biermann, D.; Shokrani, A. Future research directions in the machining of Inconel 718. *J. Mater. Process. Technol.* **2021**, *297*, 117260. [CrossRef]

120. Breidenstein, B.; Grove, T.; Krödel, A.; Sitab, R. Influence of hexagonal phase transformation in laser prepared PcBN cutting tools on tool wear in machining of Inconel 718. *Met. Powder Rep.* **2019**, *74*, 237–243. [CrossRef]

121. Rakesh, P.R.; Chakradhar, D. Machining performance comparison of Inconel 625 superalloy under sustainable machining environments. *J. Manuf. Process.* **2023**, *85*, 742–755. [CrossRef]

122. Hoier, P.; Malakizadi, A.; Krajnik, P.; Klement, U. Study of flank wear topography and surface-deformation of cemented carbide tools after turning Alloy 718. *Procedia CIRP* **2018**, *77*, 537–540. [CrossRef]

123. Sousa, V.F.C.; Silva, F.J.G. Recent Advances on Coated Milling Tool Technology—A Comprehensive Review. *Coatings* **2020**, *10*, 235. [CrossRef]

124. Sousa, V.F.C.; Da Silva, F.J.G.; Pinto, G.F.; Baptista, A.; Alexandre, R. Characteristics and Wear Mechanisms of TiAlN-Based Coatings for Machining Applications: A Comprehensive Review. *Metals* **2021**, *11*, 260. [CrossRef]

125. Jain, A.; Bajpai, V. Chapter 1—Introduction to high-speed machining (HSM). In *High Speed Machining*; Gupta, K., Paulo Davim, J., Eds.; Academic Press: Cambridge, MA, USA, 2020; pp. 1–25.

126. Montazeri, S.; Aramesh, M.; Rawal, S.; Veldhuis, S.C. Characterization and machining performance of a chipping resistant ultra-soft coating used for the machining of Inconel 718. *Wear* **2021**, *474–475*, 203759. [CrossRef]

127. Agarwal, A.; Potthoff, N.; Shah, A.M.; Mears, L.; Wiederkehr, P. Analyzing the evolution of tool wear area in trochoidal milling of Inconel 718 using image processing methodology. *Manuf. Lett.* **2022**, *33*, 373–379. [CrossRef]
128. Liu, E.; An, W.; Xu, Z.; Zhang, H. Experimental study of cutting-parameter and tool life reliability optimization in inconel 625 machining based on wear map approach. *J. Manuf. Process.* **2020**, *53*, 34–42. [CrossRef]
129. Criado, V.; Díaz-Álvarez, J.; Cantero, J.L.; Miguélez, M.H. Study of the performance of PCBN and carbide tools in finishing machining of Inconel 718 with cutting fluid at conventional pressures. *Procedia CIRP* **2018**, *77*, 634–637. [CrossRef]
130. Saleem, M.Q.; Mumtaz, S. Face milling of Inconel 625 via wiper inserts: Evaluation of tool life and workpiece surface integrity. *J. Manuf. Process.* **2020**, *56*, 322–336. [CrossRef]

*Article*

# Wear Behavior Phenomena of TiN/TiAlN HiPIMS PVD-Coated Tools on Milling Inconel 718

Vitor F. C. Sousa [1,2], Filipe Fernandes [1,3], Francisco J. G. Silva [1,2,*], Rúben D. F. S. Costa [1], Naiara Sebbe [1] and Rita C. M. Sales-Contini [4]

[1] ISEP, Polytechnic of Porto, Rua Dr. António Bernardino de Almeida, 4249-015 Porto, Portugal
[2] Associate Laboratory for Energy, Transports and Aerospace (LAETA, INEGI), Rua Dr. Roberto Frias 400, 4200-465 Porto, Portugal
[3] Department of Mechanical Engineering, CEMMPRE—Centre for Mechanical Engineering Materials and Processes, University of Coimbra, Rua Luís Reis Santos, 3030-788 Coimbra, Portugal
[4] Technological College of São José dos Campos, Avenida Cesare Mansueto Giulio Lattes, 1350 Distrito Eugênio de Melo, São José dos Campos 12247-014, SP, Brazil
* Correspondence: fgs@isep.ipp.pt; Tel.: +351-228340500

**Abstract:** Due to Inconel 718's high mechanical properties, even at higher temperatures, tendency to work-harden, and low thermal conductivity, this alloy is considered hard to machine. The machining of this alloy causes high amounts of tool wear, leading to its premature failure. There seems to be a gap in the literature, particularly regarding milling and finishing operations applied to Inconel 718 parts. In the present study, the wear behavior of multilayered PVD HiPIMS (High-power impulse magnetron sputtering)-coated TiN/TiAlN end-mills used for finishing operations on Inconel 718 is evaluated, aiming to establish/expand the understanding of the wear behavior of coated tools when machining these alloys. Different machining parameters, such as cutting speed, cutting length, and feed per tooth, are tested, evaluating the influence of these parameters' variations on tool wear. The sustained wear was evaluated using SEM (Scanning electron microscope) analysis, characterizing the tools' wear and identifying the predominant wear mechanisms. The machined surface was also evaluated after each machining test, establishing a relationship between the tools' wear and production quality. It was noticed that the feed rate parameter exerted the most influence on the tools' production quality, while the cutting speed mostly impacted the tools' wear. The main wear mechanisms identified were abrasion, material adhesion, cratering, and adhesive wear. The findings of this study might prove useful for future research conducted on this topic, either optimization studies or studies on the simulation of the milling of Inconel alloys, such as the one presented here.

**Keywords:** tool wear; machining; milling; coatings; Inconel; nickel superalloys

**Citation:** Sousa, V.F.C.; Fernandes, F.; Silva, F.J.G.; Costa, R.D.F.S.; Sebbe, N.; Sales-Contini, R.C.M. Wear Behavior Phenomena of TiN/TiAlN HiPIMS PVD-Coated Tools on Milling Inconel 718. *Metals* **2023**, *13*, 684. https://doi.org/10.3390/met13040684

Academic Editor: Badis Haddag

Received: 27 February 2023
Revised: 14 March 2023
Accepted: 28 March 2023
Published: 30 March 2023

## 1. Introduction

Inconel 718 nickel superalloys are well known for their superior mechanical properties, retaining them even at elevated temperatures of up to 650 °C [1]. In addition to nickel, these alloys also contain chromium, iron, and traces of niobium, molybdenum, titanium, and aluminum. These elements are dispersed through the nickel $\gamma$ matrix with a face-centered cubic (FCC) lattice structure [2]. These Inconel alloys exhibit high strength, resistance to thermal creep deformation, and high resistance to corrosion and oxidation phenomena. Because of this, these alloys are applied in a wide variety of industries, such as defense, food processing, automotive, and the aeronautical and aerospace industries [2]. They are heavily employed in the latter industries, particularly to produce aircraft engine components (mainly components that are subject to high service temperatures), making up for about 50% of the total weight of these components in aircraft [3], or even as explosive fasters for space shuttles [4].

Indeed, Inconel 718 has excellent mechanical properties at high temperatures when compared with other types of alloys, even other nickel-based alloys [5]. However, these high values at high temperatures, coupled with other properties, such as low thermal conductivity and work hardening, make this alloy incredibly hard to process through machining, which is the primary process used to produce engine parts and aircraft components [6]. Furthermore, the final metallurgical structure of Inconel 718 contains a significant number of hard carbides, mainly TiC and NbC, being responsible for the high abrasive behavior registered during machining, causing severe tool wear [7]. This fact, coupled with the tendency of Inconel 718 to adhere to the surface of cutting tools, only make the machining of this alloy even more challenging. The machining of this alloy generates high cutting forces and temperatures, with estimated mechanical stresses reaching peaks of 450 MPa and temperatures of up to 1100 °C, as reported by Agmell et al. [8]. These issues, coupled with high abrasive damage and material adhesion to the tools' surfaces, cause the cutting tools to fail quickly, suffering considerable damage after a reduced machining time or distance.

Machining remains a very relevant process, especially for the production of high-precision and quality parts [9]. Over the years, researchers have encountered processing problems when machining various types of materials, with this process suffering constant evolution regarding the production of new tool technologies that can attenuate these problems faced when machining hard-to-cut materials, such as (but not exclusively) Inconel 718. There have been various studies performed on the influence of machining parameters (for various cutting processes, such as drilling, turning, or milling) [9,10]. These studies have focused on the determination of the optimal parameters to produce various parts, analyzing the produced machined surface's quality, and the tool wear sustained by the employed tools in various processes [11,12]. By analyzing these factors, conclusions can be drawn regarding the best set of parameters to machine certain hard-to-cut materials, offering a basis for future works and investigation into the processing of these materials. Another important aspect when it comes to machining process control is the assessment of the cutting forces developed during this process, as these forces provide valuable information regarding the machining process's performance and stability, offering some insight into tool wear and even the machined surface's quality [13,14]. Although these empirical studies offer this basis, they are quite expensive and time-consuming. However, they offer the opportunity to develop prediction methodologies for these processes. For example, in the following study conducted by Zhao et al. [15], the authors studied the influence of cutting parameters on the milling performance of C45E. They focused on the tools' edge optimization for reducing the developed cutting forces and sustained tool wear during machining. The authors performed a simulation using DEFORM V13.0 software to predict the developed cutting forces and tool wear during the machining of this material. Practical tests were also conducted to validate the obtained results. The authors reported satisfactory results, obtaining a deviation in the measured cutting forces from the simulations performed of about 20%. They also determined the influence of the machining parameters, namely the spindle speed and feed rate, on the developed cutting forces and sustained tool wear. These simulations are quite useful, providing valuable information in a faster and more cost-efficient manner when compared with empirical tests. However, special attention is needed when employing these methods, as the configuration of these simulations is quite difficult and depends on previously acquired knowledge [16].

Regarding the mitigation of problems faced when machining hard-to-cut materials, tool coatings or even novel tool geometries can be employed to reduce these problems [10,17]. Hard coatings offer many advantages in improving the wear resistance of various surfaces, being employed not only on cutting tools, but also on other surfaces, such as injection molds [18,19]. As such, cutting tools are no different, heavily benefiting from these hard coatings. These coatings are usually obtained by CVD (chemical vapor deposition) or PVD (physical vapor deposition), conferring different properties based on the coating structure or even deposition method [17]. A recent deposition method shows some promise, which is PVD HiPIMS (High-Power Impulse Magnetron Sputtering), in which the produced coating structure is more compact and possesses improved mechanical properties. Moreover, this deposition method confers the tools a degree of compressive residual stress, which has a positive impact on the machining performance of the coated tools [20]. This happens mainly because these stresses confer the tools' edges higher strength, producing an overall better-machined surface quality. However, this makes the tools' edges more prone to wear. These novel coatings might prove useful when machining Inconel 718, as they not only confer the tool increased wear resistance and improved mechanical properties, but can also contribute to better surface roughness production. There are studies that have employed these tool coatings in the machining of Inconel 718, both PVD and CVD, as well as uncoated tools. The use of uncoated tools for machining Inconel alloys (namely Inconel 718) is not advised, especially at higher cutting speeds [21]. The main wear mechanisms that these uncoated (WC-Co) tools present are adhesive and abrasive wear, which are considerably higher when compared with those of coated tools tested at higher feed values [22]. Akhtar et al. [23] evaluated the effect of machining parameters on surface integrity in the high-speed milling of Inconel 718. In this study, the authors tested coated and uncoated tools, more precisely, uncoated SiC whisker-reinforced tools and PVD TiAlN-coated carbide tools. Despite the uncoated tools presenting overall lower wear, it was observed that the produced surface roughness for the coated tools was lower, even suggesting that the roughing operations could be conducted using these uncoated tools. This can be attributed to the fine cutting edges commonly produced by the PVD-coating process, as well as the use of TiAlN coatings, which are usually employed for high-speed machining applications [17].

Despite the use of various tool coatings and tool materials, there are still a lot of challenges related to high wear when machining Inconel 718. De Bartolomeis et al. [24] presented a review article in which there was a collection of information regarding the most desired properties for cutting tools when machining Inconel 718. These tools are preferred to have high hardness under elevated temperatures, high strength and toughness, high thermo-chemical stability, high wear resistance, and high thermal shock resistance. There is high importance regarding the thermal properties of these cutting tools, as this alloy has very low thermal conductivity. This usually results in very high cutting temperatures in small punctual areas, primarily at the tool–chip interface [25]. As such, there are some tool materials that show promising behavior when machining these types of alloys, for example, cubic boron nitrides (CBNs), ceramic materials and coatings, such as TiC, TiAlN, or ceramic tools, such as the ones presented in [23], and carbides. However, there are some disadvantages associated with the use of each of these tool materials, such as abrasion and changes in tool geometry for fine edges, or even diffusion phenomena when machining with carbides [26]. These phenomena can be attenuated by introducing a coating layer, such as AlTiN or TiAlN, to act as a chemical/thermal barrier [27]. Indeed, TiAlN-based coatings have proven to be quite promising when machining Inconel 718, as the use of these coatings under high-speed machining conditions can lead to the formation of protective $Al_2O_3$ films between the coating and the Inconel 718, as reported in the study by Grzesik et al. [28]. In short, due to the high mechanical strength and elevated temperatures developed in the machining of INCONEL® alloys, the main wear mechanism reported so far is abrasion. However, when the cutting edges are very sharp, the breakage of these edges commonly occurs, dragging the coatings with it, as the fracture occurs essentially through the substrate. Coatings that lead to the formation of oxides on the surface tend to improve chip flow, thereby reducing wear phenomena. Some studies, as previously mentioned, have shown that TiAlN coatings lead to the formation of an $Al_2O_3$ film on the surface, which has been proven to be beneficial for reducing wear phenomena.

Another important aspect to improve the machining process of Inconel 718 and the machining processes is the analysis of tool wear and wear mechanisms. The analysis of the tools' sustained wear after machining can yield relevant information regarding the process's productivity, for instance [29], which provides useful information on machining parameter adjustment and, of course, on the quality/suitability of the employed cutting tool [30,31] and even coatings [10,17]. The analysis of the tools' cutting behavior and wear enable researchers to develop new strategies to machine these difficult-to-cut alloys; for example, in the study by Fox-Rabinovich et al. [32], the authors proposed the use of a novel AlTiN/Cu coating to improve the machinability of Inconel 718 superalloy. As previously mentioned, machining tools used to cut Inconel 718 are subject to severe wear, primarily abrasive wear, resulting in early tool failure. This novel tool coating had good self-lubricating properties and reduced thermal conductivity, which resulted in an increase in tool life. The authors assessed not only the tool life and wear behavior, but also analyzed chip formation in the various studied tools. They compared the chips formed with coated and uncoated tools, verifying that the undersurface of the chips created by the coated tools exhibited a very smooth surface, indicating minimal abrasive wear. This information, obtained from analyzing the machining process using different approaches, can be used to create prediction models about tool wear, cutting forces, and even machined surface quality [33]. The configuration of these models is quite difficult and relies on practical/empirical data, either by validation after implementation or by creating a database containing information previously acquired by performing practical tests. There is a lack of these models for milling operations, particularly when considering the machining of Inconel 718 [24].

The machining of Inconel 718 is quite a relevant and popular topic, with most of the studies being performed on the turning of these alloys. However, there have been few studies regarding the milling of these alloys, particularly for finishing operations [24]. Therefore, in the present manuscript, a comparative study on the production quality of TiN/TiAlN-coated tools using different machining parameters is presented. The influence of these parameters on the sustained tool wear and machined surface quality will be assessed. Furthermore, the developed tool-wear mechanisms will be analyzed for each of the devised test conditions. This coating was chosen due to the TiN sublayer that confers high coating adhesion strength [17], which is commonly associated with improved wear behavior [34]. Furthermore, these sublayers highly contribute to the coating's crack resistance [35] and, thus, to its wear resistance. Introducing layers such as these also contributes to the coating's oxidation resistance [36,37], which is important, given that the machining of Inconel 718 causes high cutting temperatures in small areas, thus contributing to coating and tool degradation [17,38]. Test results for uncoated tools will not be presented in this study, as the use of a coated tool to machine this type of alloy is advised and is well documented in the literature. Inconel 718 tends to adhere to uncoated tools' surfaces, causing severe abrasive and adhesive wear; therefore, the authors have opted to study the mentioned tool coating.

Therefore, in this study, the authors hope that the presented findings will be used as a basis for the optimization of the milling processes of Inconel 718, particularly finishing operations. Furthermore, the findings reported in this study can be used for simulation and predictive model configuration. Although this is considered quite difficult for the milling process, as it is a very dynamic machining process, a considerable lack of studies have been conducted on this theme [24]. Moreover, the milling of Inconel 718 alloy has not been heavily explored in the recent literature, especially on finishing operations and conditions [24]. Therefore, this study aims to help fill this gap in the recent literature, as well as provide insight into the milling (finishing) operations of these alloys. Studies such as these have proven to be quite useful when assessing the machinability of various alloys, helping to adjust machining parameters and develop new strategies or even tools and tool coatings [39,40]. As mentioned, the basis to develop simulations about the milling of Inconel will prove to be quite useful. If they are properly calibrated, the simulations can yield interesting and valuable results regarding the generated cutting forces, temperature, and developed tool wear (even providing useful information regarding tool life). However, to calibrate these simulations, studies such as this one are very important as they serve as an experimental basis for these calibrations.

This paper will be divided into three main sections. The following section will be the materials and methods, where the used materials will be presented, as well as the analyses performed and the equipment used for them. Then, in Section 3, the results of the conducted tests will be presented, showing the machined surface quality results followed by the presentation of the wear sustained by the machining tools. This wear analysis will be quantitative and qualitative, showing wear measurements (regarding flank wear) as well as characterizing the sustained wear mechanisms. In Section 3, the obtained results will be discussed based on the current literature. Finally, in Section 4, the conclusions of this work will be presented.

## 2. Materials and Methods

In this section, the various materials and methods used for the conduction of this work will be presented, starting with the presentation of the machining tools and workpiece material. The adopted methods regarding sample preparation and analyses will also be presented.

### 2.1. Materials

2.1.1. Machining Tools

The employed tools are end-mills (INOVATOOLS, S.A., Leiria, Portugal) their substrate is a cemented carbide WC-Co, grade 6110. The substrate presents an average grain size of about 0.3 μm and uses 6% Co (percentual mass fraction) as a binder. Figure 1a depicts the tools' shape.

The tools' substrates (INOVATOOLS, S.A., Leiria, Portugal) were supplied with the tools' final dimensions and geometry, as they were commercial tools sold by the supplier. However, these tools were then coated with a coating developed in the laboratory. The tools were coated with a TiN/TiAlN multilayer coating. After the machining of the tools' substrates, they were subjected to a cleaning operation and then coated. The tool coating was obtained using PVD CemeCom CC800/HiPIMS equipment (CemeCon, Wuerselen, Germany) with four target holders. The adopted deposition parameters can be observed in Table 1.

**Table 1.** Deposition parameters for the TiN/TiAlN tool coatings.

| Deposition Parameters | TiN Layer | TiAlN Layer |
|---|---|---|
| Deposition time (min) | 40 | 260 |
| Reactor gases | $Ar^+ + Kr + N_2$ | $Ar^+ + Kr + N_2$ |
| Target amount/composition | 2/Ti | 2/TiAl |
| Pressure (mPa) | 580 | 580 |
| Temperature (°C) | 450 | 450 |
| Bias voltage (V) | −110 | −110 |
| Target current density (A/cm$^2$) | 20 | 20 |
| Holder rotational speed (rpm) | 1 | 1 |

Coating deposition started with the opening of the shutters of the two pure titanium targets, enabling the deposition of the first thin layer of TiN onto the tools' substrates. After this initial deposition, these shutters closed and the ones for the TiAl targets opened, thus enabling the deposition of the outer TiAlN layer, which was thicker than the former. These deposition parameters were selected based on previously conducted experiments on similar substrates. After coating deposition, the tools were carefully packed, avoiding the handling of the cutting area.

2.1.2. Workpiece Material

The selected workpiece material (Paris Saint Denis Aero Portugal, Grândola, Portugal) was an austenitic nickel–chromium-based superalloy, Inconel 718. The workpiece material was supplied as a round bar, having a 158 mm diameter. Regarding heat treatments, the material underwent a solution annealing process at 970 °C, followed by rapid cooling by water quenching. Then, it underwent precipitation hardening at 718 °C for 8 h before being cooled in a furnace until 621 °C. This temperature was maintained for 8 h, with the material then being removed from the furnace and left to cool in air. The material's mechanical properties are presented in Table 2, and its respective chemical composition (%wt) is shown in Table 3.

**Table 2.** Workpiece material's (Inconel 718) mechanical properties determined at room temperature.

| Material Property | Value |
|---|---|
| Yield strength (MPa) | 1188 |
| Tensile strength (Mpa) | 1412 |
| Elongation (%) | 15 |
| Hardness (HBW) | 441 |

**Table 3.** Workpiece material's (Inconel 718) chemical composition (%wt).

| Chemical Composition (%wt) | | | | | | | |
|---|---|---|---|---|---|---|---|
| Ni | Cr | Fe | Nb | Mo | Ti | Al | Co |
| 53.89 | 18.05 | 17.78 | 5.35 | 2.90 | 0.96 | 0.51 | 0.20 |
| Cu | Si | Mn | B | C | P | N | Mg |
| 0.10 | 0.080 | 0.078 | 0.039 | 0.023 | 0.010 | 0.0070 | 0.0017 |

In addition to the elements considered in Table 3, residual elements, such as Ag, Ca, Pb, Bi, S, and O, were present; however. only trace amounts were found; thus, they were not considered relevant to include in Table 3.

### 2.2. Methods

### 2.2.1. Sample Preparation

To characterize the tool-coating thickness and the sustained tool wear after machining, the tools were subject to a preparation procedure. The mentioned analyses were performed using SEM (FEI QUANTA 400 FEG, Field Electron and Ion Company, FEI, Hillsboro, OR, USA). Regarding the tool coating analysis, firstly, an unused tool was cut (cross-section) in a coated area using a STRUERS MINITOM (Struers, Inc., Cleveland, OH, USA) disc saw equipped with a thin disc with electroplated diamond particles in its cutting area. After cutting, the sample was mounted in thermoset resin and placed in a STRUERS PEDOPRESS hot press (Struers, Inc., Cleveland, OH, USA). This mounting step was performed to enable the sample's preparation for SEM analysis. This preparation consisted of subjecting the mounted sample to grinding and polishing processes using silicon carbide (SiC) sandpaper with different grits in the following sequence: 220, 550, 800, and 1200 grit. After the grinding process, the sample was rotated 90° to reduce undesirable abrasion marks. Following the grinding process, the samples were subject to a polishing stage using two types of diamond slurry, one with an average particle size of 3 μm and another with an average particle size of 1 μm. Each polishing step was performed for about 15 min to reduce the abrasion marks on the samples' surfaces. A total of three samples for coating analysis were produced in this manner.

The tools used in the machining tests were additionally prepared to enable the sustained wear analysis, also using SEM. After machining, the tools were cut using the same STRUERS MINITOM (Struers, Inc., Cleveland, OH, USA) disc saw equipment, this time cutting the tools in an uncoated area. All the tools were cut in the same area, guaranteeing that all samples had the same length; moreover, during the cutting of the samples, the protection of the cutting area was ensured, preventing possible damage that would negatively impact the subsequent analysis of the tool wear and wear mechanisms. This cutting procedure enabled a more detailed analysis of the top of the tools in the SEM equipment. After cutting, all of the tools were subject to a cleaning procedure, more specifically, an ultrasonic bath using acetone for 5 min.

### 2.2.2. Coating Thickness Analysis

Coating thickness was assessed by analyzing cross-sections obtained (using the process described in Section 2.2.1) from the coated end mills. Thickness was measured using an FEI QUANTA 400 FEG (Field Electron and Ion Company, FEI, Hillboro, OR, USA) scanning electron microscope, provided with an EDAX Genesys Energy Dispersive X-ray Spectroscopy microanalysis system (Edax Ametak, Mahwah, NJ, USA). The mounted samples were analyzed in different areas of the cutting area, the coating thickness was measured, and its average value was determined. The same range of magnifications was used for all images obtained from the SEM analyses, allowing a better comparison between samples. EDS analyses were also carried out to assess the coating's chemical composition (this analysis does not provide fully accurate values; however, its accuracy is sufficient to be used to ascertain the chemical composition of these coatings, avoiding the use of more expensive and time-consuming technologies). EDS analyses were performed using a beam potential of 15 kV, which was sporadically reduced to 10 kV, to decrease the interaction volume, thus reducing the noise in the obtained spectra.

### 2.2.3. Machining Tests

The tests were performed using a milling center—HAAS VF-2 CNC machining center (HAAS Automation, Oxnard, CA, USA)—with a maximum speed of 10,000 rpm and a maximum power of 20 kW. Because the workpiece material was supplied as a round bar with a 158 mm diameter, a strategy was selected in which the tool would machine the material's surface in a spiral motion. The initial bar had a considerable height (of about 1 m); as such, it was necessary to cut this bar into sections so as to enable a stiffer machining process. The bar was cut using a bandsaw, creating sections with a height of 32 mm. These sections were then face-milled, guaranteeing a constant height of 30 mm (due to possible deviations in height created by the bar's cutting process).

Further, regarding the machining strategy, it was concluded that the best solution would be for the tool to perform an initial plunge at the round bar's center, then machining from the center to the bar's periphery. By adopting this strategy, the selected radial depth of the cut would be kept constant from the beginning of the test to its end, thus preventing undesired wear phenomena related to the change in this parameter throughout the test.

The machining tests were performed using cutting fluid (water-miscible) that was projected externally (5% oil in water) relative to the tool. The machining parameters were selected based on the tool's substrate manufacturer (INOVATOOLS). The choice was made based on the optimal parameters for optimal surface roughness production. The machining tests were designed using a full factorial experiment design. The final selected machining parameters can be observed in Table 4. In this table, the values for the cutting speed (Vc), feed per tooth (fz), cutting length (Cut. Length), axial depth of cut (ap), and radial depth of cut (ae), as well as the respective tool reference for each tested condition are shown. A total of three tools per test condition were used to improve the quality of the obtained results regarding the performance and wear sustained by the tested tools.

Two cutting lengths were selected to observe how the tool wear would progress throughout the machining of the workpiece, which were 5 and 15 m. Two cutting speeds were also selected, 100 and 125 m/min, to evaluate the influence of this parameter on tool wear and production quality (machined surface roughness). Regarding feed per tooth, this parameter is also known to have an influence on tool wear and production quality; therefore, three values were tested, with the center value (100%) being 0.0700 mm/tooth. As observed in Table 4, the values for ap and ae were kept constant, with the selected ae value being 75% of the tool's cutting diameter (tool diameter was 6 mm).

**Table 4.** Selected machining parameters for the conducted tests.

| Tool Ref. | Machining Parameters | | | | |
|---|---|---|---|---|---|
| | Vc (m/min) | fz (mm/tooth) | Cut. Length (m) | ap (mm) | ae (mm) |
| S100F75L5 | 100 | 0.0525 | 5 | 0.08 | 4.5 |
| S100F75L15 | 100 | 0.0525 | 15 | 0.08 | 4.5 |
| S100F100L5 | 100 | 0.0700 | 5 | 0.08 | 4.5 |
| S100F100L15 | 100 | 0.0700 | 15 | 0.08 | 4.5 |
| S100F150L5 | 100 | 0.105 | 5 | 0.08 | 4.5 |
| S100F150L15 | 100 | 0.105 | 15 | 0.08 | 4.5 |
| S125F75L5 | 125 | 0.0525 | 5 | 0.08 | 4.5 |
| S125F75L15 | 125 | 0.0525 | 15 | 0.08 | 4.5 |
| S125F100L5 | 125 | 0.0700 | 5 | 0.08 | 4.5 |
| S125F100L15 | 125 | 0.0700 | 15 | 0.08 | 4.5 |
| S125F150L5 | 125 | 0.105 | 5 | 0.08 | 4.5 |
| S125F150L15 | 125 | 0.105 | 15 | 0.08 | 4.5 |

### 2.2.4. Machined Surface Roughness Analysis

The analysis of the machined surface roughness provides valuable information regarding machining tool performance, which can be related to tool wear and optimal machining conditions to obtain the best possible surface quality. For this analysis, a Mahr Perthometer M1 profilometer (Mahr, Gottingen, Germany) was used. This equipment was used to determine the surface roughness parameters according to the ISO 21920-1:2021 standard [41]. Each test was performed with a cut-off value of 0.8 mm and a measurement length of 5.6 mm, which was seven times the cut-off value. The first and the last measurement segments of 0.8 mm were not considered due to errors resulting from the probe acceleration and deceleration at the time of measurement. The machined material's surface roughness was evaluated in the tangential and radial directions (in relation to the machining direction). This was performed to evaluate the machining stability, as a considerable difference between these values would indicate an unstable machining process. Furthermore, surface roughness assessment was performed in various areas of the workpiece's surface, namely in the center and periphery, again to evaluate the stability of the cutting process.

Due to the workpiece's dimensions and equipment availability, it was not possible to perform surface roughness area analyses (using profilometers). As such, only the mentioned equipment was used for surface roughness assessment.

### 2.2.5. Tool Wear Analysis

After being used for machining the workpiece, the tool wear sustained by the machining tools was assessed by performing SEM analysis, identifying the main wear mechanisms as well as performing wear measurements on the tools' top views (flank wear, VB), more precisely, VB3 according to the standard ISO 8688-2:1986 [42]. This was performed in this way due to the selected machining parameters. In fact, these caused primarily localized flank wear near the tools' chamfer. EDS analyses were also carried out during tool wear assessment to characterize phenomena related to material adhesion to the tools' surfaces. The tools' rake and clearance faces were also analyzed, characterizing and comparing the sustained wear for all tested tools. Furthermore, the amount of lost tool material was assessed by comparing the change in tool geometry with that of a tool that had not been used for machining (new tool). In Figure 1, the reference for tool tooth/face identification is displayed, showing the top view of an unused end-mill, as well as the scheme used for measurement according to the standard [42].

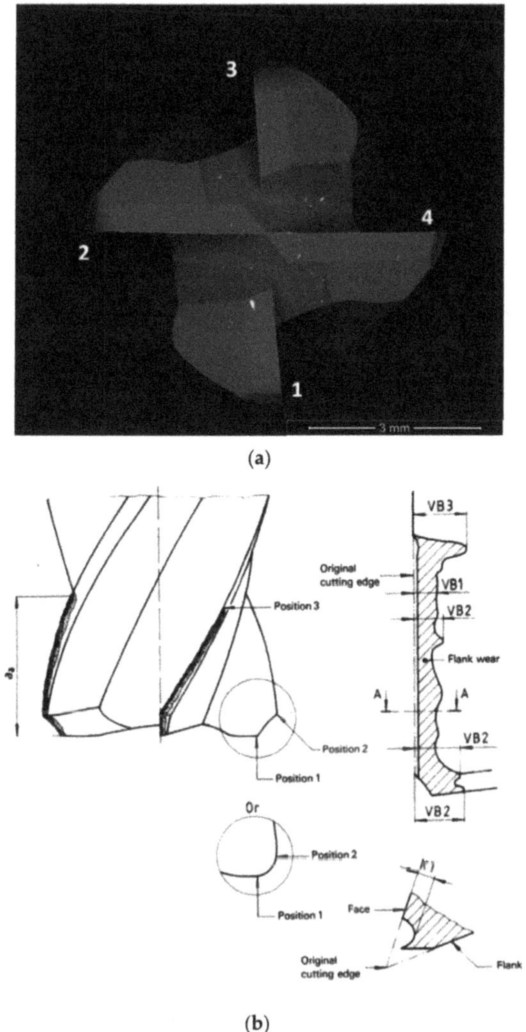

(a)

(b)

**Figure 1.** Reference used for tool wear evaluation during SEM analysis (**a**); types of wear on end-milling cutters according to ISO 8688-2:1989 (en) (**b**) [42].

As seen in Figure 1a, the numbers depicted are used to identify the tools' rake face (RF) and clearance face (CF). As mentioned, the numbers refer to the tools' teeth. Additionally, as three tools per test condition were used, an identification number ranging from 01 to 03 was added to the end of each image reference. As previously mentioned, the selected machining parameters caused a high amount of localized flank wear (VB3) and, as such, the tool wear was assessed according to the standard [42]. The wear measurements were performed in "Position 1", depicted in Figure 1b.

After each machining test, every tool was subjected to a preparation procedure to enable its analysis using the SEM equipment; this procedure was performed as described in Section 2.2.1.

## 3. Results and Discussion

### 3.1. Coating Characterization

In Figure 2, the coating's constitution can be observed, as well as the measurements performed to evaluate the total coating thickness.

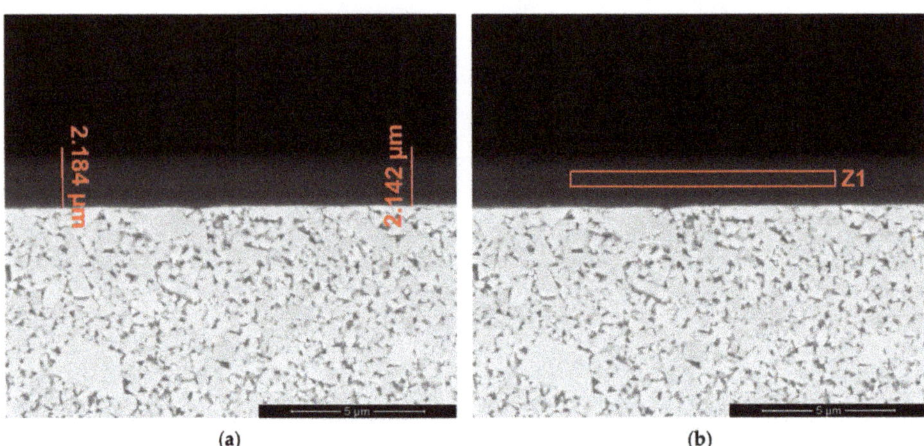


**Figure 2.** (**a**) Coating SEM analysis: thickness measurements performed on the tool's cross-section; (**b**) Z1 corresponds to the EDS evaluation zone to characterize overall coating composition.

EDS analysis was performed to confirm the coating's chemical composition in the area indicated in Figure 2b. The spectrum obtained from this analysis can be observed in Figure 3.

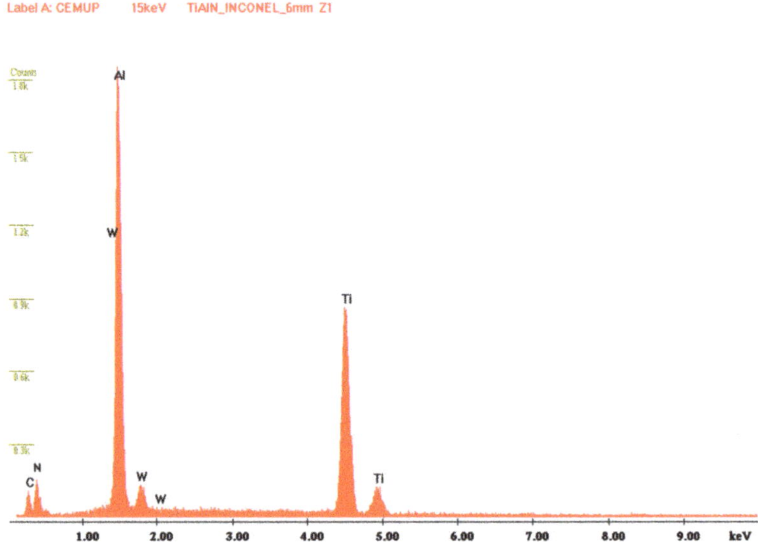


**Figure 3.** EDS spectrum obtained from the tool coating analysis.

Observing Figure 2, it can be noticed that there was a zone with a lighter tone near the tool's substrate material. This is the TiN layer deposited close to the substrate, with a view to improving its adhesion and facilitating the deposition of further layers. However, in

Figure 2, this layer is not clearly perceptible, and its composition could not be assessed by direct EDS analysis due to the interaction of the surrounding elements with the ones in this thin layer. As such, elemental mapping of the coated samples was performed, determining that this layer was in fact, the TiN layer. The base image used for the element mapping can be observed in Figure 4, while the element mapping that was performed can be observed in Figure 5.

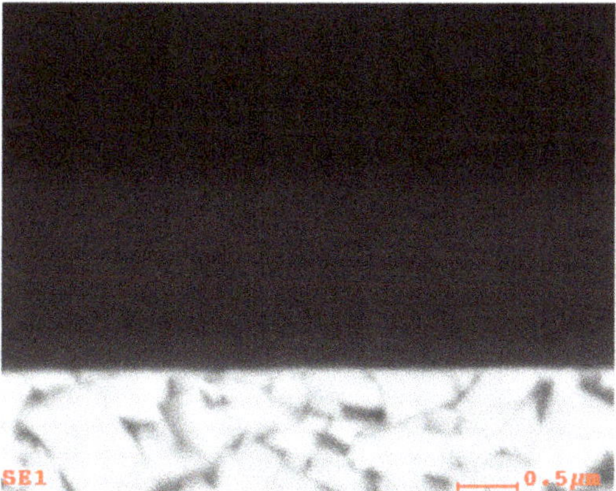

**Figure 4.** Base image of the coating's cross-section used for element mapping.

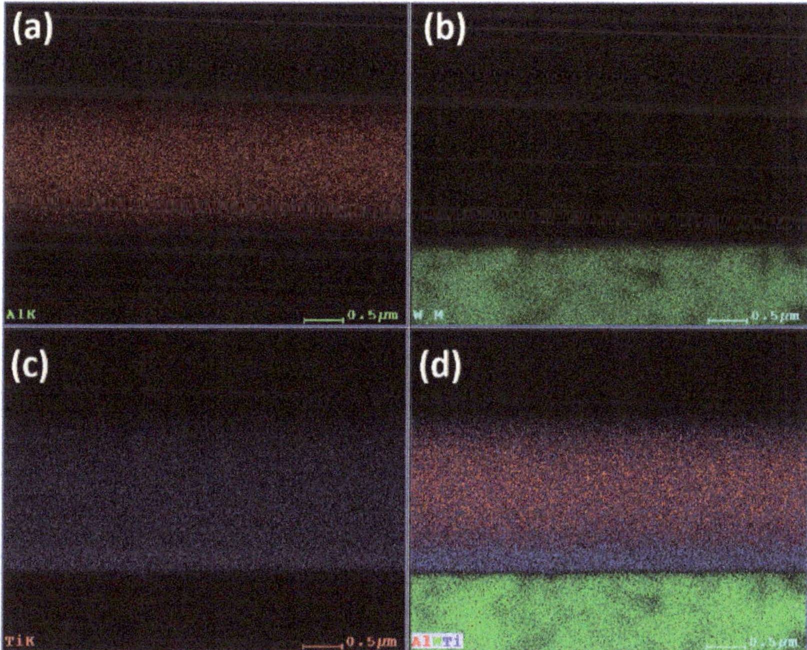

**Figure 5.** Element mapping performed on the tool coating: (**a**) Al distribution, (**b**) W distribution, (**c**) Ti distribution, and (**d**) Al, W, and Ti distribution overlapped.

As observed in Figure 5c, there seems to be a line near the substrate that has a higher content of Ti when compared with the remaining coating. This is also evident in Figure 5d, where the Al seems to be present mainly on the coating's top layer, while the bottom layer (near the substrate) is mainly composed of Ti. As such, it was determined that this was the layer composed of TiN.

Regarding the coating thickness, measurements were performed in various coated areas of the mounted tool samples, enabling the determination of the average coating thickness. These values are presented in Table 5, showing the average total coating thickness, as well as the average TiAlN and TiN layer thicknesses.

**Table 5.** Average thickness values for the complete TiAlN/TiN coating and its layers.

| Coating | Average Thickness Value (µm) |
|---|---|
| TiN/TiAlN (total coating) | $2.10 \pm 0.16$ µm |
| TiN layer | $0.25 \pm 0.023$ µm |
| TiAlN layer | $1.80 \pm 0.12$ µm |

As seen from the figures presented in this subsection, this is a multilayer coating. The TiN sublayer promoted the adhesion of the coating to the substrate, thus improving its wear resistance and behavior when machining the workpiece. The method of employing a sublayer (not exclusively TiN) brings many advantages related to tool wear behavior improvement, as reported by many authors [43,44]. This sublayer, and in multilayered coatings overall, confers the coatings improved crack resistance and adhesion strength. This increases the coatings' wear behaviors, causing overall less sustained wear [10,17]. Not only are these sublayers related to an improvement regarding the tools' wear behavior, as they confer the tools improved oxidation resistance, which might prove useful when machining certain alloys, such as Inconel 718, as its cutting generates high punctual cutting temperature [37,38], but also they result in a high temperature concentrated in a very small zone in the tool–chip interface. The sustained tool wear and wear mechanisms will be assessed in detail in Section 3.3 of the present paper.

### 3.2. Machined Surface Roughness Analysis

The average Ra values obtained for the performed tests can be observed in Table 6.

**Table 6.** Average mean surface roughness values (Ra) for the performed machining tests.

| Tool Reference | Average Ra Value (µm) |
|---|---|
| S100F75L5 | $0.409 \pm 0.0588$ |
| S100F75L15 | $0.692 \pm 0.0538$ |
| S100F100L5 | $0.452 \pm 0.0521$ |
| S100F100L15 | $0.737 \pm 0.0511$ |
| S100F150L5 | $0.517 \pm 0.0713$ |
| S100F150L15 | $0.637 \pm 0.0703$ |
| S125F75L5 | $0.429 \pm 0.0509$ |
| S125F75L15 | $0.674 \pm 0.0613$ |
| S125F100L5 | $0.502 \pm 0.0415$ |
| S125F100L15 | $0.786 \pm 0.0550$ |
| S125F150L5 | $0.596 \pm 0.0691$ |
| S125F150L15 | $0.824 \pm 0.0947$ |

As shown in Table 6, the machined surface roughness values were considerably higher for tests conducted at higher values of cutting length. This was expected, as Inconel 718 induces high levels of tool wear over a relatively low cutting length. As such, this tool wear negatively impacted the surface roughness values obtained. Regarding the performed measurements, it was noticed that, in the workpiece's periphery, the values were slightly higher than those at its center. This can be attributed, once again, to an increase in tool wear throughout the test, which was lower at the start of the machining path and, consequently, higher at its periphery (end of the toolpath). This variation was observed to be more significant for tools that exhibited higher tool wear (consequently, this variation induced higher standard deviation values).

Apart from tool wear, it was noticed that machining parameters clearly influenced the machined surface's quality, especially the feed per tooth (fz) parameter [45]. In Figure 6, the obtained results regarding the machined surface roughness for the various test parameters can be observed.

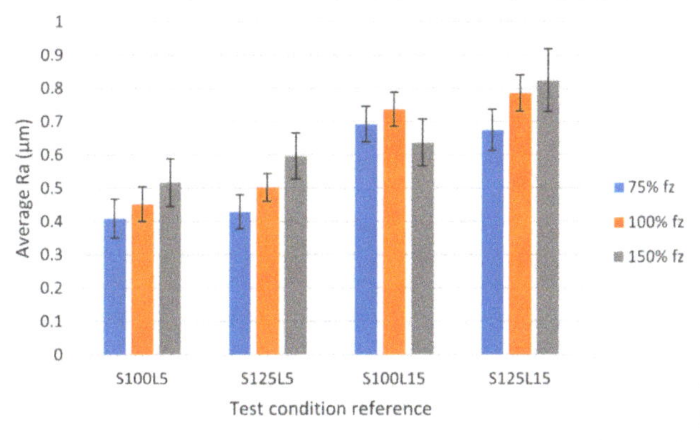


**Figure 6.** Comparison of surface roughness values for all test conditions (divided into three series based on the selected fz parameter).

As shown in Figure 6, there was a common tendency for machined surface roughness variation with the feed-per-tooth value. Usually, at lower values of this parameter, the machined surface quality is better than that obtained at higher values of feed per tooth [45,46]. This tendency was observed for both 100 m/min and 125 m/min in the tests conducted at a 5 m cutting length. However, concerning the tests conducted at a 15 m cutting length (L15), the tendency was slightly different for the case of the machined surface quality produced by tools machining at a 100 m/min cutting speed. Indeed, there was an increase in surface roughness from 75% to 100% feed per tooth in the tests conducted at a 100 m/min cutting speed. This value, however, decreased slightly from 100% to 150% feed per tooth.

In the tests conducted at a 125 m/min cutting speed and 15 m cutting length, the tendency was the same as that registered for the tests with a 5 m cutting length. This means that the machined surface quality deteriorated as the feed-per-tooth value increased. However, the machined surface roughness values obtained at a 125 m/min cutting speed tended to be slightly higher than those obtained at a 100 m/min cutting speed. This is not commonly observed, as cutting speed tends to produce a slightly better machined surface quality (although the feed per tooth is a more influential parameter in the machined surface quality) [47]. This is probably related to the fact that the tools' wear, in this case, was higher than that registered at a 100 m/min cutting speed (this will be explained in more detail in Section 3.3).

The fz parameter had a clear influence on the produced machined surface's roughness. Higher feed-per-tooth values produced higher Ra values, except in the case of S100L15 tools tested at 150% fz, which produced a better surface quality than that registered at 75% fz, which was the indicated value to obtain lower values of machined surface roughness. Regarding the cutting speed, as previously mentioned, a decrease in the surface roughness value commonly occurs when the cutting speed value increases. However, this was not the case for what was observed. This was probably due to the amount of tool-wear sustained by the tools at this cutting speed [48].

It is also worth noting that the values for the 75% fz produced at 100 m/min and 125 m/min were quite similar, only having a noticeable difference at 150% fz. Regarding the standard deviation registered for all surface roughness measurements, it seems that the value was higher for tests conducted at 150% fz, especially at 125 m/min. This was probably due to higher sustained tool wear, which resulted in a greater difference in the registered Ra from the center of the workpiece to its periphery (where the tool wear would be higher) [49,50].

### 3.3. Tool Wear Analysis

In the present section, the results regarding the tool wear analysis will be presented. This section is divided into two subsections, one focusing on the wear measurements performed according to the descriptions in Section 2, as well as the characterization of the wear sustained on the tools' rake and clearance faces. In the other subsection, the analysis of the wear mechanisms identified in the analyzed tools will be presented. The obtained results will be discussed and related to the machined surface quality achieved with the tested tools in an attempt to relate the tools' wear to these results. Furthermore, the influence of the machining parameters on the tools' wear will be discussed.

### 3.3.1. Wear Measurements and Characterization

As previously mentioned, the results obtained from the performed wear measurements will be described here, starting with the flank wear measurements conducted during SEM analysis on the tools' top surfaces (localized flank wear, VB3, according to [40]). These results can be observed in Table 7.

**Table 7.** Average flank wear (VB) values measured on the tools' top surfaces during SEM analysis.

| Tool Reference | Average VB Value ($\mu$m) |
| --- | --- |
| S100F75L5 | 92.670 $\pm$ 10.06 |
| S100F75L15 | 488.02 $\pm$ 21.80 |
| S100F100L5 | 89.370 $\pm$ 8.530 |
| S100F100L15 | 536.60 $\pm$ 34.61 |
| S100F150L5 | 75.450 $\pm$ 4.910 |
| S100F150L15 | 471.11 $\pm$ 42.13 |
| S125F75L5 | 199.16 $\pm$ 16.08 |
| S125F75L15 | 561.93 $\pm$ 39.64 |
| S125F100L5 | 190.05 $\pm$ 11.81 |
| S125F100L15 | 545.34 $\pm$ 23.65 |
| S125F150L5 | 168.03 $\pm$ 24.50 |
| S125F150L15 | 613.55 $\pm$ 19.71 |

According to the values displayed in Table 7, the values of VB3 (maximum localized flank wear) were higher for a longer cutting length. This was to be expected; however, the increase was quite significant, meaning that after only a short distance, the tools suffered a lot of flank wear. This was due to the properties of Inconel 718 causing high abrasive damage that led to tool chipping [51]. Additionally, the values that were obtained for tests conducted at 100 m/min were slightly lower than those obtained at a 125 m/min cutting speed. This indicates that an increase in cutting speed also caused an increase in the VB values.

As with the previous section, graphs displaying the values presented in Table 7 are provided for ease of comparison. As seen in Figure 7, there was a significant increase in the VB value from the tests conducted at a 5 m cutting length to those conducted at a 15 m cutting length, with the value of the latter being almost five times higher than that of the former [8]. Regarding the tests conducted at a 5 m cutting length, a common trend was identified: the VB values seemed to be higher for cutting tests carried out at 75% feed per tooth, and this value decreased with an increase in the feed per tooth parameter. This trend was identified for tests conducted at both 100 m/min and 125 m/min cutting speeds. As for the tests conducted at a 15 m cutting length, no clear trend could be identified. However, the flank wear seemed to be more intense for the tests conducted at higher feed-per-tooth values.

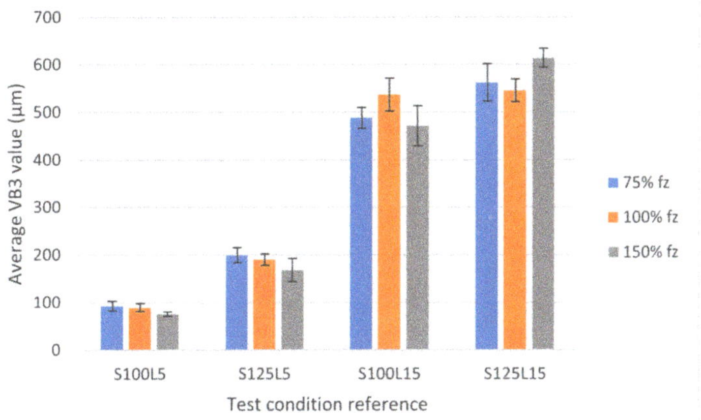

**Figure 7.** Comparison of flank wear (VB) values for all the test conditions (divided into three series regarding the selected fz parameter).

From Figure 7, it can be observed that the cutting length (as mentioned above) and cutting speed exerted clear influences. As previously mentioned, the feed per tooth parameter exerted an influence in the tests conducted at a 5 m cutting length. Regarding the influence of cutting speed on the registered flank wear values, it seems that an increase in cutting speed resulted in a more severe level of flank wear. When comparing the tests conducted at 5 m and 15 m cutting lengths, the tests carried out at 125 m/min produced higher values of flank wear. Regarding the tests conducted at a 15 m cutting length, there seemed to be no clear influence of cutting speed on the sustained flank wear. In the case of the tools tested at a 125 m/min cutting speed, the highest amount of wear was identified for the 150% fz condition, whereas in the case of the tests conducted at 100 m/min, it seems that the worst case of flank occurred under the 100% fz condition. Although an increase in cutting speed usually promotes a smoother cutting behavior for the cutting tools, it can be observed that, in this case, it produced a more accentuated flank wear.

Regarding the type of flank wear identified for tools tested at 100 m/min and 125 m/min, although the maximum localized flank wear was less intense in the case of tests conducted at 100 m/min, the wear mark was considerably wider. In this case, the sustained wear had a more prominent area than that in the tests conducted at 125 m/min. For these tested tools, the wear appeared to be more localized. An example can be observed in Figures 8 and 9, where a top view and a rake face view can be observed for tools tested at 100 m/min and 125 m/min.

In Figure 8, two images obtained through SEM analysis of two tools' top surfaces are displayed. In Figure 8a, a tool that was tested at 100 m/min with a 5 m cutting length and 0.07 mm/tooth (100% fz) is displayed. In Figure 8b, a tool that was tested at 125 m/min is shown (with the remaining machining parameters staying the same as those for the tool depicted in Figure 8a). In the case of Figure 8b, the wear is slightly more accentuated when compared with Figure 8a.

Figure 9 displays two rake faces of tools tested under the same conditions as those shown in Figure 8. Although the wear is quite similar, it seems that the rake face wear sustained by the tools tested at 125 m/min was more intense. Of course, these are the tools tested at a 5 m cutting length. As previously mentioned, the impact of the cutting length seemed to increase the wear significantly. This is shown in Figure 10 (15 m cutting length), where the tools' top views (Figure 10a) and rake faces (Figure 10b) are shown.

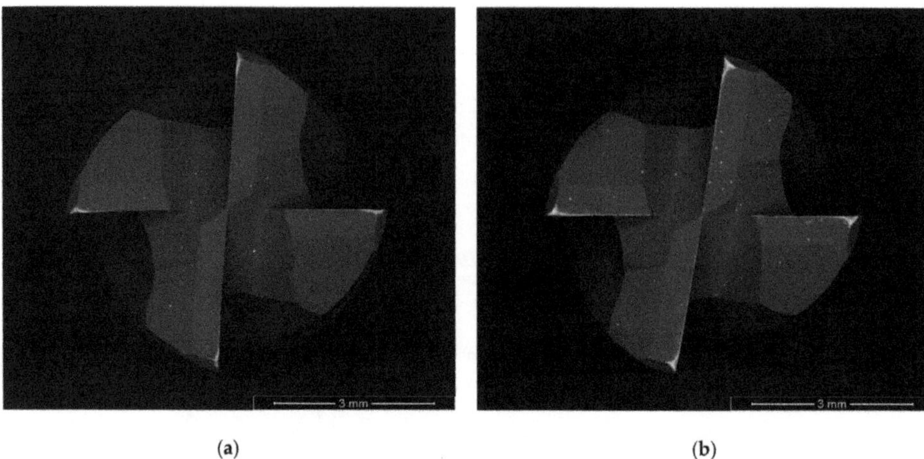

(a)    (b)

**Figure 8.** Top view of tools tested at S100F100L5 (**a**) and S125F100L5 (**b**).

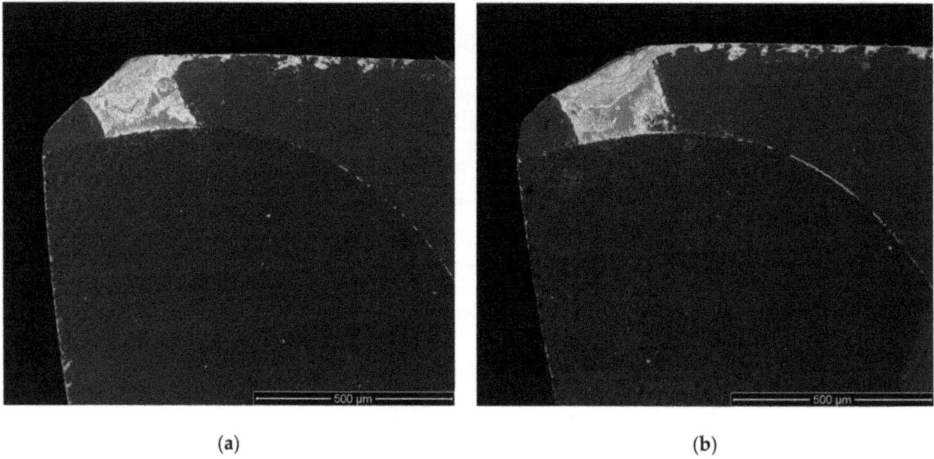

(a)    (b)

**Figure 9.** Rake face view (RF1) of the tools tested at S100F100L5 (**a**) and S125F100L5 (**b**).

Figure 10 clearly shows the impact of cutting length on tool wear. Longer cutting lengths contributed to higher levels of tool wear, which was mainly because the tools would machine more material and be subject to adverse cutting conditions for longer periods of time (when compared with shorter cutting lengths). It was observed that the wear mark was wider for tools tested at lower feed rates and cutting speed values, and deeper for tools tested at higher feed rates and cutting speeds. This was because the tool was subjected to higher attrition wear at lower cutting speeds and feed-per-tooth values, causing more severe wear and altering the tools' geometry. This, in turn, promoted a different chip formation mechanism that seemed to accentuate the tool wear even more.

Regarding the influence of tool wear on the machined surface's quality, it seems that the tools with more severe wear produced lower machined surface quality. However, this was particularly observed for the tools tested at 125 m/min and tested at a 15 m cutting length, although the value of fz usually influenced the machined surface quality, with a higher value resulting in worse machined surface quality and better results in terms of tool wear (measured flank wear). However, this tendency was only observed for tools tested at a 5 m cutting length, with the fz parameter not greatly influencing the measured tool wear.

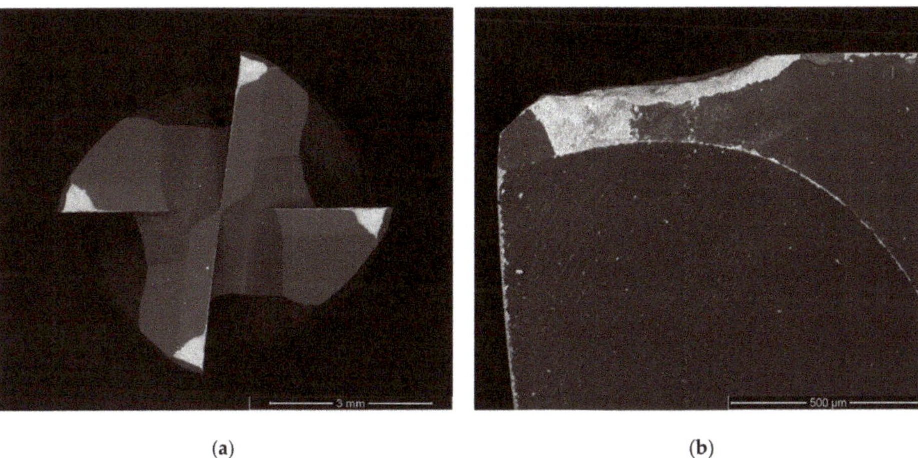

(a)                                (b)

**Figure 10.** Top view of tools tested at S100F100L15 (**a**); rake face view (RF1) of tools tested at S100F100L15 (**b**).

### 3.3.2. Tool Wear Mechanism Analysis

In this subsection, the identified wear mechanisms will be presented, evaluated, and discussed. They were common for all tested tools, with the ones tested at a 15 m cutting length having more severe levels of these mechanisms and overall wear (as expected) [8]. However, these wear mechanisms, as mentioned, were the same for all tested conditions.

The predominant wear mechanism was abrasion; in Figure 11, this wear is evident due to the smooth appearance caused by material abrasion on the tool's surface. Abrasive wear is also evident in Figure 10, where the loss of material caused by Inconel 718 abrasion during testing can be seen. Abrasive wear usually leads to the loss of tool material, even causing some alterations regarding geometry. This can also be seen in the tools' coating, as presented later in the manuscript. Abrasive wear is usually observed when machining Inconel 718 [7], which was observed in all the tested tools, albeit at different intensities. As previously mentioned, the tools tested at a 15 m cutting length exhibited considerably higher levels of wear than those tested at 5 m. The former tools exhibited more evidence of abrasive wear, as can be observed in Figure 11.

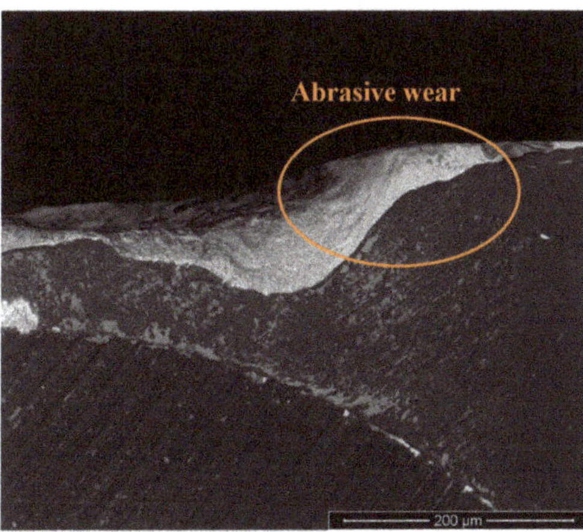


**Figure 11.** Rake face (RF2) of a tested S125F100L15 tool exhibiting abrasive wear.

Although Inconel 718 is known to adhere to cutting tools [23], the registered material adhesion was not abundant. That is, light workpiece material adhesion was registered on the tools' flanks, edges, and rake faces, as shown in Figure 12. Furthermore, there seemed to be a higher level of adhesion to the tools' uncoated areas (substrate) than to the tool coatings themselves (as shown in the marked area of Figure 12). Further, regarding material adhesion, the formation of a built-up edge was also registered, albeit in a primordial stage (not developed), which can be observed in Figure 13.

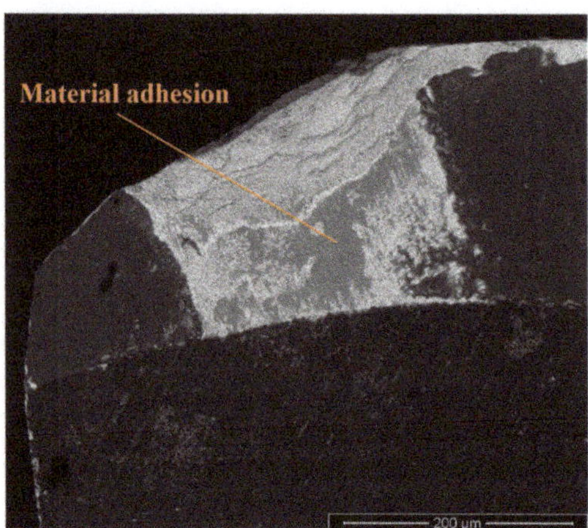


**Figure 12.** Material adhesion on the rake face surface (RF1) of a tested S125F100L5 tool.

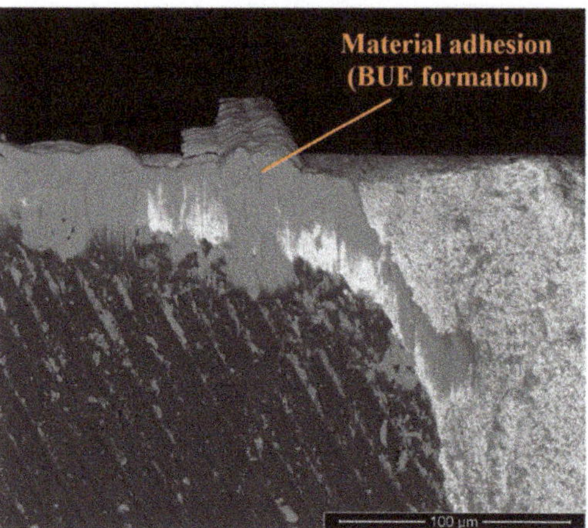


**Figure 13.** Initiation of the formation of a built-up edge (BUE) on a tool's top surface, exhibiting material adhesion on the exposed substrate (shown on a tested S100F75L15 tool).

The adhered material's composition was assessed by performing EDS analyses, proving that it was indeed Inconel 718. This adhesion may eventually lead to adhesive wear. It was detected on some of the tools' faces, such as the rake face depicted in Figure 12. Adhesive wear can lead to premature tool failure, as it promotes the removal of tool coatings and substrates, with are removed with the adhered material on the tools' surface.

A considerable amount of chipping, mainly of the tools' substrate was registered in the tested tools, particularly those tested at a 15 m cutting length (as seen in Figure 10). Indeed, machining Inconel 718 can be quite aggressive to the cutting tools, promoting high levels of wear [7,24]. This chipping hinders the production quality that can be achieved with these tools, as well as negatively impacts the overall machining performance [10,17]. This is due to the change in the cutting tools' geometry, altering the chip formation mechanism and, thus, promoting a higher amount of wear, cutting forces, and machined surface roughness values [9,10].

The tool coatings seemed to hold quite well. The TiN layer improved the coatings' crack resistance and, as such, no major cracking was identified in the tested coated tools [17]. Furthermore, the coating's adhesion was deemed to be quite good, as there was no major evidence of coating delamination, with the main coating wear mechanism being abrasion, as shown in Figure 14. As shown in Figure 11, the abrasion smoothed out and marked the tool's surface. In this case, the coating was eroded due to the machined material being "dragged" across the tools' surface.

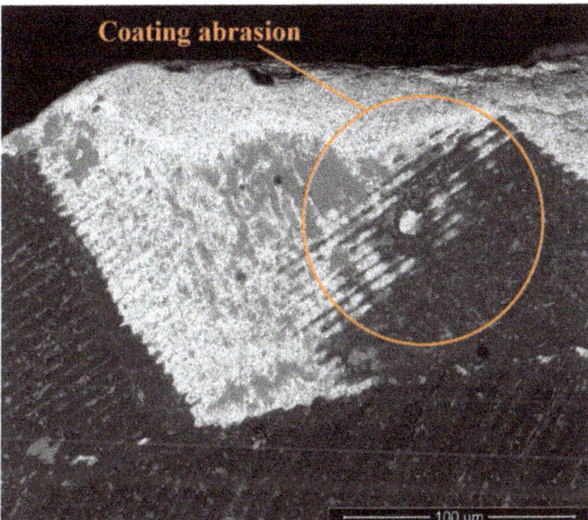


**Figure 14.** Coating abrasion shown on a tested S100F75L15 tool rake face (RF1).

## 4. Conclusions

The present work presents a comparative study of coating end-mills used to machine Inconel 718 alloy, known for its low machinability and excellent mechanical properties. These cause elevated levels of tool wear, even after only a few meters of cutting.

From the conducted and presented study, the following conclusions can be drawn:

- The machining parameters have an effect on the machined surface's quality, particularly the feed per tooth parameter;
- The optimal machined surface roughness value is obtained for the S100F75L5 test conditions;
- Tool wear impacted the machined surface's quality, however, this was mostly noticeable for longer cutting lengths due to the tools machining a higher amount of material;
- Less flank wear was observed for the S100F150L5 test conditions, with a decrease in the fz value increasing the amount of registered tool wear, although only slightly;
- An increase in cutting speed promoted an increase in tool wear. However, the most impactful parameter on tool wear was indeed the cutting length, as expected;
- Regarding the tool wear mechanisms, the predominant one was the abrasive wear mechanism, followed by tool chipping (substrate) and slight material adhesion to the tools' surfaces;
- Material adhesion was found to be more common in uncoated areas of the tested tool; furthermore, the uncoated regions were subject to more considerable abrasive wear;
- The resistance to cracking and high adhesive strength of the tool coating could be attributed to the TiN sublayer, which enhanced the coatings' crack resistance and adhesion to the substrate materials;

- The obtained results, in terms of machined surface roughness, can be considered quite satisfactory, even when tool wear is quite high. However, the tools' coatings suffered a considerable amount of wear only after a few meters of cutting, not being recommended for longer operations that require many meters of cutting.

These results highlight the advantages of using TiAlN coatings to obtain a satisfactory surface roughness quality for finishing operations applied to Inconel 718. Furthermore, the inclusion of coating sublayers can bring many advantages regarding the wear behavior of the cutting tools, especially when machining a material such as Inconel 718, known for its low machinability and capability of inducing high levels of wear on cutting tools. These results can be used as a basis for future work, especially when aiming to optimize/improve the outcomes of finishing operations applied to Inconel 718, particularly for milling applications.

**Author Contributions:** Investigation, V.F.C.S. and F.J.G.S.; Conceptualization, F.J.G.S. and F.F.; Supervision, F.J.G.S., F.F. and R.C.M.S.-C.; Data collection, V.F.C.S., R.C.M.S.-C., F.F. and N.S.; Formal analysis, F.J.G.S., F.F., R.C.M.S.-C., N.S. and R.D.F.S.C.; Methodology, V.F.C.S., R.C.M.S.-C. and N.S.; Visualization, R.C.M.S.-C., R.D.F.S.C., F.F. and N.S.; Data curation: F.J.G.S., N.S. and R.D.F.S.C.; Writing—draft, V.F.C.S. and F.J.G.S.; Writing—reviewing, R.C.M.S.-C., R.D.F.S.C., F.F. and N.S.; Resources, V.F.C.S., F.J.G.S. and R.C.M.S.-C.; Main research, V.F.C.S. All authors have read and agreed to the published version of the manuscript.

**Funding:** This research received no external funding.

**Data Availability Statement:** Not applicable.

**Acknowledgments:** The authors would like to thank Rui Rocha from CEMUP for his crucial collaboration in taking SEM pictures and commenting on some phenomena. The authors would also like to thank CETRIB/INEGI/LAETA, namely Jorge Seabra and Carlos Fernandes, for their continuing support in experimental activities and work dissemination. Moreover, Luís Durão and Victor Moreira from Mechanical Technology Lab of ISEP are acknowledged for their commitment and support in all machining experiments.

**Conflicts of Interest:** The authors declare no conflict of interest.

## References

1. Thakur, A.; Gangopadhyay, S. State-of-the-art in surface integrity in machining of nickel-based super alloys. *Int. J. Mach. Tools Manuf.* **2016**, *100*, 25–54. [CrossRef]
2. Buddaraju, K.M.; Sastry, G.R.K.; Kosaraju, S. A review on turning of Inconel alloys. *Mater. Today Proc.* **2021**, *44*, 2645–2652. [CrossRef]
3. Ulutan, D.; Ozel, T. Machining induced surface integrity in titanium and nickel alloys: A review. *Int. J. Mach. Tools Manuf.* **2011**, *51*, 250–280. [CrossRef]
4. Asala, G.; Andersson, J.; Ojo, O.A. A study of the dynamic impact behaviour of IN 718 and ATI 718Plus® superalloys. *Philos. Mag.* **2019**, *99*, 419–437. [CrossRef]
5. Hong, S.J.; Chen, W.P.; Wang, T.W. A diffraction study of the $\gamma''$ phase in Inconel 718 superalloy. *Metall. Mater. Trans. A* **2001**, *32*, 1887–1901. [CrossRef]
6. Paturi, U.M.R.; Darshini, B.V.; Reddy, N.S. Progress of machinability on the machining of Inconel 718: A comprehensive review on the perception of cleaner machining. *Clean. Eng. Technol.* **2021**, *5*, 100323. [CrossRef]
7. Zhou, J.; Bushlya, V.; Avdovic, P.; Ståhl, J. Study of surface quality in high-speed turning of Inconel 718 with uncoated and coated CBN tools. *Int. J. Adv. Manuf. Technol.* **2012**, *58*, 141–151. [CrossRef]
8. Agmell, M.; Bushlya, V.; M'Saoubi, R.; Gutnichenko, O.; Zaporozhets, O.; Laakso, S.V.A.; Ståhl, J. Investigation of mechanical and thermal loads in pcBN tooling during machining of Inconel 718. *Int. J. Adv. Manuf. Technol.* **2020**, *107*, 1451–1462. [CrossRef]
9. Sousa, V.F.C.; Silva, F.J.G. Recent Advances in Turning Processes Using Coated Tools—A Comprehensive Review. *Metals* **2020**, *10*, 170. [CrossRef]
10. Sousa, V.F.C.; Silva, F.J.G. Recent Advances on Coated Milling Tool Technology—A Comprehensive Review. *Coatings* **2020**, *10*, 235. [CrossRef]
11. Silva, F.J.G.; Martinho, R.P.; Martins, C.; Lopes, H.; Gouveia, R.M. Machining GX2CrNiMoN26-7-4 DSS Alloy: Wear Analysis of TiAlN and TiCN/Al$_2$O$_3$/TiN Coated Carbide Tools Behavior in Rough End Milling Operations. *Coatings* **2019**, *9*, 392. [CrossRef]
12. Gouveia, R.M.; Silva, F.J.G.; Reis, P.; Baptista, A.P.M. Machining Duplex Stainless Steel: Comparative Study Regarding End Mill Coated Tools. *Coatings* **2016**, *6*, 51. [CrossRef]

13. Sousa, V.F.C.; Silva, F.J.G.; Fecheira, J.S.; Lopes, H.M.; Martinho, R.P.; Casais, R.B.; Ferreira, L.P. Cutting Forces Assessment in CNC Machining Processes: A Critical Review. *Sensors* **2020**, *20*, 4536. [CrossRef]

14. Sousa, V.F.C.; Silva, F.J.G.; Alexandre, R.; Fecheira, J.S.; Silva, F.P.N. Study of the wear behaviour of TiAlSiN and TiAlN PVD coated tools on milling operations of pre-hardened tool steel. *Wear* **2021**, *476*, 203695. [CrossRef]

15. Zhao, X.; Qin, H.; Feng, Z. Influence of tool edge form factor and cutting parameters on milling performance. *Adv. Mech. Eng.* **2021**, *13*, 1–12. [CrossRef]

16. Niyas, S.; Jappes, J.T.W.; Adamkhan, M.; Brintha, N.C. An effective approach to predict the minimum tool wear of machining process of Inconel 718. *Mater. Today Proc.* **2022**, *60*, 1819–1834. [CrossRef]

17. Sousa, V.F.C.; Silva, F.J.G.; Pinto, G.F.; Baptista, A.; Alexandre, R. Characteristics and Wear Mechanisms of TiAlN-Based Coatings for Machining Applications: A Comprehensive Review. *Metals* **2021**, *11*, 260. [CrossRef]

18. Silva, F.; Martinho, R.; Andradre, M.; Baptista, A.; Alexandre, R. Improving the Wear Resistance of Moulds for the Injection of Glass Fibre–Reinforced Plastics Using PVD Coatings: A Comparative Study. *Coatings* **2017**, *7*, 28. [CrossRef]

19. Nunes, V.; Silva, F.J.G.; Andrade, M.F.; Alexandre, R.; Baptista, A.P.M. Increasing the lifespan of high-pressure die cast molds subjected to severe wear. *Surf. Coat. Technol.* **2017**, *332*, 319–331. [CrossRef]

20. Sousa, V.F.C.; Silva, F.J.G.; Lopes, H.; Casais, R.C.B.; Baptista, A.; Pinto, G.; Alexandre, R. Wear Behavior and Machining Performance of TiAlSiN-Coated Tools Obtained by dc MS and HiPIMS: A Comparative Study. *Materials* **2021**, *14*, 5122. [CrossRef]

21. Luo, M.; Li, H. On the Machinability and Surface Finish of Superalloy GH909 Under Dry Cutting Conditions. *Mater. Res.* **2019**, *21*, e20171086. [CrossRef]

22. Bhatt, A.; Attia, H.; Vargas, R.; Thomson, V. Wear mechanisms of WC coated and uncoated tools in finish turning of Inconel 718. *Tribol. Int.* **2010**, *43*, 1113–1121. [CrossRef]

23. Akhtar, W.; Sun, J.; Chen, W. Effect of Machining Parameters on Surface Integrity in High Speed Milling of Super Alloy GH4169/Inconel 718. *Mater. Manuf. Process.* **2016**, *31*, 620–627. [CrossRef]

24. De Bartolomeis, A.; Newman, S.T.; Jawahir, I.S.; Biermann, D.; Shokrani, A. Future research directions in the machining of Inconel 718. *J. Mater. Process. Technol.* **2021**, *297*, 117260. [CrossRef]

25. Astakhov, V.P. Tribology of Cutting Tools. In *Tribology in Manufacturing Technology*; Springer: Berlin/Heidelberg, Germany, 2012; pp. 1–66. [CrossRef]

26. Grzesik, W.; Niestony, P.; Habrat, W.; Sieniawasky, J.; Laskowski, P. Investigation of tool wear in the turning of Inconel 718 superalloy in terms of process performance and productivity enhancement. *Tribol. Int.* **2018**, *118*, 337–346. [CrossRef]

27. Kim, E.J.; Lee, C.M. A Study on the Optimal Machining Parameters of the Induction Assisted Milling with Inconel 718. *Materials* **2019**, *12*, 233. [CrossRef]

28. Grzesik, W.; Malecka, J.; Kwasny, W. Identification of oxidation process of TiALN coatings versus heat resistant aerospace alloys based on diffusion couples and tool wear tests. *CIRP Ann.* **2020**, *69*, 41–44. [CrossRef]

29. Gueli, M.; Ma, J.; Cococcetta, N.; Pearl, D.; Jahan, M.P. Experimental investigation into tool wear, cutting forces, and resulting surface finish during dry and flood coolant slot milling of Inconel 718. *Procedia Manuf.* **2021**, *53*, 236–245. [CrossRef]

30. Finkeldei, D.; Sexauer, M.; Bleicher, F. End milling of Inconel 718 using solid Si3N4 ceramic cutting tools. *Procedia CIRP* **2019**, *81*, 1131–1135. [CrossRef]

31. Li, H.Z.; Zeng, H.; Chen, X.Q. An experimental study of tool wear and cutting force variation in the end milling of Inconel 718 with coated carbide inserts. *J. Mater. Process. Technol.* **2006**, *180*, 296–304. [CrossRef]

32. Fox-Rabinovich, G.S.; Yamamoto, K.; Aguirre, M.H.; Cahill, D.G.; Veldhuis, S.C.; Biksa, A.; Dosbaeva, G.; Shuster, L.S. Multi-functional nano-multilayered AlTiN/Cu PVD coating for machining of Inconel 718 superalloy. *Surf. Coat. Technol.* **2010**, *204*, 2465–2471. [CrossRef]

33. Ducroux, E.; Fromentin, G.; Viprey, F.; Prat, D.; D'Acunto, A. New mechanistic cutting force model for milling additive manufactured Inconel 718 considering effects of tool wear evolution and actual tool geometry. *J. Manuf. Process.* **2021**, *64*, 67–80. [CrossRef]

34. Ruan, H.; Wang, Z.; Wang, L.; Sun, L.; Peng, H.; Ke, P.; Wang, A. Designed Ti/TiN sub-layers suppressing the crack and erosion of TiAlN coatings. *Surf. Coat. Technol.* **2022**, *438*, 128419. [CrossRef]

35. Shuai, J.; Zuo, X.; Wang, Z.; Guo, P.; Xu, B.; Zhou, J.; Wang, A.; Ke, P. Comparative study on crack resistance of TiAlN monolithic and Ti/TiAlN multilayer coatings. *Ceram. Int.* **2020**, *46*, 6672–6681. [CrossRef]

36. Çomakli, O. Improved structural, mechanical, corrosion and tribocorrosion properties of Ti45Nb alloys by TiN, TiAlN monolayers, and TiAlN/TiN multilayer ceramic films. *Ceram. Int.* **2021**, *47*, 4149–4156. [CrossRef]

37. Ananthakumar, R.; Subramanian, B.; Kobayashi, A.; Jayachandran, M. Electrochemical corrosion and materials properties of reactively sputtered TiN/TiAlN multilayer coatings. *Ceram. Int.* **2012**, *38*, 477–485. [CrossRef]

38. Zhang, M.; Cheng, Y.; Xin, L.; Su, J.; Li, Y.; Zhu, S.; Wang, F. Cyclic oxidation behaviour of Ti/TiAlN composite multilayer coatings deposited on titanium alloy. *Corros. Sci.* **2020**, *166*, 108476. [CrossRef]

39. Sousa, V.F.C.; Castanheira, J.; Silva, F.J.G.; Fecheira, J.S.; Pinto, G.; Baptista, A. Wear Behavior of Uncoated and Coated Tools in Milling Operations of AMPCO (Cu-Be) Alloy. *Appl. Sci.* **2021**, *11*, 7762. [CrossRef]

40. Byrghardt, A.; Szybicki, D.; Kurc, K.; Muszynska, M.; Mucha, J. Experimental Study of Inconel 718 Surface Treatment by Edge of Robotic Deburring with Force Control. *Strength Mater.* **2017**, *49*, 594–604. [CrossRef]

41. *ISO 21920-1:2021*; Geometrical Product Specifications (GPS)—Surface Texture: Profile Part 1: Indication of Surface Texture. International Organization for Standardization: Geneva, Switzerland, 2021.
42. *ISO 8688-2:1986*; Tool Life Testing in Milling—Part 2: End Milling. International Organization for Standardization: Geneva, Switzerland, 1986.
43. Silva, F.J.G.; Fernandes, A.J.S.; Costa, F.M.; Baptista, A.P.M.; Pereira, E. A new interlayer approach for CVD diamond coating of steel substrates. *Diam. Relat. Mater.* **2004**, *13*, 828–833. [CrossRef]
44. Silva, F.J.G.; Fernandes, A.J.S.; Costa, F.M.; Teixeira, V.; Baptista, A.P.M.; Pereira, E. Tribological behaviour of CVD diamond films on steel substrates. *Wear* **2003**, *255*, 846–853. [CrossRef]
45. Marakini, V.; Pai, S.P.; Bhat, U.K.; Thakur, D.S.; Achar, B.P. High-speed face milling of AZ91 Mg alloy: Surface integrity investigations. *Int. J. Lightweight Mater. Manuf.* **2022**, *5*, 528–542. [CrossRef]
46. Kumar, S.; Saravanan, I.; Patnaik, L. Optimization of surface roughness and material removal rate in milling of AISI 1005 carbon steel using Taguchi approach. *Mater. Today Proc.* **2019**, *22*, 654–658. [CrossRef]
47. Korkut, I.; Donertas, M.A. The influence of feed rate and cutting speed on the cutting forces, surface roughness and tool-chip contact length during face milling. *Mater. Des.* **2007**, *28*, 308–312. [CrossRef]
48. Natasha, A.R.; Ghani, J.A.; Haron, C.H.C.; Syarif, J. The influence of machining condition and cutting tool wear on surface roughness of AISI 4340 steel. *IOP Conf. Ser. Mater. Sci. Eng.* **2017**, *290*, 012017. [CrossRef]
49. Liang, X.; Liu, Z. Tool wear behaviors and corresponding machined surface topography during high-speed machining of Ti-6Al-4V with fine grain tools. *Tribol. Int.* **2018**, *121*, 321–332. [CrossRef]
50. Usca, U.A.; Uzun, M.; Sap, S.; Kuntoglu, M.; Giasin, K.; Pimenov, D.Y.; Wojciechowski, S. Tool wear, surface roughness, cutting temperature and chips morphology evaluation of Al/TiN coated carbide cutting tools in milling of Cu–B–CrC based ceramic matrix composites. *J. Mater. Res. Technol.* **2022**, *16*, 1243–1259. [CrossRef]
51. Zimmermann, R.; Welling, D.; Venek, T.; Ganser, P.; Bergs, T. Tool wear progression of SiAlON ceramic end mills in five-axis high-feed rough machining of an Inconel 718 BLISK. *Procedia CIRP* **2021**, *101*, 13–16. [CrossRef]

*Article*

# On the Influence of Binder Material in PCBN Cutting Tools for Turning Operations of Inconel 718

Francisco Matos [1], Tiago E. F. Silva [2], Vitor F. C. Sousa [2,*], Francisco Marques [3], Daniel Figueiredo [3], Francisco J. G. Silva [4] and Abílio M. P. de Jesus [1,2]

[1] DEMec, Department of Mechanical Engineering, Faculty of Engineering, University of Porto, 4200-465 Porto, Portugal
[2] INEGI, Institute of Science and Innovation in Mechanical Engineering, University of Porto, 4200-465 Porto, Portugal
[3] Palbit, S.A., Product Development Department, 3854-908 Branca, Portugal
[4] ISEP, School of Engineering, Polytechnic of Porto, 4200-072 Porto, Portugal
* Correspondence: vsousa@inegi.up.pt

**Abstract:** Inconel 718 is a highly valued material in the aerospace and nuclear industries due to the fact of its exceptional properties. However, the processing of this material is quite difficult, especially through machining processes. Machining this material results in rapid tool wear, even when low material removal rates are considered. In this study, instrumented turning experiments were employed to evaluate the machinability of Inconel 718 alloy using PCBN tools while assessing the usage of two distinct binder phases, TiN and TiC, for those cutting tools. It was found that the tool life was highly sensitive to the cutting speeds but also affected by the workpiece mechanical properties. At lower cutting speeds, notch wear significantly impacted the tool integrity, whereas at higher cutting speeds, flank wear was the primary failure mode of the tool. The flank wear of the tools with TiN-based binder outperformed TiC by almost 30%, presenting a more consistent behavior when machining.

**Keywords:** Inconel 718; PCBN cutting tools; tool binder material; machinability assessment; tool wear

**Citation:** Matos, F.; Silva, T.E.F.; Sousa, V.F.C.; Marques, F.; Figueiredo, D.; Silva, F.J.G.; Jesus, A.M.P.d. On the Influence of Binder Material in PCBN Cutting Tools for Turning Operations of Inconel 718. *Metals* **2023**, *13*, 934. https://doi.org/10.3390/met13050934

Academic Editor: George A. Pantazopoulos

Received: 13 March 2023
Revised: 17 April 2023
Accepted: 6 May 2023
Published: 11 May 2023

## 1. Introduction

Nickel superalloys, such as Inconel 718, are well known for their excellent mechanical properties, even at elevated temperatures [1,2]. This makes these superalloys the choice material for applications that are quite aggressive and require high mechanical strength at higher temperatures. For example, these alloys are employed in the making of turbine components that operate at temperatures above 800 °C. In addition to their high mechanical properties, these alloys also exhibit high resistance to corrosion and oxidation phenomena [3], making them suited for a wide variety of industries, such as the defense, food processing, automotive, and aeronautical and aerospace industries, in which more Inconel 718 is used [4,5].

Inconel 718 is also known for being a hard material to process, especially with machining operations, and it is classified as a hard-to-cut metal. This is due to the fact of their high mechanical property values coupled with the fact that this alloy has low thermal conductivity and a tendency to work harden [6]. Furthermore, the metallurgical structure of Inconel 718 has a significant number of hard carbides, namely, TiC and NbC, which causes the material to exhibit highly abrasive behavior while being cut [7]. As such, the main problems encountered when machining Inconel 718 is the short tool life caused by the high abrasive wear, high cutting temperatures [8], and tendency of the metal to adhere to the tools' surface. Moreover, the machining of this material damages the workpiece itself at a microstructural level due to the very high cutting forces generated during the process, as well as the surface tearing and distortion of the final machined components [9]. This is a problem as most aircraft engine components are obtained via machining.

A wide variety of strategies can be employed to mitigate the problems faced when machining hard-to-cut materials, such as the use of tool coatings or novel tool geometries [10]. Alternative machining strategies are also employed to mitigate problems that arise from machining Inconel 718, such as laser-assisted machining aimed at improving surface quality [11] or even different machining strategies using robotics with cutting force control to improve the surface quality of Inconel components, including turbines [12]. Hard coating offers many advantages, especially by improving the wear resistance of cutting tools [13]. However, there are some machining strategies that can be employed in the machining of these hard-to-cut materials, such as high-speed machining (cutting speeds over 120 m/min). This strategy has shown some potential in the machining of hardened steels [14,15], titanium alloys [16,17], and super alloys [18,19]. By employing this machining strategy, higher material removal rates can be achieved (compared to conventional machining). Nonetheless, employing this strategy using carbide tools (uncoated or coated) causes severe tool wear and coating delamination, primarily due to the high abrasive wear caused by the Inconel 718. As such, some alternative tool materials can be employed, such as cBN (cubic boron nitride) or PCBN (polycrystalline cubic boron nitride), which has excellent chemical inertness, high thermal stability, and very high hardness (being second to diamond) [20]. CBN can better maintain its mechanical strength values when subjected to higher temperatures compared to other conventional tool materials; moreover, due to the fact of its high hardness values, it has excellent abrasion resistance [21]. Regarding its resistance to diffusion phenomena, PCBN can resist interactions with the iron present in Inconel 718 up to temperatures of 1300 °C, both for low and high contents of CBN [22].

PCBN tools can be divided into three categories: low content CBN (50–65%); high content CBN (80–90%); and binderless sintered CBN (no binder) [23]. Several studies have been conducted to determine the most appropriate cutting tools for turning Ni-based alloys at higher cutting speeds. Criado et al. [21] compared the performance of low-content PCBN tools and carbide tools during machining operations of Inconel 718. In this study, it was found that the PCBN tools could handle higher cutting speeds, however, at lower values of depth of cut compared to the uncoated WC cutting tools. However, the authors concluded that even with this parameter's adjustment, the PCBN tools had an increased value for machined volume compared to the WC cutting tools while retaining a very similar level of tool wear. Regarding the machined surface roughness, the best quality was obtained using the PCBN tools, which outperformed the WC cutting tools by up to five times in this regard. Studies such as these highlight the potential of using these PCBN tools, particularly for finish-turning operations of Inconel 718. Dudzinski et al. [19] stated that although carbide tools are suited for machining Inconel 718 at lower cutting speeds (between 20 and 30 m/min), for enhanced productivity, ceramic tools are a more appropriate choice, even when considering that WC tools are less prone to crater wear compared to ceramic cutting tools.

Costes et al. [24] analyzed the influence of CBN content when finishing Inconel 718. The authors found that percentages of CBN content in the range of 45–60% in the cutting tool and used at cutting speeds between 250 m/min and 300 m/min led to the best behavior in terms of the best machined surface quality and least amount of sustained wear. Bushlya et al. [25] compared the performance of coated and uncoated PCBN (50% CBN content) with whisker-reinforced alumina tools in high-speed turning operations of Inconel 718. It was concluded that the tool life was highly sensitive to the variations in the cutting speed. The whisker-reinforced alumina tool life was lower than that of the coated and uncoated PCBN tools. The authors also studied the developed cutting forces during the machining operations, finding that these force values were higher for the whisker-reinforced alumina tools compared to the PCBN cutting tools. However, the cutting forces were slightly higher for the coated PCBN cutting tools.

Still, regarding studies conducted on the use of PCBN tools, Khan et al. [26] reported that a 0.2 mm/rev feed rate and a cutting speed value of 300 m/min offered a good productivity/material removal rate when turning Inconel 718. However, in TiN-coated PCBN inserts, as the cutting speed value increased from 300 m/min to 350 m/min, no significant improvement in the performance of the tool life was noticed. This is due to the rapid oxidation of the coating layer [26,27]. Bushlya et al. [28] also tested TiN-coated PCBN tools, comparing them with uncoated PCBN tools. Again, the coated tools produced higher cutting forces during the turning than those observed for the uncoated PCBN tools. In addition, the coated PCBN tools did not produce the best results in terms of the machined surface quality. Some experimental tests showed that low-content CBN tools with TiN binder correspond to excellent wear resistance because the ceramic binder phase has better chemical stability with respect to nickel [29,30].

Understanding tool wear and the developed wear mechanisms during machining is crucial, especially when optimizing a certain process. Wear generally occurs over time, and this failure is a gradual, cumulative process that affects the tool life [31]. As such, studies evaluating the wear behavior of cutting tools can provide useful information about the optimization of the cutting process. Understanding and predicting the effect of the selected cutting parameters on the tool life and wear behavior are also very important. Cores et al. [24] reported adhesion and diffusion wear as the predominant wear mechanisms sustained by CBN tools when turning Inconel 718. The authors evaluated speeds of 250 m/min, 350 m/min, and 450 m/min. The workpiece was subjected to high temperatures and stresses, suffering superficial plasticization. As such, the alloy spread over the contact area between the insert and the workpiece. Regarding diffusion wear mechanisms, Bushlya et al. [29] also reported that, under both high pressure and temperature, boron and nitrogen diffuse from the CBN into the Inconel 718 while forming solid solutions. A chemical reaction triggers the formation of Ti and Nb nitrides and Cr, Mo, and Nb borides. The authors also found that the presence of TiC binder in PCBN is an obstacle to further diffusional loss of the CBN phase. Despite the diffusional loss of carbon, TiC remains more stable than CBN and acts as an inert obstacle. These results made it possible to explain the significantly lower wear rate of PCBN tools with low CBN content and ceramic binder compared to the high CBN content grades.

In the present study, instrumented turning experiments were devised to evaluate the machinability of Inconel 718 alloy using PCBN tools. Two PCBN tool types were used that had distinct binder phases: TiN and TiC. The effect of the binder on the generated cutting loads and the sustained tool wear is presented and discussed. Although the use of PCBN tools for machining Inconel alloys has been researched to some extent, the influence of the binder phases could use additional exploration, as this may bring some advantages in the use of these tools, especially when aiming to mitigate common problems associated with the machining of Inconel 718.

## 2. Materials and Methods

Longitudinal cylindrical turning tests were performed using a Mazak Integrex i-200ST multitasking CNC machine instrumented with a load cell (Kistler 9129A piezoelectric dynamometer) coupled to a multichannel charge amplifier (Kistler 5070A) and data acquisition system (Kistler 5697A). In addition to the acquisition of the cutting forces, tool flank wear measurements and image collection of the rake and flank face were conducted using a DinoLite digital microscope. SEM images were taken, and a spectroscopy analysis was also performed. The SEM/EDS analysis was conducted using a high-resolution (Schottky) environmental scanning electron microscope with X-ray microanalysis and electron backscattered diffraction analysis (FEI Quanta 400 FEG ESEM/EDAX Genesis X4M). To avoid contaminants, all samples were submitted to an ultrasonic cleaning bath prior to the SEM analysis. The samples were immersed in ethanol and subjected to ultrasonic waves for 2 min to remove any debris or unwanted particles that may have accumulated on the surface of the samples. Table 1 displays the mechanical properties at both room temperature

(RT) and 649 °C, as specified by the alloy manufacturer and determined through tensile testing according to ASTM E8/E8M-16a and ASTM E21.

**Table 1.** Mechanical properties of the tested IN718 alloy as a function of temperature.

| Temp. | Yield S. (MPa) | Tensile S. (Mpa) | Elongation (%) | Area Reduction (%) |
|---|---|---|---|---|
| RT | 52.70 | 19.06 | 18.43 | 4.73 |
| 649 °C | 50.00 | rem | 17.00 | 4.75 |

A schematic representation of the described experimental setup is shown in Figure 1. A tool holder (ISO DCLNL 2525M 12) was clamped to the dynamometer, resulting in an insert (ISO CNGA 120408) setting angle of 95°, a rake angle of −6.5, and a clearance angle of 6.5°. A sampling rate of 1000 Hz was defined for the three components, $F_x$, $F_y$, and $F_z$, which corresponded to the passive, cutting, and feed forces, respectively. Concerning the lubrication conditions, a soluble oil emulsion was used with a pressure of 3.7 bar pointed towards the cutting edge. The tested levels of each cutting parameter are presented in Table 2. The full combination of the presented parameters was used for the two different cutting inserts (with different binder phases). Experimental tests usually vary the cutting speed ($v_c$) from 150 to 400 m/s, so five cutting speed levels within this range were selected. A feed ($f$) of 0.1 and 0.2 mm/rev and a depth of cut ($a_p$) of 0.15 mm were selected for the experimental runs. This combination of cutting conditions resulted in a matrix of 10 tests for each tool grade, adding up to 20 different experimental runs.

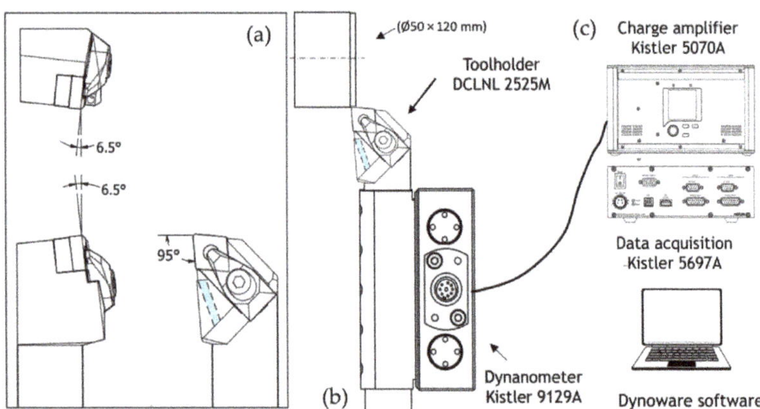


**Figure 1.** Instrumented setup of the longitudinal, cylindrical turning machining operations: (**a**) tool holder and effective rake, clearance, and cutting-edge angles; (**b**) cutting tool and tool holder relative to the position of the workpiece and load cell mount; (**c**) charge amplifier and data acquisition system.

**Table 2.** Tested levels for each cutting parameter.

| Cutting Parameter | Levels | | | | |
|---|---|---|---|---|---|
| Depth of cut, $a_p$ (mm) | | | 0.15 | | |
| Feed, $f$ (mm) | | 0.1 | | 0.2 | |
| Cutting speed, $v_c$ (m/min) | 150 | 200 | 250 | 300 | 350 |

Two identical cutting inserts were selected that varied the ceramic binder: (i) titanium carbide (TiC) for the PBY603 grade and (ii) titanium nitride (TiN) for the PBY602 grade. These 50% CBN content grades are suitable for the continuous and lightly interrupted cutting of hardened steel and the finishing of abrasive high-strength cast irons, and they can also be used to machine heat-resistant superalloys. It is worth noting that the inserts were composed of a PCBN tip brazed on a cemented carbide substrate, as shown in Figure 2. The inserts had an AlTiN/TiSiAlN double-compound coating (1.5 μm thickness) applied using the physical vapor deposition (PVD) process. Both inserts had the same negative 80° rhombic-shaped cutting insert and a honed-edge preparation with a 0.020 mm radius, which was measured (for thorough control of the cutting conditions) using the Brucker Alicona apparatus based on 3D optical measurement technology (at Palbit S.A.), as shown in Figure 2.

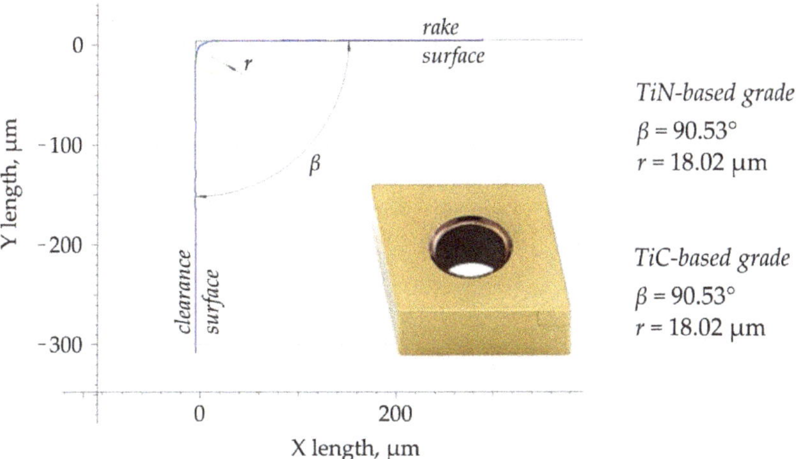


**Figure 2.** Overall geometry of the cutting insert (brazed tip construction) and edge radius measurement (at <0.05 mm from tool tip) of TiN-based and TiC-based PCBN inserts.

The Inconel 718 samples were equally pre-machined (facing and turning) into cylindrical shapes (50 mm in diameter), and a free length of 55 mm was defined for each longitudinal turning pass, ensuring the repeatability of the process. Prior to the machining operations, hardness tests were conducted (using an Emco M4U 075 hardness testing machine, according to ISO 6508) on the cylindrical rods. The following steps were successively repeated for each condition of the turning experiment: (i) placing a virgin (i.e., new), turning the insert on the tool holder; (ii) starting the data acquisition system without the lathe rotation; (iii) starting the turning test while acquiring data; (iv) stopping the test after a determined number of passes; (v) inspecting the tool and measuring the flank wear with a digital microscope; (vi) collecting the chip; (vii) repeating steps (i) to (iv) until a flank wear ($VB_{max}$) of 0.2 mm is achieved or 5 min of cutting time have been completed.

An example of the as-recorded cutting force data is shown in Figure 3. Distinct stages can be noticed. During the first stage (I), the force noise came from the lathe since the spindle was already rotating, but the cutting tool was motionless, whereas during the second stage (II) the cutting tool started its movement, entering the workpiece and beginning to cut. This second stage was characterized by a sudden increase in the forces, which stabilized less than half a second later into the (III) cutting stage itself. This third stage corresponded to the steady-state cutting conditions that were considered for this analysis. The last stage (IV) was characterized by a peak in the forces, possibly due to the fact of inertial effects and occasional chip tangling and winding around the workpiece.

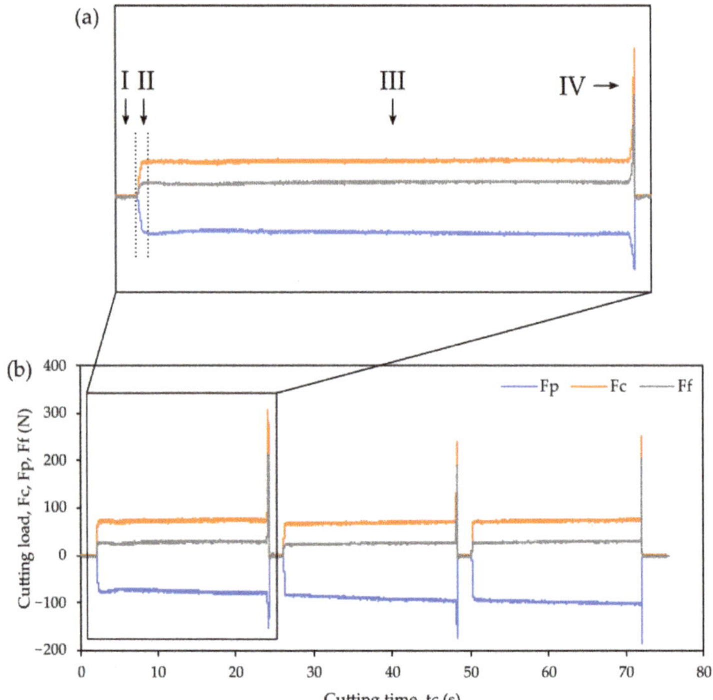



**Figure 3.** Load evolution of the Inconel 718 turning operation using the TiC binder cutting tool ($v_c$ = 200 m/min, $f$ = 0.10 mm/rev, and $a_p$ = 0.15 mm) while recording the cutting force signals: continuous cutting of (**a**) multiple (3) longitudinal turning passages; (**b**) distinct stages in each passage.

## 3. Results and Discussion

### 3.1. Workpiece Material Inspection Analysis

A chemical composition analysis enabled a comparison with material standards, confirming its compliance, as depicted in Table 3. Moreover, small samples of Inconel 718 were prepared for microstructural observation. After polishing (as seen in Figure 4a), these samples were chemically etched with oxalic acid, enabling different phases and grain structures, as shown in Figure 4b. Large carbide particles (1 to 7 μm) can be seen in the unetched sample, which are responsible for the hardened condition along with the very fine precipitates [32] dispersed along the matrix. The microstructure shows equiaxed grains and annealing twins, suggesting a recrystallized structure [33,34]. In terms of the hardness, different measurements were performed for different radial distances to the center of the cylindrical samples. No significant changes in hardness were noticed along the diameter of the samples (44.1 ± 0.1 HRC).

**Table 3.** Chemical composition (wt.%) of the Inconel 718 alloy processed in the current study.

| Element | Ni | Fe | Cr | Nb | Mo | Ti | Al | W | Si | Co |
|---|---|---|---|---|---|---|---|---|---|---|
| Measured | 52.70 | 19.06 | 18.43 | 4.73 | 3.00 | 1.02 | 0.541 | 0.114 | 0.114 | 0.109 |
| Standard min. | 50.00 | rem | 17.00 | 4.75 | 2.80 | 0.65 | 0.20 | - | - | - |
| Standard max. | 55.00 | - | 21.00 | 5.50 | 3.30 | 1.15 | 0.80 | 0.35 | 0.35 | 1.00 |

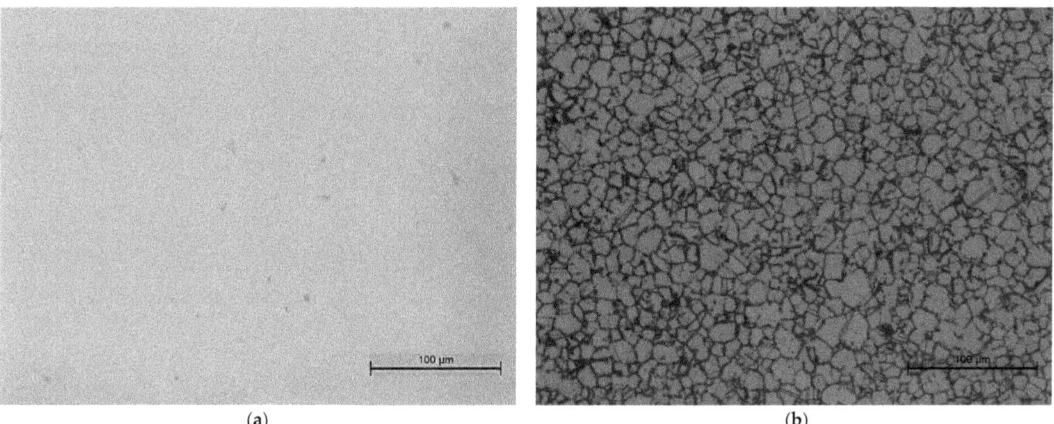

**Figure 4.** Optical microscopic images of Inconel 718: (**a**) polished condition evidencing the occurrence of large precipitates; (**b**) chemically etched (oxalic acid) condition, evidencing the microstructure.

### 3.2. Cutting Load

Figure 5 allows for the observation of the cutting loads according to each selected operational condition. The significant evolution of these loads can be noticed for each selection of cutting parameters due to the considerable wear rates promoted by the low machinability of the alloy. The initial force values, which were minimum due to the fact of insignificant tool wear, are represented by "min.", whereas "max." illustrates the last passage load values, which were maximum due to the fact of tool degradation. The average force values over the entire test are represented by "avg.". Overall, the highest values were achieved by the passive force component (Fp) followed by the cutting force (Fc) and feed force (Ff), in descending order, given that when the depth of cut is smaller than the nose radius of the tool, the passive or radial component of the tool forces is found to be most dominant [24,25,29,35,36]. Moreover, passive force had a large evolution over time, often amounting to triple and even quadruple the initial (i.e., virgin tool) value, when reaching the end of the tool's life. Monitoring the evolution of this force can provide relevant information on the tool wear evolution. The force appears to be highly sensitive to the evolution of the tool wear and degradation of the tool edge surface, which is a tendency also observed by other researchers [24,35,37]. In addition, the passive force seems to be associated to the ploughing force, which is caused by the flow of material ploughed by the clearance face of the tool due to the elastic deformation of the workpiece material under the clearance surface [35].

The first pass cutting forces for each tool, cutting speed, and feed are shown in Figure 6 for each distinct cutting tool. This first pass values are important, since at this point the cutting tools have the most similar wear conditions, enabling a more suitable comparison of the cutting forces. As expected, higher cutting forces were achieved when machining with higher feeds for both tools, given that maintaining the depth of cut and increasing the feed results in a higher chip cross-sectional area along with higher cutting forces. It is important to note the high stability of the cutting forces in the steady-state regime (standard deviation < 5 N). When varying the feed from 0.1 mm/rev to 0.2 mm/rev and cutting with the TiC binder inserts, this resulted in an average increase of 52.1% in the cutting forces, while under the same conditions, an increase of 55.5% was observed when machining with the TiN binder inserts. The cutting forces proved to have a slightly decreasing tendency when increasing the cutting speed. As the cutting speed increased, there was a larger energy input in the machining process, which combined with the lower heat diffusion (near-adiabatic process), promoted a temperature increase that resulted in thermal softening of the workpiece material. For the tests, it was not possible to perceive a significant difference in the cutting forces among the distinct CBN binders. Regardless of the employed cutting insert, the measured load values oscillated at approximately 120 N when machining with a feed of 0.2 mm/rev and 80 N with a feed of 0.1 mm/rev.

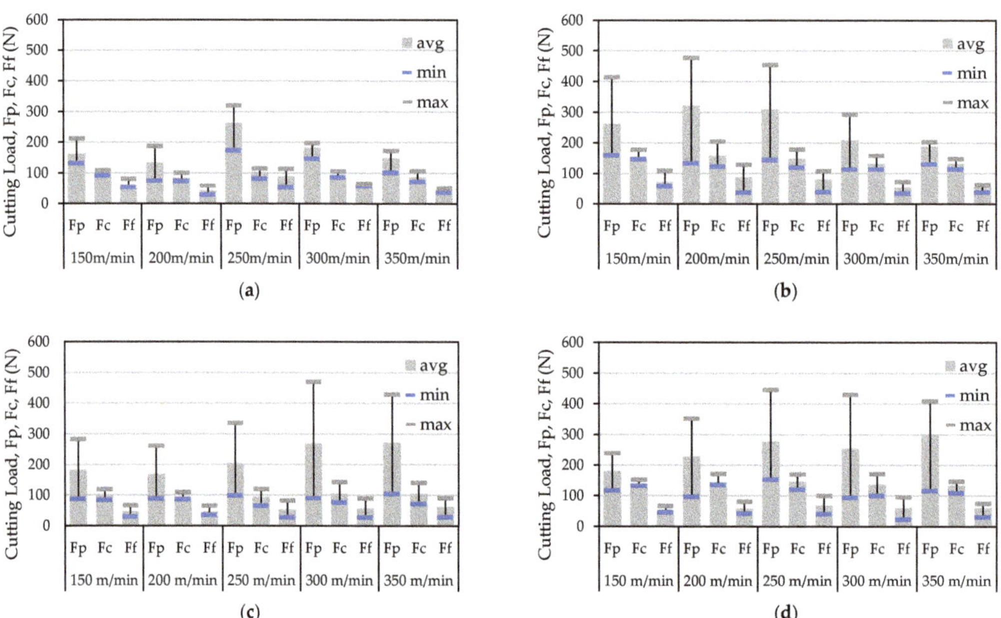

**Figure 5.** Cutting load evolution according to the selected cutting parameters and depending on the first (min.) and last (max.) cutting tool passages: (**a**) TiC binder cutting tool and $f$ = 0.1 mm/rev; (**b**) TiC binder cutting tool and $f$ = 0.2 mm/rev; (**c**) TiN binder cutting tool and $f$ = 0.1 mm/rev; (**d**) TiN binder cutting tool and $f$ = 0.1 mm/rev.

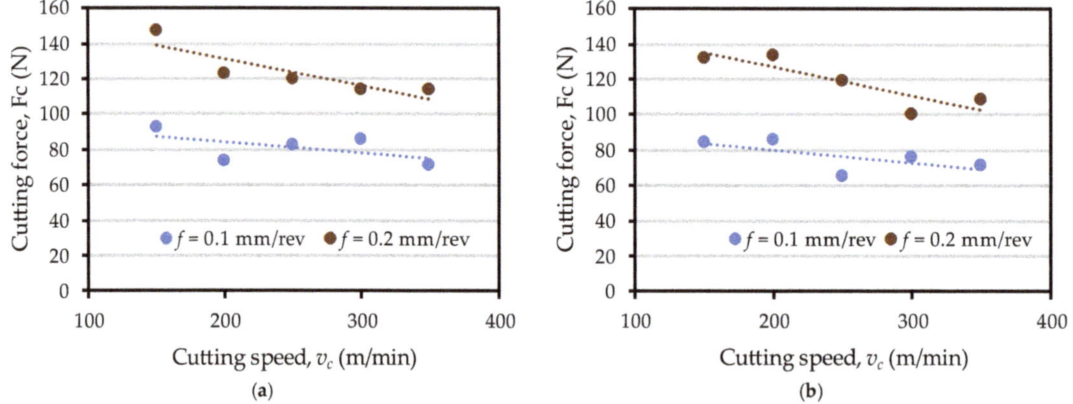

**Figure 6.** Measured cutting forces on the initial passage of each tool for each tested configuration of cutting parameters: (**a**) TiC binder cutting tool; (**b**) TiN binder cutting tool.

The measurement of the first pass cutting forces enabled the calculation of the specific cutting pressure, $K_c$, which is a highly relevant parameter used for the evaluation of a material's machinability and enables an estimate of the cutting force for scenarios with distinct operational conditions. The specific cutting pressure can be defined as the ratio between the main cutting force and cross-sectional area of an undeformed chip, $A_0$. The cross-sectional area of the undeformed chip can be calculated using Equation (1), and the specific cutting pressure, $K_c$, can be calculated using Equation (2).

$$A = a_p \cdot f \tag{1}$$

$$K_c = \frac{F_C}{A} \tag{2}$$

Figure 7 exhibits the evolution of the specific cutting pressure according to the feed conditions. Typically, $K_c$ exponentially increases for a smaller chip section while stabilizing when increasing the chip section, which is a phenomenon associated with size effects in metal cutting, as widely observed by researchers [35,38,39]. The occurrence of defects in metals, such as grain boundaries or missing atoms, is generally acknowledged, and the chance of encountering such stress-reducing defects increases for larger sizes of removed material [39]. In addition, the occurrence of ploughing for increasingly smaller feed rates explains the size effect. When the uncut chip thickness becomes smaller or similar to the tool edge radius, the cutting loads did not decrease proportionally due to the considerable relative contribution of workpiece deformation (compression), leading to higher $K_c$ values.

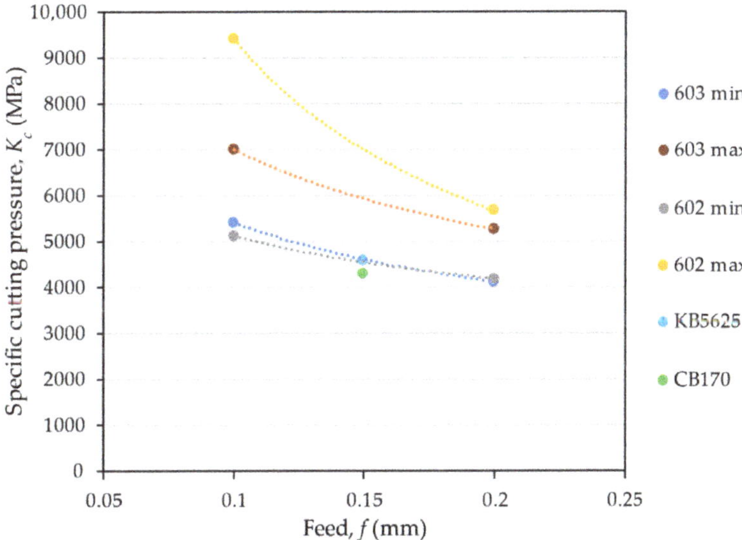


**Figure 7.** Specific cutting pressure values considering new and used tools and their comparison with values in the literature for similar cutting insert geometries.

The turning tests comprised two different chip cross-sectional areas, with the highest values of cutting pressure being achieved for the lowest feed, which corresponded to the smaller chip section, as explained by the size effect in the metal cutting. Values between 4000 and 5500 N/mm$^2$ were obtained for a feed of 0.2 and 0.1 mm/rev, respectively, at a cutting speed of 300 m/min and depth of cut of 0.15 mm. These are typical values for Inconel 718 turning processes with negative rake angles (commonly used for Ni-based alloys), hence, inducing high values of specific cutting forces. Furthermore, Cantero et al. [35] obtained similar $K_c$ values when turning Inconel 718 (45 HRC) with commercially available KB5625 and CBN170 PCBN inserts. These inserts have a medium CBN content and ceramic binder, with a honed edge preparation (25 μm) similar to the tested tools in this work. Whereas in Figure 7 the specific cutting pressure labeled with "min." corresponds to the first passage of the longitudinal turning (i.e., virgin cutting insert), "max." corresponds to the last passage of that insert. Significantly higher specific cutting pressures (>30%) occur when reaching the tool's end of life, proving that the wear affects the cutting power, leading to a less efficient machining along with a higher surface roughness of the part, thus affecting the process sustainability and overall quality of the generated surface.

### 3.3. Tool Wear

The adopted experimental procedure made possible the tracking of the evolution of the tool wear. It is an inevitable, gradual, and complex phenomenon affecting workpiece surface accuracy and integrity and contributing to the cost of the machining process. It occurs due to the fact of geometrical damage, frictional force, and temperature increases at the tool–workpiece interface region. Studying the influence of the cutting parameters and their effect on tool wear allows to optimize the process, achieving both better results on the final product and on the economical side of the operation.

To establish the tool flank wear progression depending on the machining time, wear data were acquired approximately every 30 to 60 s of the machining time. To do so, the test was stopped and the flank wears ($VB_{max}$ and $VB_{notch}$) were measured according to ISO 3685. Herein, flank wear refers to $VB$ for simplicity. Figure 8 shows the monotonic linear wear increase for the CBN cutting tool according to the cutting time ($t_c$). These data were then computed into a graph, where the ratio of $VB$ and cutting time represents the tool flank wear rate ($TW_f$) in mm/s, as presented in Figure 8.

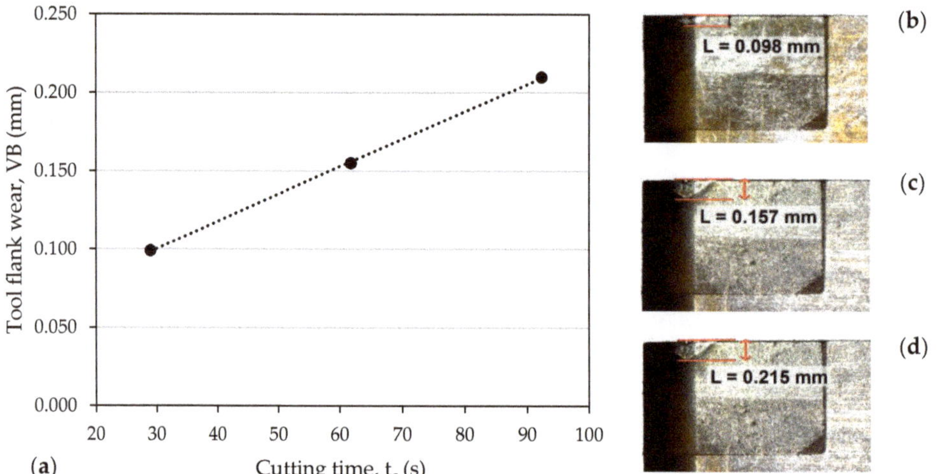


**Figure 8.** Tool flank wear evolution (**a**) as a function of the cutting time for a cutting speed ($v_c$) of 300 m/min and feed (*f*) of 0.2 mm using a TiC binder insert: (**b**) flank face for a cutting time of 30 s; (**c**) flank face for a cutting time of 62 s; (**d**) flank face for a cutting time of 92 s.

As the cutting speed increased, the tool wear rate tended to increase as well, which may be explained by the increase in the cutting temperature at the cutting edge for increasing cutting speeds. Those high-speed (and, thus, high-temperature) conditions may lead to softening of the tool material which, in turn, may result in an accelerated tool wear rate (monotonic trend of the curves shown in Figure 9). Still, there seems to be a tendency towards the stabilization of the wear rates for longer cutting speeds. This might be associated with the obtained cutting loads that were more stable for faster cutting speeds (refer to Figure 5), especially for TiC binders. Thus, it is noticeable that within cutting speeds between 250 m/min and 350 m/min, the wear rate tended to a steadier rate when machining with TiC binder inserts. However, when machining the same alloy with the

CBN tool with TiN binder, this effect was less evident, since the tool flank wear rate kept increasing with an increasing cutting speed, while the cutting loads also showed lower stability (Fp >> Fc, Ff), as illustrated in Figure 5.

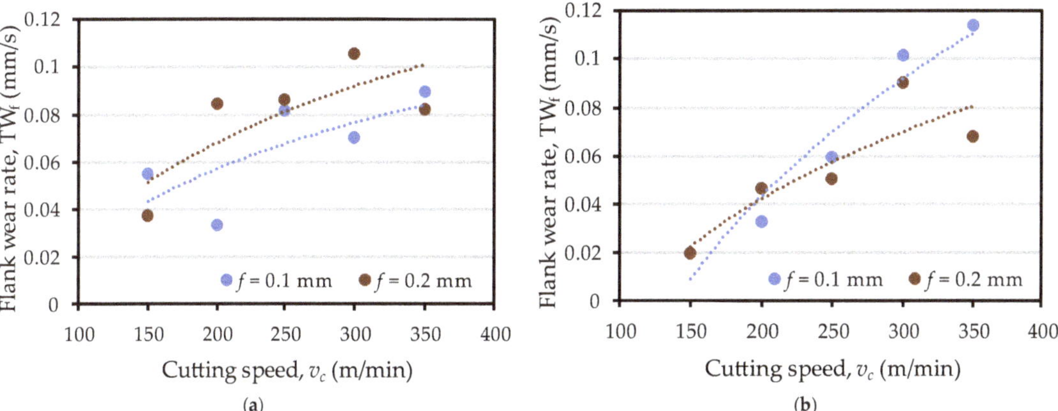

Figure 9. Influence of the cutting speed ($v_c$) on the tool flank wear rate for each distinct tested insert: (a) TiC binder; (b) TiN binder.

Throughout the tests, all inserts experienced noticeable crater wear on the rake face and flank wear due to the fact of abrasion on the flank face. In most of the experimental runs, an increased cutting speed improved the stability of the cut. In Figure 10, two distinguishable wear patterns that were often found in the current experiment are shown. At lower cutting speeds (i.e., between 150 and 200 m/min), the flank wear was extremely irregular with the length variations of the flank itself in addition to the formation of build-up edge and notching. At higher cutting speeds, despite the rapid wear of the tool, this phenomenon was less discernible. Regarding the influence of the feed on flank wear, the results do not show a noticeable influence for the different levels of tested feeds (0.1 and 0.2 mm/rev), which is supported by Cantero et al. [35], who found similar results when machining 45 HRC Inconel 718.

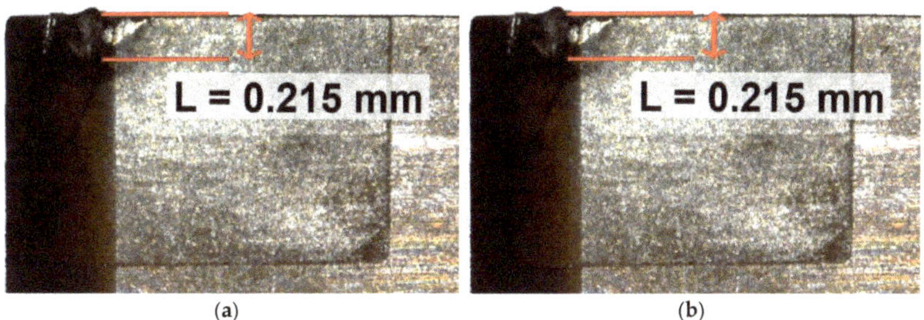


Figure 10. Example of cutting inserts (TiC binder) with exceeded flank wear ($V_B > 0.2$ mm) for different operational cutting conditions: (a) cutting speed ($v_c$) of 200 m/min and feed ($f$) of 0.2 mm; (b) cutting speed ($v_c$) of 350 m/min and feed ($f$) of 0.2 mm.

Knowing the speed at which each tool flank degrades, the flank wear rate, as shown in Figure 9, enables the calculation of the expected tool life, as presented in Figure 11, using a flank wear of 0.2 mm as the wear criterion. It was chosen to display the average tool life of each tool for both feeds, as no minor differences were found for the flank wear with the two tested feeds (0.1 and 0.2 mm/rev). It is also important to note that the tool life at a cutting speed of 150 m/min was extrapolated and could hardly be achieved due to the chipping and notching phenomena, which is further addressed. Moreover, it is important to remark that even though the cutting speed greatly affects the tool life (which shows considerable differences regarding the CBN binder at low cutting speed), it plateaus at 300 m/min. This could be explained by a change in the phenomena, leading the wear to change from a tribological type (i.e., abrasion and adhesion) to a chemical (i.e., diffusion and oxidation). Within the selected range of the tested parameters, the cutting speeds between 250 and 350 m/min resulted in smaller notching, better cutting stability, and chip control despite a short tool life.

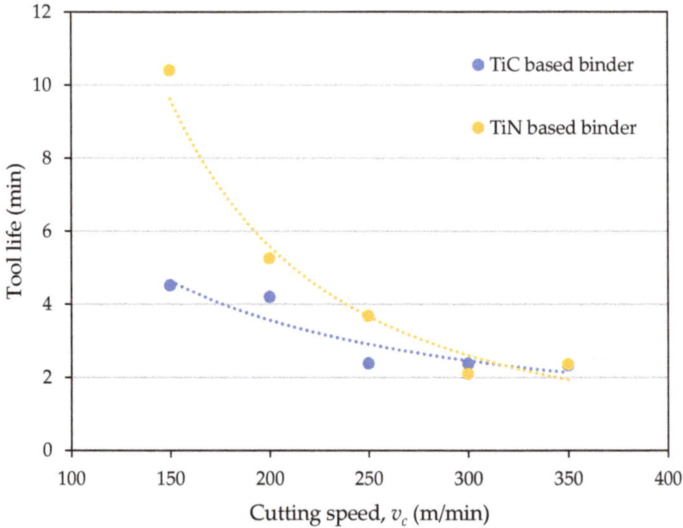


**Figure 11.** Influence of cutting speed on tool life prediction based on a maximum flank wear of 0.2 mm for TiC and TiN based binders.

Productivity should not only be assessed by the geometric parameters of the operation (which allow to calculate the MRR) but also in terms of the tool's life, particularly at an industrial level where tool replacement and set-up times are detrimental. Figure 12 exhibits the removed material volume as a function of the feed and cutting speed for the maximum tool life ($VB > 0.2$ mm), enabling an assessment of the tool's performance.

Figure 12 allows for the observation of two distinct conclusions: (i) the highest machined volume was achieved with a feed of 0.2 mm and (ii) the high material removal rates did not necessarily correspond to a lower machined volume (due to the shorter life), constituting improved productivity scenarios.

Notch wear induced by chipping, contrary to flank wear, decreased with an increase in the cutting speed and became indistinguishable from flank wear at $v_c = 250$ m/min, which was also reported by several other authors, for the same tool–workpiece material pairs [35,40]. The notching and built-up edge were particularly intense at lower cutting speeds for both tested feeds with the TiC binder insert, as seen in Figure 13, whereas with the TiN binder inserts under the same cutting conditions, less notching occurred. This excessive notch wear at lower cutting speeds led to an accelerated deterioration of the cutting edge and its ability to maintain strength properties, culminating in either (i) premature failure of the tool or (ii) flank wear homogenization covering the initial notching. The second occurrence is a more convenient scenario, resulting in a more gradual tool wear and, thus, the higher stability of the cutting operation. This is consistent with the fact that the tool flank wear pattern is more homogenous at higher cutting speeds, becoming the failure mode for most tools at speeds above 200 m/min. Unexpected tool failure in addition to an increasing frequency of tool change may also destroy the previously machined surface, causing an inherent impact on the economical side of the process.

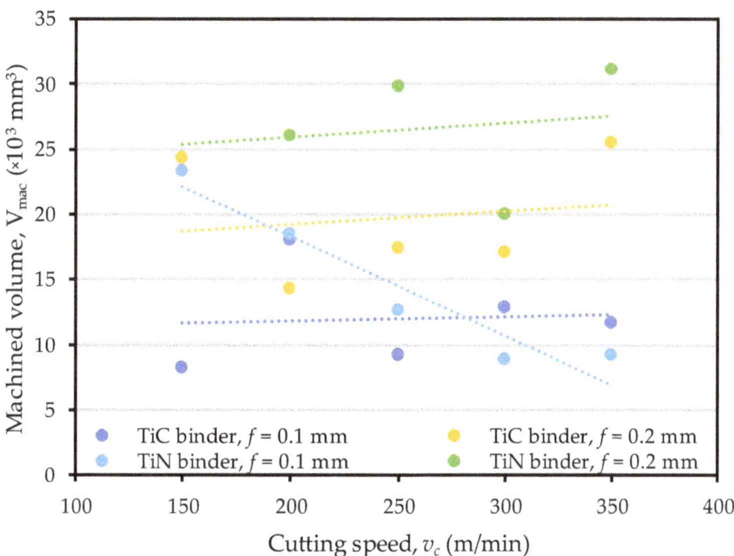


**Figure 12.** Influence of the cutting speed, feed, and PCBN binder on the machined volume.

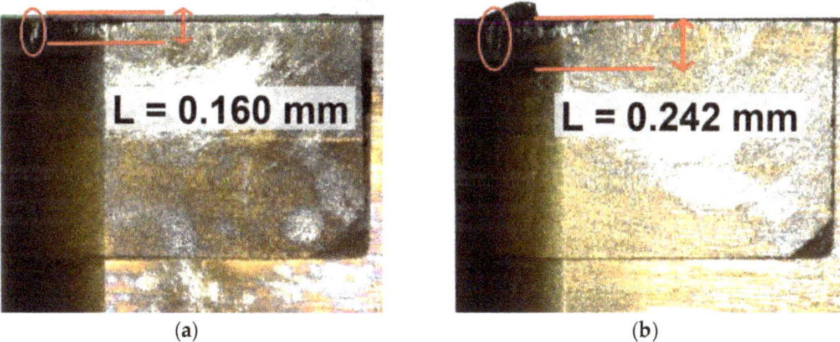


**Figure 13.** Flank surface of the TiC binder insert exhibiting notching under a cutting speed ($v_c$) of 150 m/min and (**a**) feed (*f*) of 0.1 mm after 105 s of cutting; (**b**) feed (*f*) of 0.2 mm after 65 s of cutting time.

Tool failure due to the fact of notching located at the depth of cut (see Figure 14c) at cutting speeds within the range of 150–200 m/min and a reduction of notching as the cutting speed increases has also been reported by various authors [26,41]. This tool nose notching at the depth of cut is located at the intersection between the cutting edge and the machined surface, which is a common failure mode when machining nickel-based alloys. Notch formation results from a combination of aggressive cutting conditions involved during the machining process of Ni-based alloys: high temperature, elevated strength, and strain hardening of the workpiece maintained at high temperature and the abrasive chips [42].

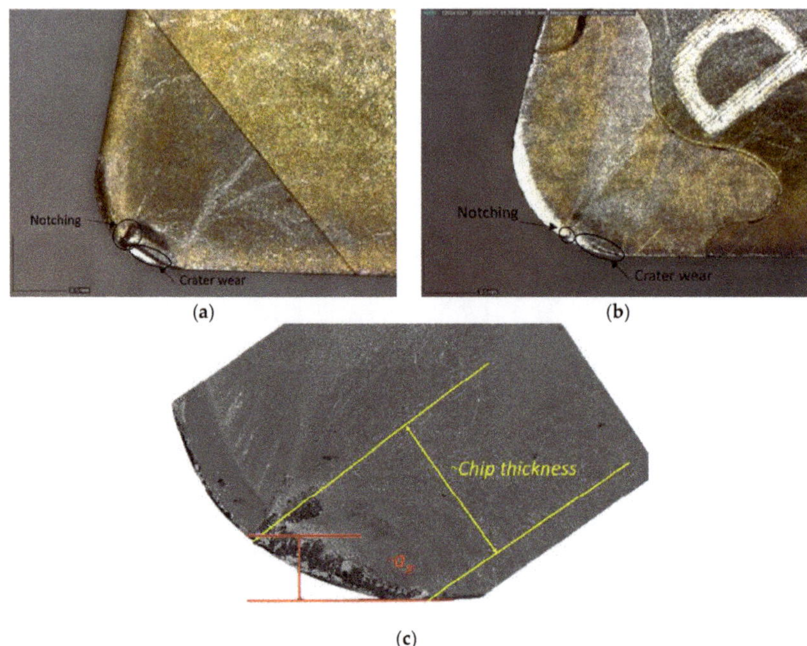



**Figure 14.** Rake face after 1.5 min of machining time: (**a**) insert 603; (**b**) insert 602; (**c**) notching occurrence located at depth of cut dimension.

Comparing the tool wear with the TiN binder insert and the TiC binder insert, during the previous turning experiments, it seemed that the first one appeared to have a more homogenous and predictable flank wear. The flank wear rate (TWf) of insert 602 showed less variation than insert 603 for the two tested feeds, except at a cutting speed of 350 m/min, so a further investigation and comparison of these two inserts was performed.

Table 4 shows the measured flank wear (*VB*) for cutting speeds of 250 m/min and feed of 0.1 mm, confirming that TiN binder inserts outperform TiC binder inserts, achieving less flank wear. The rake face images shown in Figure 14 also support the initial premise that the TiC binder insert has a tendency to notch at the depth-of-cut line. The TiN binder inserts outperformed TiC by almost 30%, a value that is in line with the 20% difference reported in the studies by Bushlya et al. [40] when machining Inconel 718 under flood coolant lubrication cutting conditions at a cutting speed of 300 m/min, feed of 0.1 mm/rev, and depth of cut of 0.3 mm. Curiously, it was reported that the oxidation of TiN and TiC start at 700 °C and 800 °C, respectively, forming $TiO_2$ [43].

**Table 4.** Flank wear after machining for 1.5 min for $v_c$ of 250 m/min and $f$ of 0.1 mm.

| Insert | VB | | | Average |
|---|---|---|---|---|
| | #1 | #2 | #3 | |
| TiC | 0.099 | 0.082 | 0.085 | 0.089 |
| TiN | 0.089 | 0.061 | 0.058 | 0.069 |

Electron backscatter diffraction (EBSD) imaging was used to further investigate and distinguish compositional changes in the tools tested. Different zones usually have different light intensities that correspond to distinct chemical compositions. EDS analyses and SEM images allowed to identify compositional changes on the tool's surface, such as the build-up edge, and to recognize that, as consequence of the abrasion at the tool–workpiece interface, small portions of the workpiece tended to become attached to the tool while the coating was removed.

Figures 15 and 16 exhibit SEM images of the three TiC and TiN binder inserts used to machine alloy 718. All tested inserts exhibited a curvilinear, elongated branded zone where the chip primarily exited at the rake face, resulting in both a loss of the tool coating due to the tool–chip interaction, diffusion, and adhesion of workpiece elements. Regarding the TiC and TiN inserts, there was a noticeable difference among them, which is visible as prominent abrasion wear roughly in the same place on the three TiC inserts, and this is coincident with the notching present on the flank face, as seen in Figure 17. This excessive notching tendency of the TiC binder insert compared to the TiN binder insert has previously been noticed in tool wear rate analyses and further confirmed with SEM analysis.

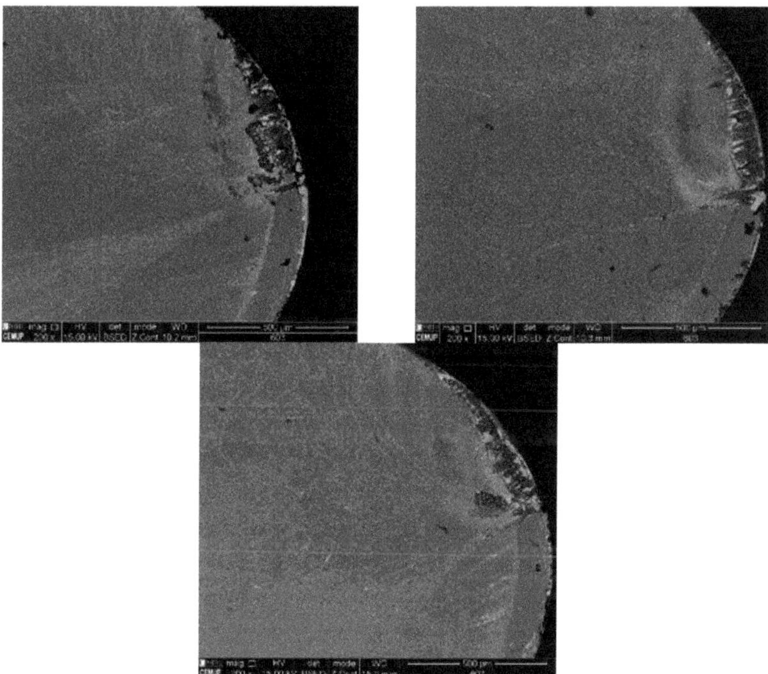

**Figure 15.** SEM images of the TiC binder insert rake face after 1.5 min of machining.

In addition, the flank wear on the TiN binder inserts was generally more homogeneous than for the TiC binder inserts mainly due to the absence of notching. Moreover, SEM images made it possible to understand that the cutting edge of the TiC binder inserts underwent more workpiece adhesion than the TiN binder inserts. While the workpiece material adhered to the tool, a new edge was created resulting in the loss of geometric and mechanical characteristics. This newly generated layer made the cutting edge less sharp, thus increasing the forces during the cutting, particularly the passive or radial force.

The TiN binder inserts seemed to have a geometrically consistent wear pattern on the rake face and tool, and the TiC binder insert's crater wear appeared to be darker in the images, indicating the presence of greater erosion, consequently reaching the tool's substrate. Both inserts show groove occurrence roughly at the same location, which seems to be related with the cutting depth; although for the TiC binder inserts, the cutting edge appeared to be more affected by the chipping, for the TiN binder inserts, the cutting edge seemed to better withstand its geometrical characteristics.

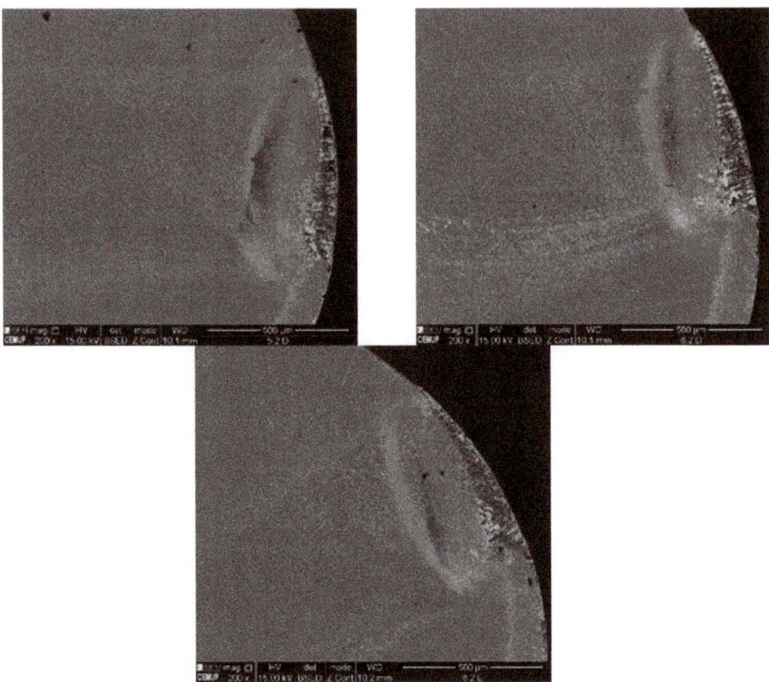

**Figure 16.** SEM images of the TiN binder insert rake face after 1.5 min of machining.

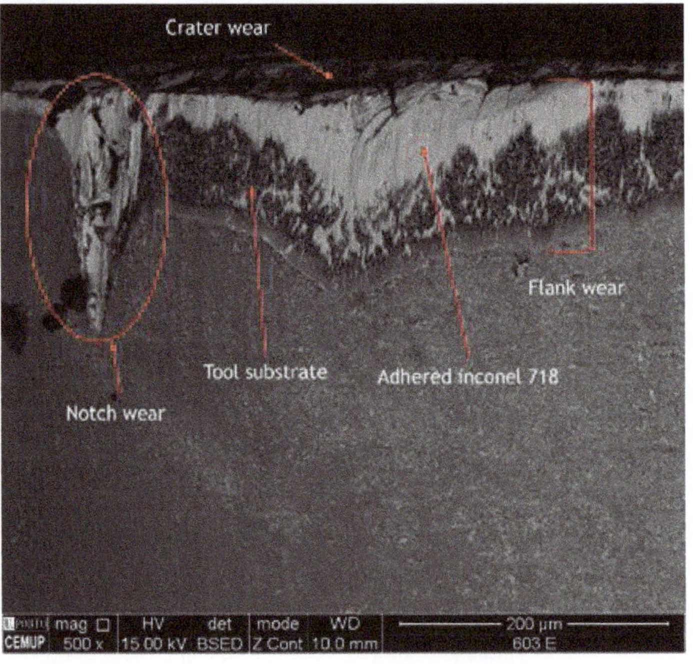


**Figure 17.** SEM image of a worn TiC binder insert tool flank.

## 4. Conclusions

In this study, instrumented turning experiments to assess the machinability of Inconel 718 alloy using PCBN cutting tools and two different binder phases, TiN and TiC, were selected. The primary cutting force displayed a consistent and gradual increase in magnitude as the depth of cut and feed rate increased, confirming the expected relationship between the machining process and operational conditions. In other words, the observed increase in the cutting force with an increasing depth of cut and feed rate was consistent with the predicted geometric relationship between these factors and the machining process. The following are the key findings:

- In terms of the cutting forces, there is a clear dependence between the obtained values and the adopted cutting parameters. The main cutting force increases for higher feed rates and decreases for higher cutting speeds;
- Increasing the cutting speed ($v_c$ > 300 m/min) seems to promote a change in the dominant wear mode, from mechanical abrasion to chemical diffusion, as evidenced by the type of tool wear noticed in the cutting tools;
- Low-content PCBN tools are suitable to machine Inconel 718 at high cutting speeds ($250 < v_c < -350$ m/min) and low depth of cut (0.15 mm). At lower cutting speeds ($150 < v_c < 200$ m/min), notching and chipping are the main causes of wear, resulting in instability of the cutting operation and leading to unpredictable tool failure. At high cutting speeds, adhesion and diffusion of Inconel 718 into the tool surface deteriorate the cutting edge's geometry and toughness;
- Lower cutting speeds and the lack of chip breaker resulted in poor chip control at low cutting speeds, and there was a tendency of chips to entangle around the workpiece. Increasing the cutting speed was favorable to the appearance of the chip segmentation;
- Of all forces involved in the turning operation, passive force was the greatest, which is typical when the depth of cut is smaller than the nose radius of the tool, and showed an evolution that was close with the increase in tool wear;
- The results of the experiment indicate that a higher removal rate does not always lead to a corresponding decrease in the durability of the cutting tool, which may ultimately enable more productive cutting conditions in terms of productivity. This highlights the potential of using PCBN tools when turning Inconel 718;
- TiN binder seemed to improve the tool resistance when machining this superalloy; yet, more experiments should be conducted, particularly focusing on the evaluation of the tool's thermal evolution through combined approaches of in situ temperature measurement and numerical modeling allowing for an assessment of the wear performance and stability for different ranges of temperature (promoted by distinct operational conditions). The chemical stability of the tested tools can also be assessed under distinct lubrication conditions.

**Author Contributions:** Conceptualization, A.M.P.d.J. and T.E.F.S.; methodology, T.E.F.S.; software, F.M. (Francisco Matos); validation, D.F., F.M. (Francisco Marques) and A.M.P.d.J.; formal analysis, A.M.P.d.J.; investigation, F.M. (Francisco Matos); resources, D.F. and F.M. (Francisco Marques); data curation, F.M. (Francisco Matos); writing—original draft preparation, T.E.F.S., V.F.C.S. and F.M. (Francisco Matos); writing—review and editing, T.E.F.S., V.F.C.S. and F.J.G.S.; visualization, F.M. (Francisco Marques), A.M.P.d.J. and F.J.G.S.; supervision, A.M.P.d.J. and F.J.G.S. All authors have read and agreed to the published version of the manuscript.

**Funding:** The authors gratefully acknowledge the funding from Project Hi-rEV—Recuperação do Setor de Componentes Automóveis (C644864375-00000002), cofinanced by Plano de Recuperação e Resiliência (PRR), República Portuguesa, through NextGeneration EU.

**Institutional Review Board Statement:** Not applicable.

**Informed Consent Statement:** Not applicable.

**Data Availability Statement:** The data presented in this study are available on request from the corresponding author.

**Conflicts of Interest:** The authors declare no conflict of interest.

## References


1. Mouritz, A.P. Introduction to aerospace materials. In *Introduction to Aerospace Materials*; Woodhead Publishing: Sawston, UK, 2012; pp. 1–14.
2. Choudhury, I.A.; El-Baradie, M.A. Machinability of nickel-base super alloys: A general review. *J. Mater. Process. Technol.* **1998**, *77*, 278–284. [CrossRef]
3. Buddaraju, K.M.; Sastry, G.R.K.; Kosaraju, S. A review on turning of Inconel alloys. *Mater. Today Proc.* **2021**, *44*, 2645–2652. [CrossRef]
4. Ulutan, D.; Ozel, T. Machining induced surface integrity in titanium and nickel alloys: A review. *Int. J. Mach. Tools Manuf.* **2011**, *51*, 250–280. [CrossRef]
5. Asala, G.; Andersson, J.; Ojo, O.A. A study of the dynamic impact behaviour of IN 718 and ATI 718Plus® superalloys. *Philos. Mag.* **2019**, *99*, 419–437. [CrossRef]
6. Shalaby, M.A.; Veldhuis, S.C. Tool wear and chip formation during dry high speed turning of direct aged Inconel 718 aerospace superalloy using different ceramic tools. *J. Eng. Tribol.* **2019**, *233*, 1127–1136. [CrossRef]
7. Zhou, J.; Bushlya, V.; Avdovic, P.; Ståhl, J.E. Study of surface quality in high speed turning of Inconel 718 with uncoated and coated CBN tools. *Int. J. Adv. Manuf. Technol.* **2012**, *58*, 141–151. [CrossRef]
8. Agmell, M.; Bushlya, V.; M'saoubi, R.; Gutnichenko, O.; Zaporozhets, O.; Laakso, S.V.; Ståhl, J.-E. Investigation of mechanical and thermal loads in pcBN tooling during machining of Inconel 718. *Int. J. Adv. Manuf. Technol.* **2020**, *107*, 1451–1462. [CrossRef]
9. Rahman, M.; Seah, W.K.H.; Teo, T.T. The machinability of Inconel 718. *J. Mater. Process. Technol.* **1997**, *63*, 199–204. [CrossRef]
10. Sousa, V.F.C.; Silva, F.J.G. Recent Advances on Coated Milling Tool Technology—A Comprehensive Review. *Coatings* **2020**, *10*, 235. [CrossRef]
11. Wojciechowski, S.; Przestacki, D.; Chwalczuk, T. The Evaluation of Surface Integrity During Machining of Inconel 718 with Various Laser Assistance Strategies. *MATEC Web Conf.* **2017**, *136*, 01006. [CrossRef]
12. Burghardt, A.; Szybicki, D.; Kurc, K.; Muszyñska, M.; Mucha, J. Experimental Study of Inconel 718 Surface Treatment by Edge Robotic Deburring with Force Control. *Strength Mater.* **2017**, *49*, 594–604. [CrossRef]
13. Sousa, V.F.C.; Da Silva, F.J.G.; Pinto, G.F.; Baptista, A.; Alexandre, R. Characteristics and Wear Mechanisms of TiAlN-Based Coatings for Machining Applications: A Comprehensive Review. *Metals* **2021**, *11*, 260. [CrossRef]
14. Fallbohmer, P.; Rodrigues, C.A.; Ozel, T.; Altan, T. High-speed machining of cast iron and alloy steels for die and mold man-ufacturing. *J. Mater. Process. Technol.* **2000**, *98*, 104–115. [CrossRef]
15. Liao, Y.; Lin, H. Mechanism of minimum quantity lubrication in high-speed milling of hardened steel. *Int. J. Mach. Tools Manuf.* **2007**, *47*, 1660–1666. [CrossRef]
16. Wang, B.; Liu, Z.Q.; Cai, Y.K.; Luo, X.C.; Ma, H.F.; Song, Q.H.; Xiong, Z.H. Advancements in material removal mechanism and surface integrity of high speed metal cutting: A review. *Int. J. Mach. Tools Manuf.* **2021**, *166*, 103744. [CrossRef]
17. Ezugwu, E.O. High speed machining of aero-engine alloys. *J. Braz. Soc. Mech. Sci. Eng.* **2004**, *26*, 1–11. [CrossRef]
18. Tu, L.Q.; Ming, W.W.; Xu, X.W.; Cai, C.Y.; Chen, J.; An, Q.L.; Xu, J.Y.; Chen, M. Wear and failure mechanisms of SiAlON ceramic tools during high-speed turning of nickel-based superalloys. *Wear* **2022**, *488–489*, 204171. [CrossRef]
19. Dudzinski, D.; Devillez, A.; Moufki, A.; Larrouquère, D.; Zerrouki, V.; Vigneau, J. A review of developments towards dry and high speed machining of Inconel 718 alloy. *Int. J. Mach. Tools Manuf.* **2004**, *44*, 439–456. [CrossRef]
20. Toenshoff, H.K.; Denkena, B. *Basics of Cutting and Abrasive Processes*; Springer: Berlin/Heidelberg, Germany, 2013. [CrossRef]
21. Criado, V.; Alvarez, J.D.; Cantero, J.L.; Miguélez, M.H. Study of the performance of PCBN and carbide tools in finishing ma-chining of Inconel 718 with cutting fluid at conventional pressures. *Procedia CIRP* **2018**, *77*, 634–637. [CrossRef]
22. Giménez, S.; Van der Biest, O.; Vleugels, J. The role of chemical wear in machining iron based materials by PCD and PCBN super-hard tool materials. *Diam. Relat. Mater.* **2007**, *16*, 435–445. [CrossRef]
23. Zhangiang, L.; Xing, A. Cutting tool materials for high speed machining. *Prog. Nat. Sci.* **2005**, *15*, 777–783. [CrossRef]
24. Costes, J.; Guillet, Y.; Poulachon, G.; Dessoly, M. Tool-life and wear mechanisms of CBN tools in machining of Inconel 718. *Int. J. Mach. Tools Manuf.* **2007**, *47*, 1081–1087. [CrossRef]
25. Bushlya, V.; Zhou, J.; Avdovic, P.; Ståhl, J.-E. Performance and wear mechanisms of whisker-reinforced alumina, coated and uncoated PCBN tools when high-speed turning aged Inconel 718. *Int. J. Adv. Manuf. Technol.* **2013**, *66*, 2013–2021. [CrossRef]
26. Khan, S.; Soo, S.; Aspinwall, D.; Sage, C.; Harden, P.; Fleming, M.; White, A.; M'Saoubi, R. Tool wear/life evaluation when finish turning Inconel 718 using PCBN tooling. *Procedia CIRP* **2012**, *1*, 283–288. [CrossRef]
27. Soo, S.L.; Khan, S.A.; Aspinwall, D.K.; Harden, P.; Mantle, A.L.; Kappmeyer, G.; Pearson, D.; M'Saoubi, R. High speed turning of Inconel 718 using PVD-coated PCBN tools. *CIRP Ann.* **2016**, *65*, 89–92. [CrossRef]
28. Bushlya, V.; Zhou, J.; Ståhl, J. Effect of Cutting Conditions on Machinability of Superalloy Inconel 718 During High Speed Turning with Coated and Uncoated PCBN Tools. *Procedia CIRP* **2012**, *3*, 370–375. [CrossRef]
29. Bushlya, V.; Bjerke, A.; Turkevich, V.Z.; Lenrick, F.; Petrusha, I.A.; Cherednichenko, K.A.; Stahl, J.E. On chemical and diffu-sional interactions between PCBN and superalloy Inconel 718: Imitational experiments. *J. Europ. Ceramic Soc.* **2019**, *39*, 2658–2665. [CrossRef]

30. Sugihara, T.; Tanaka, H.; Enomoto, T. Development of Novel CBN Cutting Tool for High Speed Machining of Inconel 718 Focusing on Coolant Behaviors. *Procedia Manuf.* **2017**, *10*, 436–442. [CrossRef]

31. Sousa, V.F.C.; Silva, F.J.G. Recent Advances in Turning Processes Using Coated Tools—A Comprehensive Review. *Metals* **2020**, *10*, 170. [CrossRef]

32. Slama, C.; Servant, C.; Cizeron, G. Aging of Inconel 718 alloy between 500 and 750 °C. *J. Mater. Res.* **1997**, *12*, 2298–2316. [CrossRef]

33. Azarbarmas, M.; Aghaie-Khafri, M.; Cabrera, J.; Calvo, J. Dynamic recrystallization mechanisms and twining evolution during hot deformation of Inconel 718. *Mater. Sci. Eng. A* **2016**, *678*, 137–152. [CrossRef]

34. Wu, Z.; Parish, C.; Bei, H. Nano-twin mediated plasticity in carbon-containing FeNiCoCrMn high entropy alloys. *J. Alloy. Compd.* **2015**, *647*, 815–822. [CrossRef]

35. Cantero, J.L.; Díaz-Álvarez, J.; Infante-García, D.; Rodríguez, M.; Criado, V. High Speed Finish Turning of Inconel 718 Using PCBN Tools under Dry Conditions. *Metals* **2018**, *8*, 192. [CrossRef]

36. Huang, Y.; Liang, S.Y. Modeling of Cutting Forces Under Hard Turning Conditions Considering Tool Wear Effect. *J. Manuf. Sci. Eng.* **2005**, *127*, 262–270. [CrossRef]

37. Bouacha, K.; Yallese, M.A.; Mabrouki, T.; Rigal, J.-F. Statistical analysis of surface roughness and cutting forces using response surface methodology in hard turning of AISI 52100 bearing steel with CBN tool. *Int. J. Refract. Met. Hard Mater.* **2010**, *28*, 349–361. [CrossRef]

38. Wanner, B.; Eynian, M.; Beno, T.; Pejryd, L. Process stability strategies in milling of thin-walled Inconel 718. *AIP Conf. Procc.* **2012**, *1431*, 465. [CrossRef]

39. Silva, T.E.; Amaral, A.; Couto, A.; Coelho, J.; Reis, A.; Rosa, P.A.; de Jesus, A.M. Comparison of the machinability of the 316L and 18Ni300 additively manufactured steels based on turning tests. *J. Mater. Des. Appl.* **2021**, *235*, 2207–2226. [CrossRef]

40. Bushlya, V.; Lenrick, F.; Bjerke, A.; Aboulfadl, H.; Thuvander, M.; Ståhl, J.-E.; M'Saoubi, R. Tool wear mechanisms of PcBN in machining Inconel 718: Analysis across multiple length scale. *CIRP Ann.* **2021**, *70*, 73–78. [CrossRef]

41. Bhattacharyya, S.K.; Jawaid, A.; Lewis, M.H.; Wallbank, J. Wear mechanisms of Syalon ceramic tools when machining nick-el-bases materials. *Metals Technol.* **1983**, *10*, 482–489. [CrossRef]

42. Xue, C.; Wang, D.; Zhang, J. Wear Mechanisms and Notch Formation of Whisker-Reinforced Alumina and Sialon Ceramic Tools during High-Speed Turning of Inconel 718. *Materials* **2022**, *15*, 3860. [CrossRef]

43. Tampieri, A.; Bellosi, A. Oxidation Resistance of Alumina-Titanium Nitride and Alumina-Titanium Carbide Composites. *J. Am. Ceram. Soc.* **1992**, *75*, 1688–1690. [CrossRef]

*Article*

# Investigations on the Wear Performance of Coated Tools in Machining UNS S32101 Duplex Stainless Steel

Vitor F. C. Sousa [1], Francisco J. G. Silva [1,2,*], Ricardo Alexandre [3], Gustavo Pinto [1,2], Andresa Baptista [1,2] and José S. Fecheira [1]

[1] ISEP—School of Engineering, Polytechnic of Porto, Rua Dr. António Bernardino de Almeida 431, 4200-072 Porto, Portugal; vcris@isep.ipp.pt (V.F.C.S.); gflp@isep.ipp.pt (G.P.); absa@isep.ipp.pt (A.B.); jsf@isep.ipp.pt (J.S.F.)

[2] INEGI—Instituto de Ciência e Inovação em Engenharia Mecânica e Engenharia Industrial, Rua Dr. Roberto Frias 400, 4200-465 Porto, Portugal

[3] TeandM—Tecnologia, Engenharia e Materiais, S.A., Parque Industrial do Taveiro, Lotes 41–42, 3045-504 Taveiro, Portugal; ricardo@teandm.pt

* Correspondence: fgs@isep.ipp.pt

**Abstract:** Due to their high mechanical property values and corrosion resistance, duplex stainless steels (DSSs) are used for a wide variety of industrial applications. DSSs are also selected for applications that require, especially, high corrosion resistance and overall good mechanical properties, such as in the naval and oil-gas exploration industries. The obtention of components made from these materials is quite problematic, as DSSs are considered difficult-to-machine alloys. In this work, the developed wear during milling of the UNS S32101 DSS alloy is presented, employing four types of milling tools with different geometries and coatings. The influence of feed rate and cutting length variations on the tools' wear and their performance was evaluated. The used tools had two and four flutes with different coatings: TiAlN, TiAlSiN and AlCrN. The cutting behavior of these tools was analyzed by collecting data regarding the cutting forces developed during machining and evaluating the machined surface quality for each tool. After testing, the tools were submitted to SEM analysis, enabling the identification of the wear mechanisms and quantification of flank wear, as well as identifying the early stages of the development of these mechanisms. A comparison of all the tested tools was made, determining that the TiAlSiN-coated tools produced highly satisfactory results, especially in terms of sustained flank wear.

**Keywords:** duplex stainless steel; milling; tool coatings; surface roughness; wear mechanisms

**Citation:** Sousa, V.F.C.; Silva, F.J.G.; Alexandre, R.; Pinto, G.; Baptista, A.; Fecheira, J.S. Investigations on the Wear Performance of Coated Tools in Machining UNS S32101 Duplex Stainless Steel. *Metals* **2022**, *12*, 896. https://doi.org/10.3390/met12060896

Academic Editor: Pavel Krakhmalev

Received: 26 March 2022
Accepted: 20 May 2022
Published: 25 May 2022

**Publisher's Note:** MDPI stays neutral with regard to jurisdictional claims in published maps and institutional affiliations.



## 1. Introduction

The use of duplex stainless steels (DSSs) in industries that benefit from the set of proprieties usually presented by these alloys, namely, high corrosion resistance and good mechanical properties, has deeply increased [1]. Thus, DSSs became appealing for many different applications in a wide range of industrial sectors [2–4], with these having better overall properties than most stainless steels, and even sometimes being presented as a viable alternative to some Ni-based alloys [5,6].

The machining process is heavily employed in the fabrication of high-quality and precision parts. Thus, the machining optimization of these alloys is quite appealing; however, the research about this topic is quite scarce. There are some authors that use techniques such as the Taguchi method, which is commonly employed in the optimization of machining processes [7–10]. These studies enable the identification of the machining parameters on the process, enabling optimization of production quality [11], material removal rate [12] and even generated cutting forces [13]. Another common method that can be employed to obtain information regarding the impact of a certain parameter is a multiple regression analysis [14,15], or response surface methodology (RSM), coupled with an analysis of variance (ANOVA) [6]. These methods allow for the creation of predictive techniques that can be applied not only to surface roughness, as seen in [14], but to other parameters, such as the lowering of cutting forces or improving the material removal rate. This can be advantageous, as it avoids the execution of expensive and time-consuming machining tests.

Although parameter optimization is very important, from a process improvement standpoint, the tool being used is equally important, as it directly impacts process efficiency. Most machining tools for turning and milling, nowadays, are coated tools or coated carbide inserts [16,17]. These coatings have a wide range of benefits for machining applications that are well documented, such as an improved surface finish, tool longevity, reduction in cutting forces [18], lowering the coefficient of friction [19] and greatly improving the wear resistance of coated tools and surfaces [20–22]. Tool coatings are usually obtained either by physical vapor deposition (PVD) or by chemical vapor deposition (CVD). Each of the processes produces different kinds of coatings with different properties and structures, as they use different methods to achieve the deposited films [23–26]. There are many studies on the comparison of coatings for the machining of various alloys, especially hard-to-machine alloys, with comparisons between PVD and CVD coatings being commonly made as researchers try to find advantages of one type of coating over the other and, also, try to optimize/improve the machining process by using coated tools [27–29]. These studies are very useful, as the coating's properties greatly affect the machining performance, with the coating's mechanical properties, structure, microstructure and residual stresses having a great influence on the machining process, stressing the importance of correct coating selection [30–33].

Cutting forces that are generated during the process provide valuable information regarding the overall process' state, thus enabling machining process optimization. This means that cutting force data provide a way to monitor the process or identify certain aspects that can be improved in the machining process. The knowledge of these cutting forces can be used to monitor tool behavior [34,35] and give information on optimal machining parameters [36,37], thus enabling the improvement of the machining process, either by parameter optimization, by improving tool life or even by helping with the development of new machining tool technology [38,39]. There are different methods for determining cutting forces, either by using a direct or indirect approach [40–43]. Further regarding machining process optimization, the analysis of the wear mechanisms that cutting tools present also brings many advantages, enabling machining process optimization by improving tool life, creating new geometries more adequate for the machining of certain materials and even optimizing the coatings of these tools by providing information on the right coating to use for a certain application. Analysis of these wear mechanisms provides valuable information that can be used for parameter adjustment as well [44]. There are studies made in this regard, focused on DSS, with common wear mechanisms being registered, such as abrasion and adhesive wear, due to the high strength of the material and high friction values reached during machining of these alloys [45–47]. Knowledge about tool wear behavior is highly important [48], as it provides information on what is occurring during machining, enabling for a decision-making process that is focused on improving the process. Either by implementing a new cooling method or adjusting machining parameters, this knowledge also enables the selection of the right tool for the job (prioritizing tool life, surface roughness and overall process efficiency), considering tool geometry and coating [45,49–51]

In this work, the wear behavior and machining performance of milling tools used in finishing operations of the UNS S32101 DSS alloy is presented. Four tools with differing geometries and coatings were used, namely, TiAlN, TiAlSiN and AlCrN coatings. Two of the tested tools were coated with AlCrN, one having two flutes and the other four flutes, as this coating is recommended for the machining of DSS alloys. The machining parameters were also varied, testing two values of cutting length and three values of feed rate. The influence of these coatings, tool geometries and machining parameter variations on the wear behavior and production quality of the tools was assessed. This assessment was performed by subjecting the tools to SEM analysis, measuring flank wear (VB) and identifying the sustained wear mechanisms. The machined surface roughness was also evaluated to characterize this influence. The obtained results are presented in this manuscript, with one section dedicated to each of the performed analyses (surface roughness assessment, flank wear measurement and wear mechanism analysis). The authors hope that with these results, the gap regarding the machining of these alloys can be somewhat filled, contributing information that could be relevant when it comes to the optimization of the machining of these alloys.

## 2. Materials and Methods

In this section, the various materials and methods employed during this work will be presented in various subsections, as to provide a clear understanding of what was used throughout the various work phases.

### 2.1. Materials

#### 2.1.1. Employed Tools

In this work, four tool types were used to machine the DSS alloy. The tools' substrate was cemented carbide WC-Co grade 6110, provided by INOVATOOLS, S.A. (Leiria, Portugal). This cemented carbide grade presented as micro-grain, with a measurement of about 0.3 μm, and used 6% Co (wt.) as a binder. The tools' substrate was ground, yet the patterns left by this process were seen in the tools' surfaces, even after the deposition process. Regarding the tools' dimensions, all the produced tools were 4 mm in diameter and had a total length of 68 mm. As this study aimed to compare the influence of different coatings as well as different tool geometries in the wear mechanism suffered during machining tests, four tools with different geometries were produced and were coated with three different coatings, TiAlN, TiAlSiN and AlCrN. After production, the tools were subjected to an ultrasonic bath using acetone for 10 min; after this first bath, the tools underwent a second one for 5 min. The acetone was changed between the baths. These tools were coated with three types of PVD coatings, the TiAlN, TiAlSiN and AlCrN coatings. This enabled the comparison of the wear behavior and machining performance of these coatings by creating four different tools, as seen in Table 1.

**Table 1.** Tools created for the machining of UNS S32101.

| Tool Ref. | Tool Geometry | Number of Flutes/Edges | Rake Angle | Relief Angle |
|---|---|---|---|---|
| T1 | Flat end-mill | 2 | 30° | 10° |
| T2 | End-mill with a 45° chamfer, 0.08 mm from the cutting edge | 4 | 35° | 10° |
| T3 | Flat end-mill | 4 | 40° | 5° |
| T4 | End-mill with a 0.2 mm corner radius | 4 | 35° | 10° |

Regarding the deposition of these coatings, CemeCon CC800/9ML PVD unbalanced magnetron sputtering equipment provided with four target holders was used. After the cleaning process, the tools were assembled in a holder and placed inside the equipment. To ensure that the coatings were deposited onto the tools' surfaces as homogenously as possible, the tool holders were rotated at a speed of 1 rpm during the process. The coating process was similar for the TiAlN and TiAlSiN coatings, with the use of four similar targets. In the case of the AlCrN coating, four targets were also used, two being composed of 100% Cr and the other two of 100% Al. These targets were distributed alternately inside the reactor, resulting in the deposition of alternating layers of Cr and Al. Thin, alternating layers of Al and Cr were achieved on the T1 and T3 tools. The set of parameters used for the deposition of the mentioned coatings can be observed in Table 2. During the deposition of the various tool sets, flat substrates were placed inside the deposition chamber to create a sample that could be used to determine the coatings' mechanical properties.

**Table 2.** TiAlN, TiAlSiN and AlCrN PVD coatings' deposition parameters.

| Parameters | TiAlN Coating | TiAlSiN Coating | AlCrN Coating |
|---|---|---|---|
| Deposition time (min.) | 240 | 240 | 240 |
| Reactor gases | $Ar^+ + Kr + N_2$ | $Ar^+ + Kr + N_2$ | $Ar^+ + Kr + N_2$ |
| Target material | 4TiAl 40/60 | 4TiAlSi 38/57/5 | 2Al + 2Cr |
| Pressure (mPa) | 580 | 580 | 580 |
| Temperature (°C) | 450 | 450 | 450 |
| Bias (V) | $-110$ | $-110$ | $-110$ |
| Target current density (A/cm$^2$) | 20 | 20 | 20 |
| Holder rotational speed (rpm) | 1 | 1 | 1 |

These parameters have been selected based on previous successful experiments carried out in similar substrates. After deposition, all the tools were packed carefully, avoiding the handling of the tools' cutting areas.

### 2.1.2. Machined Material

The machined material is a duplex stainless steel, UNS S32101, which was provided as a round bar that was 75 mm in diameter. Regarding this material's mechanical properties, it has a yield strength of 450 MPa and ultimate tensile strength of 650 MPa, following the information provided by the material's manufacturer (Outokumpu). This steel has an average hardness of 280 ± 20 HV5, and its chemical composition was provided by the manufacturer and can be observed in Table 3. The hardness values were confirmed by performing ten measurements with universal hardness testing equipment, the EMCO M4U, using a 5 kgf and a dwell time of 30 s. The hardness values were in accordance with those present in the material's data sheet (290 HV5), exhibiting a slight deviation.

**Table 3.** Chemical composition (wt%) of the DSS, UNS S32101.

| C | Mn | Cu | Cr | Ni | Mo | N |
|---|---|---|---|---|---|---|
| 0.03 | 5.0 | 0.3 | 21.5 | 1.5 | 0.3 | 0.22 |

### 2.2. Methods

### 2.2.1. Coatings' Thickness Analysis

The coating thickness analysis was carried out on cross-sections of samples, obtained as described in the previous section. The coating thickness measurement was performed using a FEI QUANTA 400 FEG scanning electron microscope (SEM), provided with an EDAX Genesis energy dispersive X-ray spectroscopy microanalysis system. The same range of magnifications were used in all the pictures obtained by the SEM, allowing for a greater ease of comparison of phenomena between different samples. The values for coating thickness were obtained by calculating the mean value of the measurements registered during SEM analysis. A total of six different measurements across different zones for each of the coated tools were performed. EDS analyses were carried out with a beam potential of 15 kV; however, this was sporadically reduced to 10 kV to decrease the volume of interaction and reduce noise in the spectra. Although the reading accuracy of the EDS analyses as a quantitative evaluation is not the best, it was also used to confirm the chemical composition of the coatings, which was sufficient for this purpose and avoided employment of more costly technologies.

### 2.2.2. Machining Tests

Machining Strategy

The machining tests were performed in a HAAS VF2 CNC milling center, capable of reaching 10,000 rpm and having a maximum power of 20 kW. A strategy was devised to machine the material: as its geometry was round, the tools would machine the material in a spiral motion, going from the outside of the raw material to its center. The machining tests were performed using a water-miscible cutting fluid projected externally (5% oil in water); furthermore, a set of parameters were chosen by considering the optimal values for machining operations that were suggested by the tools' substrate provider. The machining parameter values were determined and applied to all the tools to enable the comparison of their wear behavior and cutting performance. These parameters can be observed in Table 4. To ensure the repeatability of the tests, each milling test was performed three times for each set of cutting conditions to produce a greater number of results regarding wear mechanisms and the surface roughness of the machined surface, thus improving the consistency of the obtained data. Regarding the machining parameters, a constant radial depth of cut and vertical depth of cut were used, these being 3 mm and 0.08 mm, respectively. Furthermore, a constant cutting speed of 60 m/min was used. To evaluate the influence of feed on the machined surface roughness and on the tools' wear, this parameter was varied by 25% by creating test conditions that employed 75% and 125% of the recommended feed value, which was 479 mm/min. To determine the best cutting length to carry out this comparative experimental work, the cutting force analysis was used. For this purpose, one of each fabricated tool was used for the machining of the material, using the strategy and parameters described above. For this preliminary test, a cutting length of 25 m was used in the procedure described in detail (mentioning the equipment) in the following subsection.

**Table 4.** Parameters used in the machining tests.

| Sample | Coating | Cutting Speed | Number of Edges | Feed Rate | Depth of Cut | Radial Depth of Cut | Cutting Length |
|---|---|---|---|---|---|---|---|
| Reference | Type | mm/min | | mm/min | mm | mm | m |
| T1L4F75 | AlCrN | 60 | 2 | 359.25 | 0.08 | 3 | 4 |
| T1L2F75 | AlCrN | 60 | 2 | 359.25 | 0.08 | 3 | 2 |
| T1L4F100 | AlCrN | 60 | 2 | 479 | 0.08 | 3 | 4 |
| T1L2F100 | AlCrN | 60 | 2 | 479 | 0.08 | 3 | 2 |
| T1L4F125 | AlCrN | 60 | 2 | 598.25 | 0.08 | 3 | 4 |
| T1L2F125 | AlCrN | 60 | 2 | 598.25 | 0.08 | 3 | 2 |
| T2L4F75 | TiAlN | 60 | 4 | 359.25 | 0.08 | 3 | 4 |
| T2L2F75 | TiAlN | 60 | 4 | 359.25 | 0.08 | 3 | 2 |
| T2L4F100 | TiAlN | 60 | 4 | 479 | 0.08 | 3 | 4 |
| T2L2F100 | TiAlN | 60 | 4 | 479 | 0.08 | 3 | 2 |
| T2L4F125 | TiAlN | 60 | 4 | 598.25 | 0.08 | 3 | 4 |
| T2L2F125 | TiAlN | 60 | 4 | 598.25 | 0.08 | 3 | 2 |
| T3L4F75 | AlCrN | 60 | 4 | 359.25 | 0.08 | 3 | 4 |
| T3L2F75 | AlCrN | 60 | 4 | 359.25 | 0.08 | 3 | 2 |
| T3L4F100 | AlCrN | 60 | 4 | 479 | 0.08 | 3 | 4 |
| T3L2F100 | AlCrN | 60 | 4 | 479 | 0.08 | 3 | 2 |
| T3L4F125 | AlCrN | 60 | 4 | 598.25 | 0.08 | 3 | 4 |
| T3L2F125 | AlCrN | 60 | 4 | 598.25 | 0.08 | 3 | 2 |
| T4L4F75 | TiAlSiN | 60 | 4 | 359.25 | 0.08 | 3 | 4 |
| T4L2F75 | TiAlSiN | 60 | 4 | 359.25 | 0.08 | 3 | 2 |
| T4L4F100 | TiAlSiN | 60 | 4 | 479 | 0.08 | 3 | 4 |
| T4L2F100 | TiAlSiN | 60 | 4 | 479 | 0.08 | 3 | 2 |
| T4L4F125 | TiAlSiN | 60 | 4 | 598.25 | 0.08 | 3 | 4 |
| T4L2F125 | TiAlSiN | 60 | 4 | 598.25 | 0.08 | 3 | 2 |

Cutting Force Analysis

Cutting forces developed during machining can provide valuable information regarding the tools' wear. To acquire these cutting forces, a 4-component KISTLER 9171A dynamometer was used, coupled to a KISTLER 5697A1 data acquisition system, which allowed to record the cutting forces developed in X, Y and Z axes, as well as the developed torque (Mz), through appropriate software supplied with the dynamometer. The equipment was coupled to the spindle of the CNC machining center, having then the appropriate clamping system for placing the tool. The acquisition rate of the developed forces was selected according to the spindle's rotation speed, and the force of all the cutting edges of the tool at each rotation was registered. The results were collected and analyzed, enabling the identification of any increase or abnormal behavior of the cutting forces that could be caused by wear phenomena on the tool.

Machining Test Parameters

With the determined cutting lengths, the wear mechanisms that these tools suffer during machining could be evaluated. The main concern was to identify the early wear phenomena as well as its evolution under these cutting conditions, and not to determine the lifespan of the tools. The determined parameters for each test can be observed in Table 4, including the sample reference used for each of the tested tools.

Three tools were used for each of the conditions presented above. The use of lubricant during the machining tests caused difficulties in the analysis of the tools' surfaces. Thus, all the tools underwent one ultrasonic bath using acetone for 5 min to remove the presence of lubricant on the areas where SEM analysis was to be performed. This bath had a short duration to prevent the eventual removal of adhered material or sections of coating (near detachment). The surface roughness of the material was also assessed after each machining test.

### 2.2.3. Surface Roughness Test

Surface roughness evaluation is very important in the machining process, as it is closely related to machining performance, tool wear and process stability. In this work, the material's machined surface roughness was characterized in two directions: radial and tangential (machining direction). To assess this parameter, surface roughness tests were conducted using a MAHR PERTHOMETER M2 profilometer, following the standard procedure described in DIN EN ISO 4288/ASME b461. Each test covered a length of 5.6 mm, corresponding to seven segments of the cut-off value (0.8 mm), with the first and last value being ignored due to the acceleration and deceleration of the probe arm. For the surface roughness analysis, the following parameters were considered: arithmetic mean roughness (Ra) and the maximum roughness (Rmax). The R profile was also analyzed to identify any sharp peaks in the surface roughness values and other phenomena.

### 2.2.4. Tool Wear Analysis

After machining, all the tools were subjected to SEM analyses to access the amount of wear that these tools sustained and to identify the wear mechanisms developed on the various tools during machining. To access the wear, the values for flank wear were measured (VB) according to the ISO 8688-2:1986 standard [52]. Furthermore, the values of VB that were presented were the mean values for each tool for each of the test conditions.

A reference for the tool analysis was created and can be observed in Figure 1; both the rake face and clearance face were analyzed, identifying the wear mechanisms that were present and measuring the flank wear (VB) on the clearance face of the tested tools.

The numerations observed in Figure 1 were adopted in the identification of the SEM images, with the rake face and clearance face being identified with RF and CF, respectively, identifying the edge under analysis as presented in Figure 1. Additionally, since three tools were used for each of the test conditions, an identification number (ranging from 01 to 03) was added at the end of the image reference.

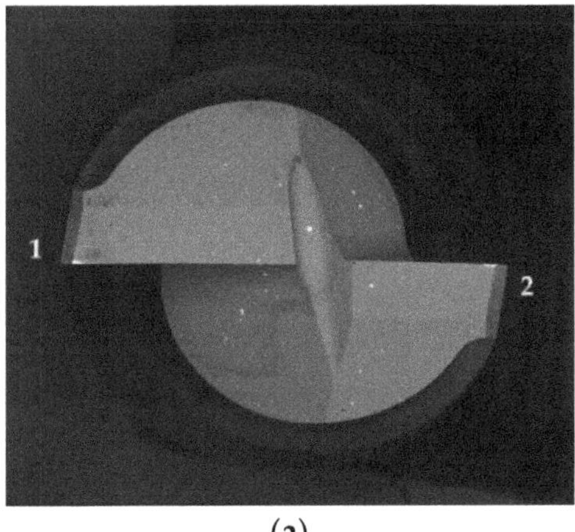

(a)

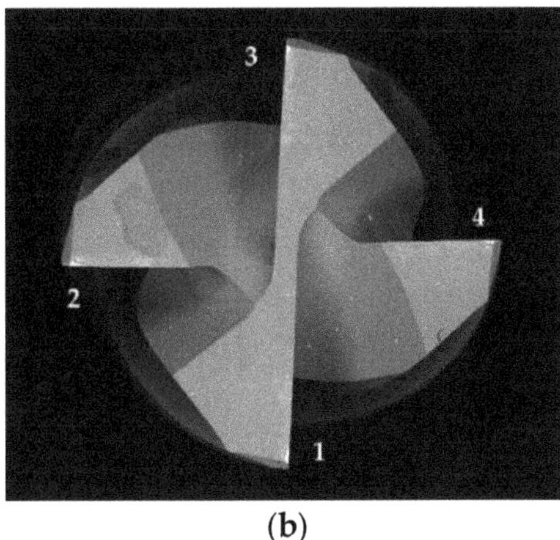

(b)

**Figure 1.** Reference used for the SEM analysis of the end-mills with (**a**) two cutting edges, and (**b**) four cutting edges.

## 3. Results and Discussion

In this section, the results obtained from all the tests and analyses are presented, organized and divided into various subchapters.

### 3.1. Coatings' Characterization

For the coatings' characterization, the thickness of the three applied coatings was analyzed by preparing the unused tools for SEM analysis. Both the T1 and T3 tools were considered in the characterization of the AlCrN coating, as they were coated with the same coating using the same deposition parameters. Regarding the thickness measurements, these were performed in the same area for all the tested tools. Thickness values were obtained near each tool's rake and clearance face (obtained from the tools' cross-sections). The average thickness value was then determined. It is worthy to note that there were no significant changes detected for the analyzed coatings' thicknesses.

The average thickness values registered for each of the coatings can be observed in Table 5.

**Table 5.** Average thickness values for the analyzed coatings.

| Coating | Thickness (μm) |
|---------|----------------|
| TiAlN | 2.812 ± 0.121 |
| TiAlSiN | 2.799 ± 0.163 |
| AlCrN | 2.965 ± 0.227 |

Although the AlCrN coating is a multi-layered coating, in the polished samples the interface between layers was not observed. This was due to the polishing process itself, as it smoothed the sample's cross-section, homogenizing the coating and making it look like a monolithic coating. However, if the coating was fractured, this multi-layered structure was observed, as seen in Figure 2a. Also in Figure 2, SEM measurements of the TiAlSiN coating thickness can be observed. This is an example of a measurement performed for all the coatings' thicknesses. In Figure 2a, the Al layers are the lighter ones, whereas the Cr-rich layers are the darker ones. Regarding their thickness, due to the SEM equipment's limitations, this was not determined.

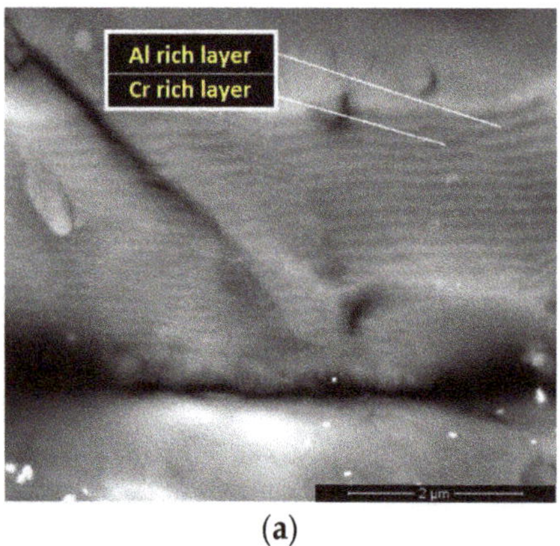


(a)

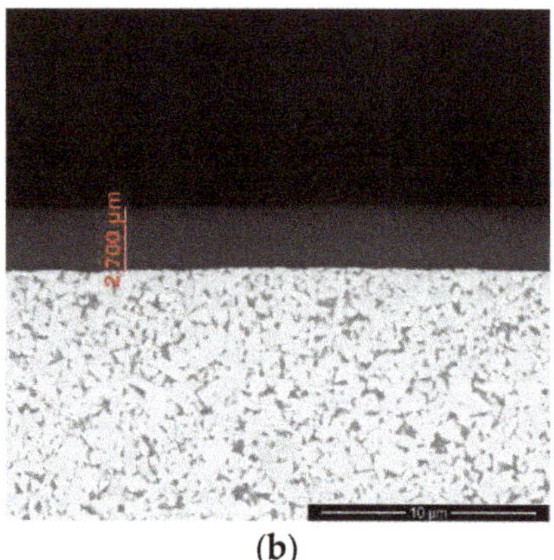


(b)

**Figure 2.** Coatings' thickness analysis: multi-layered structure of the AlCrN coating (**a**); measurement of TiAlSiN (**b**).

The coatings' mechanical properties were also evaluated, namely, hardness and the Young's modulus, by conducting ultra-micro hardness tests on the TiAlN, TiAlSiN and AlCrN coatings using a dynamic ultra-micro hardness tester, the Fischerscope H100. As previously mentioned, the coatings were deposited onto flat substrate samples that were coated simultaneously with the tools. Ten tests were conducted on each of the deposited coatings, determining values for hardness and the reduced Young's modulus. The average of these values was calculated and is presented in Table 6. In Figure 3, the hardness values of the coatings are presented as a column graph, comparing these with the hardness value of the base material.

**Table 6.** Average hardness, Young's modulus, H/E and $H^3/E^2$ ratio values obtained for the produced coatings.

| Coating | Hardness (H)—GPa | Reduced Young's Modulus (Er)—GPa | H/E | $H^3/E^2$ |
|---|---|---|---|---|
| TiAlN | $22.9 \pm 1.2$ | $312 \pm 9$ | 0.073 | 0.123 |
| TiAlSiN | $21.6 \pm 0.9$ | $266 \pm 7$ | 0.081 | 0.142 |
| AlCrN | $36.8 \pm 2.1$ | $335 \pm 5$ | 0.11 | 0.395 |

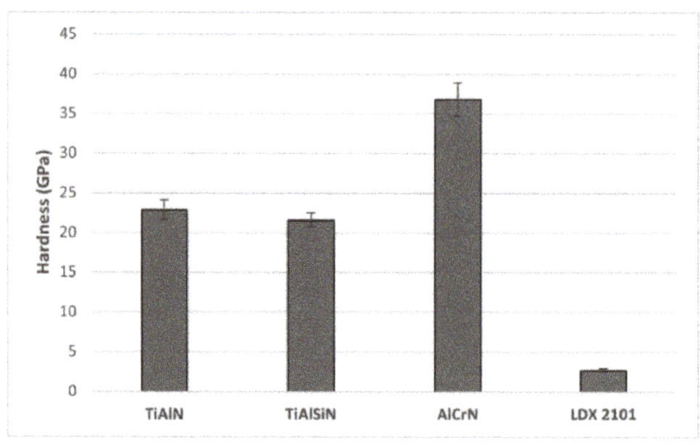


**Figure 3.** Hardness values of the tested coatings and base material (UNS S32101).

As seen in Table 6, the $H/E$ and $H^3/E^2$ ratios are displayed. These ratios provide valuable information regarding the coatings' wear behavior [53], with the $H/E$ ratio being a good indicator of the coating's wear resistance, with higher values being preferred (closer to 1; however, the presented values are quite satisfactory for all analyzed coatings). The $H^3/E^2$ ratio is related to the coating's ability to resist plastic deformation, which impacts the coating's wear behavior as well [54] (for this ratio, the higher value is also preferred, with ratios closer to 1 indicating very good plastic deformation resistance).

Still regarding the coating's wear behavior, another indicator of this is the ratio between the Young's modulus value of the coating and the substrate. Usually, a higher E value in the substrate is preferred (when compared to the coating's E value), as this can delay the onset of plastic deformation in the substrate and avoid subsequent coating cracking and chipping [55]; as such, lower values of this ratio are preferred. Substrate/coating systems that have this ratio closer to 1 are more susceptible to phenomena such as cracking and coating delamination. The substrate's Young's modulus was determined to be 611 GPa (average value). In Table 7, the ratio between the coatings ($E_c$) and the substrate's ($E_s$) Young's modulus value can be observed.

**Table 7.** $E_c/E_s$ ratio determined for each of the tested coatings.

| Coating | $E_c/E_s$ |
| --- | --- |
| TiAlN | 0.511 |
| TiAlSiN | 0.435 |
| AlCrN | 0.581 |

### 3.2. Cutting Force Analysis

Cutting forces provide valuable information regarding the machining process, primarily about process stability and tool wear. Using the methodology presented in the previous chapter, cutting forces were also registered during all the conducted tests; however, due to the used parameters, primarily a low depth of cut, the cutting forces produced did not present high values.

Analyzing the cutting force graphs obtained from each of the conducted tests, a common trend was noticed. Independent of feed rate, the cutting forces registered an increase in value for the Fx, Fy and Fz components, with the increase in the Fz component being quite significant for all the cases. This increase was registered at around 2 m of the cutting length for all the tested tools. It was also registered that the values for these components increased until the end of the test, at 4 m of the cutting length.

This increase in cutting force value may be attributed to the wear sustained by the cutting tools; however, this cutting force value variation was quite low (as previously mentioned). These low values were attributed to the selected cutting parameters, and more concretely, the low value of the axial depth of cut. Due to this fact, these values were omitted in further analyses.

### 3.3. Surface Roughness Analysis

In this section, the surface roughness analysis results are presented, showing the average Ra values for the tangential and radial measurement directions, as any deviation between these values of roughness indicates instability during machining, usually presenting as a higher level of vibration during the cutting operation. These results are divided into subsections, one for each of the tested tools, with an extra subsection where the comparison between the tools, regarding the roughness values, is presented. The adopted procedure is described in detail in Section 2.

#### 3.3.1. T1—Produced Surface Roughness

The average surface roughness values that were measured in the material machined with the T1 tools are presented in Table 8. These were used to create surface graphs to depict the surface roughness variation for all the tested conditions. Table 8 adopted the tool reference described in Section 2.

**Table 8.** Average surface roughness values, measured in the radial and tangential directions for the tests conducted with T1.

| Tool Ref. | Ra (Radial Direction) (μm) | Ra (Tangential Direction) (μm) |
| --- | --- | --- |
| T1L2F75 | 0.193 ± 0.015 | 0.182 ± 0.019 |
| T1L4F75 | 0.325 ± 0.021 | 0.345 ± 0.031 |
| T1L2F100 | 0.219 ± 0.019 | 0.223 ± 0.011 |
| T1L4F100 | 0.394 ± 0.033 | 0.389 ± 0.016 |
| T1L2F125 | 0.254 ± 0.012 | 0.327 ± 0.044 |
| T1L4F125 | 0.481 ± 0.036 | 0.492 ± 0.056 |

From the values presented in Table 8, it was observed that the values measured in the tangential and radial directions are very similar. However, this deviation was not significant. Thus, it can be concluded that in this case, the machining process itself was stable.

By analyzing the surface roughness values measured for both directions, it was observed that they follow a similar trend of rising with the cutting length; i.e., tests conducted with the tool with the 2-m cutting length produced the best results in terms of surface roughness. This was due to the tool's wear, which was aggravated for higher cutting lengths. Another parameter that seemed to influence the surface roughness of the material was the feed rate, as it was observed for lower feed rates that the tools produce a better surface quality. From the lowest feed rate value to the original, a slight increase in surface roughness was registered; however, for feed rates of 125%, the surface roughness that was produced was the highest registered, with the maximum produced under the condition of 4 m of cutting length while using 125% of the original feed rate (T1L4F125).

### 3.3.2. T2—Produced Surface Roughness

Regarding the Ra values obtained for the tests conducted using T2 tools, these can be observed in Table 9.

**Table 9.** Average surface roughness values, measured in the radial and tangential directions for the tests conducted with T2.

| Tool Ref. | Ra (Radial Direction) (μm) | Ra (Tangential Direction) (μm) |
| --- | --- | --- |
| T2L2F75 | 0.291 ± 0.018 | 0.260 ± 0.012 |
| T2L4F75 | 0.401 ± 0.026 | 0.436 ± 0.021 |
| T2L2F100 | 0.325 ± 0.017 | 0.301 ± 0.015 |
| T2L4F100 | 0.413 ± 0.031 | 0.405 ± 0.026 |
| T2L2F125 | 0.409 ± 0.019 | 0.409 ± 0.014 |
| T2L4F125 | 0.505 ± 0.037 | 0.477 ± 0.022 |

As seen from Table 9, the Ra values measured in the tangential and radial directions were very similar, meaning that there were no excessive vibrations produced during the machining tests. From the surface roughness values analysis, it was observed that, as seen from the results of the tests using T1, the surface roughness was lower for lower feed values. Furthermore, the cutting length also had a clear influence in the machined material's surface quality, registering a clear increase in surface roughness for the tests conducted at a 4 m cutting length. Once again, the highest surface roughness value was obtained under the condition of a 4-m cutting length and higher feed rate value (125%). It is also worth noting that the increase in surface roughness value from a 75% to 100% feed rate was less accentuated than the increase registered from a 100% to 125% feed rate. This was a similar behavior to that observed in the tests conducted with T1.

### 3.3.3. T3—Produced Surface Roughness

In Table 10, the average calculated values of Ra from the surface roughness measurements conducted on the machined material (using T3 tools) can be observed.

**Table 10.** Average surface roughness values, measured in the radial and tangential directions for the tests conducted with T3.

| Tool Ref. | Ra (Radial Direction) (μm) | Ra (Tangential Direction) (μm) |
|---|---|---|
| T3L2F75 | $0.234 \pm 0.016$ | $0.206 \pm 0.013$ |
| T3L4F75 | $0.270 \pm 0.019$ | $0.273 \pm 0.029$ |
| T3L2F100 | $0.209 \pm 0.009$ | $0.211 \pm 0.021$ |
| T3L4F100 | $0.299 \pm 0.011$ | $0.295 \pm 0.031$ |
| T3L2F125 | $0.172 \pm 0.012$ | $0.190 \pm 0.012$ |
| T3L4F125 | $0.228 \pm 0.022$ | $0.228 \pm 0.025$ |

Comparing the values shown in Table 10 to those obtained from the tests conducted with the previous tools (T1 and T2), it was observed that the Ra values were quite lower for the tests performed with the T3 tools. Analyzing the variation of Ra values with the different test conditions, similarly to the previously presented results, showed that the surface roughness values tended to increase for higher cutting lengths. However, there was a different trend noticed with the feed variation. Unlike what was observed for the other test conditions, a significant decrease in surface roughness values were registered for higher feed rate test conditions (125%), with these producing, by far, the best surface quality registered to this point. In this case, the worst surface quality was obtained under the test condition of 4 m of cutting length while using the original feed rate (T3L4F100). The lowest value of Ra was obtained under the condition of a 2-m cutting length, using 125% of the original feed rate (T3L2F125).

### 3.3.4. T4—Produced Surface Roughness

In this subsection, the surface roughness results for the machining tests conducted with T4 tools are presented. The average values of the measured Ra (in both radial and tangential directions) parameter can be observed in Table 11. As seen in the previous subsections regarding the surface roughness results obtained using the different tools, surface roughness varies with the cutting length and feed rate values.

**Table 11.** Average surface roughness values, measured in the radial and tangential directions for the tests conducted with T4.

| Tool Ref. | Ra (Radial Direction) (μm) | Ra (Tangential Direction) (μm) |
|---|---|---|
| T4L2F75 | $0.194 \pm 0.018$ | $0.175 \pm 0.011$ |
| T4L4F75 | $0.253 \pm 0.025$ | $0.253 \pm 0.018$ |
| T4L2F100 | $0.259 \pm 0.016$ | $0.247 \pm 0.011$ |
| T4L4F100 | $0.274 \pm 0.019$ | $0.284 \pm 0.016$ |
| T4L2F125 | $0.250 \pm 0.022$ | $0.270 \pm 0.014$ |
| T4L4F125 | $0.284 \pm 0.033$ | $0.323 \pm 0.021$ |

From the analysis of Table 11, it was concluded that the process was stable, as the values for Ra measured in the tangential and radial directions did not deviate significantly from each other. The variation in Ra values for different parameters was like the variation observed in the tests conducted with T1 and T2 tools. Once again, the 2-m cutting length test condition produced the best surface roughness quality, with this deteriorating with the increase in this parameter, as observed in all the analyzed tools. Regarding the variation of Ra with the feed rate, as previously mentioned, it was observed that the lowest Ra value was obtained by using lower feed rate values, with the surface quality quickly deteriorating with the increase in feed rate.

### 3.4. Flank Wear Measurements

The wear measured in the tools' flanks for each test condition is presented in this section in numerical form, presenting the average values of flank wear, as well as in graphical form, enabling for a clear perception on how this wear varies with the machining parameters. These values were measured on the clearance faces of all the tested tools, following the standard ISO 8688-2:1989 [52], as mentioned in Section 2. As in the previous Section 3.3., these values are in different subsections, one for each of the analyzed tools, with an extra section dedicated to the comparison of the flank wear values obtained.

#### 3.4.1. T1—Flank Wear Measurement

In this subsection, the wear measurement values, performed on the clearance faces of all the tested T1 tools, are presented in Table 12.

**Table 12.** VB values obtained from measurements performed on the clearance faces of T1 tools.

| Tool Ref. | Flank Wear, VB ($\mu$m) |
|---|---|
| T1L2F75 | $14.22 \pm 1.05$ |
| T1L4F75 | $18.34 \pm 1.16$ |
| T1L2F100 | $9.670 \pm 0.78$ |
| T1L4F100 | $12.27 \pm 0.98$ |
| T1L2F125 | $23.32 \pm 1.84$ |
| T1L4F125 | $25.20 \pm 1.76$ |

From the values presented in Table 12, it was concluded that the increase in cutting length caused a slight increase in flank wear. However, it was a very small increase, especially when compared to the variation in flank wear that was observed with feed variation. For the lower feed rate, the tools presented slightly more flank wear than those that were tested at a 100% feed rate. There was also an increase in flank wear for higher feed rates; however, these feed rates produced the most severe wear, with high values of flank wear being registered at these values of feed rate when 2- and 4-m cutting lengths were used. In terms of a less amount of wear, this was obtained under the original cutting condition, with the lowest amount of wear being registered for T1L2F100 and followed by T1L4F100, indicating that the machining parameters were well defined for this tool in terms of wear behavior.

#### 3.4.2. T2—Flank Wear Measurement

In Table 13, the average flank wear values obtained from measurements performed on the clearance faces of the T2 tools are presented.

**Table 13.** VB values obtained from measurements performed on the clearance faces of T2 tools.

| Tool Ref. | Flank Wear, VB ($\mu$m) |
|---|---|
| T2L2F75 | $8.380 \pm 0.69$ |
| T2L4F75 | $13.18 \pm 1.12$ |
| T2L2F100 | $7.040 \pm 0.24$ |
| T2L4F100 | $10.01 \pm 0.98$ |
| T2L2F125 | $13.21 \pm 1.12$ |
| T2L4F125 | $25.12 \pm 1.05$ |

As seen in the analysis of T1, the variation of the flank wear for the T2 tools was deeply influenced by the feed rate, increasing slightly under the 75% feed rate condition, and registering a significant increase under the 125% feed rate condition. Once again, the original values of feed rate produced better results in terms of tool wear, even at 4 m of cutting length. The variation in cutting length induced a higher tool wear for the 125% feed rate condition, registering a significant increase, whereas in the case of the 75% and 100% feed rates, the cutting length did not appear to influence the tools' wear so severely.

### 3.4.3. T3—Flank Wear Measurement

In Table 14, the average VB values that were measured on the clearance faces of the T2 tools are presented.

**Table 14.** VB values obtained from measurements performed on the clearance faces of T3 tools.

| Tool Ref. | Flank Wear, VB ($\mu$m) |
|---|---|
| T3L2F75 | 11.72 ± 1.06 |
| T3L4F75 | 12.54 ± 0.85 |
| T3L2F100 | 14.17 ± 1.26 |
| T3L4F100 | 16.26 ± 1.55 |
| T3L2F125 | 6.820 ± 0.55 |
| T3L4F125 | 8.570 ± 0.19 |

Observing the values presented in Table 13, it was concluded that T3 tools did not suffer wear in the same manner as the T1 and T2 tools. In fact, the behavior of the wear was like the variation of the Ra values of the studied tools, where T1 and T2 tools produced better results at lower feed rates in terms of surface roughness and even wear (as the wear sustained at a 75% feed rate is very similar to that sustained at 100%). However, in the case of T3 tools, it was observed that the best results in terms of wear were obtained at higher feed rates (125%). This can be compared to the results obtained from the surface roughness analysis of this tool, as the best surface quality was also achieved at higher feed rates [11,14].

In terms of amount of wear, a slight decrease in wear for the lowest feed rate value was observed, albeit very similar to the wear sustained at the original feed rate. However, for higher feed rate values, the wear was significantly lower. This fact, coupled with the results obtained from the surface roughness analysis, indicates that this tool has a high-performance for higher feed values, producing low surface roughness values while sustaining less flank wear [11]. Regarding the flank wear variation with the increasing cutting length, as expected, these values rose slightly for higher cutting lengths; however, the main influence in flank wear for this case seemed to be the feed rate.

### 3.4.4. T4—Flank Wear Measurement

The flank wear analysis for the final tool type, T4, is presented in this subsection. Average flank wear values, measured on the clearance faces of these tools and tested for all different conditions, can be observed in Table 15.

**Table 15.** VB values obtained from measurements performed on the clearance faces of T4 tools.

| Tool Ref. | Flank Wear, VB ($\mu$m) |
|---|---|
| T4L2F75 | 6.710 ± 0.36 |
| T4L4F75 | 13.71 ± 0.74 |
| T4L2F100 | 10.47 ± 0.66 |
| T4L4F100 | 18.80 ± 1.16 |
| T4L2F125 | 4.770 ± 0.22 |
| T4L4F125 | 5.830 ± 0.65 |

In this case, the flank wear increased with cutting length, especially at 75% and 100% feed rate values. There was an influence in wear registered for the 125% feed rate; however, the registered increase was very slight. In terms of flank wear variation coupled with the feed rate, as seen in the previously analyzed tool (T3), the lowest values of flank wear were registered for higher feed rates; however, the wear values registered at the lowest feed rate condition were like those obtained at the 125% feed rate, especially for the test conducted with 2 m of cutting length. The highest amount of wear was produced at a 100% feed rate value at 4 m of cutting length, which impacted the overall quality of the machined surface [11] as seen in the previous section, and was followed by T4L2F100. It is also worth noting that the produced values of flank wear were low when compared to those obtained in T1 and T2, indicating that this tool has good wear performance when machining this type of material, even at higher feed rates.

### 3.5. Analysis of the Wear Mechanisms

In the present section, a detailed analysis of the wear mechanisms sustained by the different tools is presented. Following the same method of presentation seen in the previous section, the different wear mechanisms are presented for each tool, presenting a comparison as well between the wear observed in the SEM images and the values presented in the previous subsection. Furthermore, at the end of this section, an additional subsection is presented that gives a summary of the wear mechanisms that were analyzed for each tool, identifying common trends in terms of wear mechanisms and wear behavior.

#### 3.5.1. T1—Wear Mechanism Analysis

Here, the images taken from the SEM analysis regarding the T1 tools are presented. The main types of wear that these tools sustained were abrasive wear, coating cracking and delamination, exhibiting some adhesion of machined material. As expected, there was an influence on wear severity caused by the increase in cutting length, as can be observed in Figure 4.

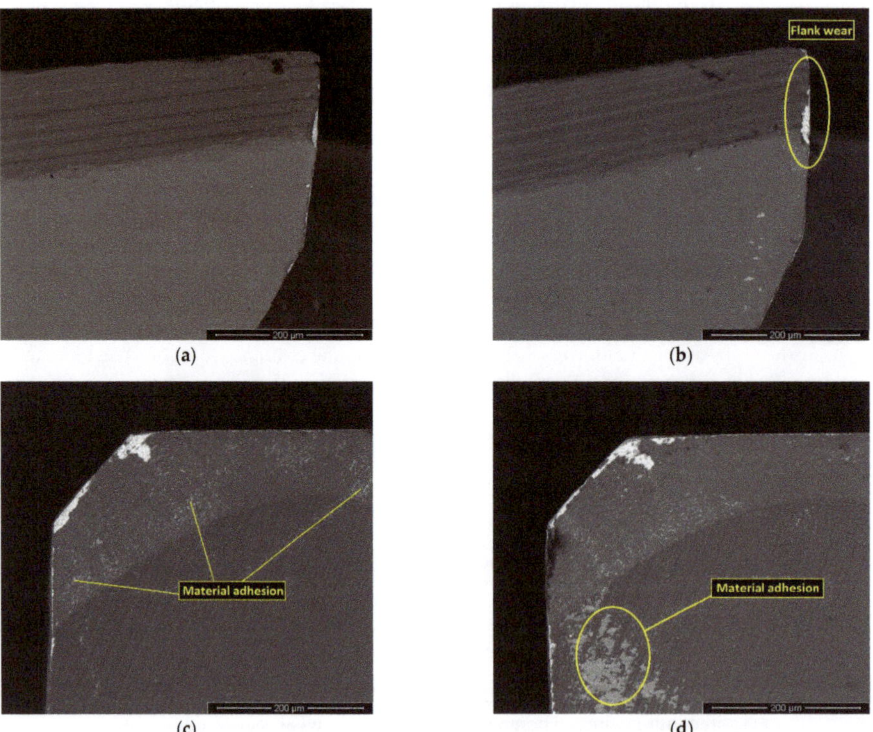


**Figure 4.** Clearance face, with a 500× magnification, of the T1L2F100 (**a**) and T1L4F100 (**b**) tools, and rake face, with a 500× magnification, of the T1L2F100 (**c**) and T1L4F100 (**d**) tools.

As seen in Figure 4, the wear was more severe for the 4-m cutting length machining tests, with these exhibiting more damage to the flank and the presence of well-developed wear mechanisms, such as machined material adhesion. The machined material adheres to the tool's surface overtime, promoting the development of wear mechanisms such as abrasion, coating delamination and adhesive wear. The adhesion present in the rake face of the T1L4F100 tool is presented with larger magnification in Figure 5.

**Figure 5.** Adhered workpiece material present in the T1L4F100 tool's rake face, with a magnification of 2500×.

The adhesive damage observed in Figure 5 was analyzed to determine if the material present on the surface was the one that was being machined. Thus, EDS analyses were performed to determine the chemical composition of the damage, as observed in Figure 6. It was indeed confirmed that the adhered material was the DSS, as its chemical composition was very close to that supplied by the manufacturer. Therefore, it can be concluded that the increase in cutting length promotes the adhesion of machined material to the tool's surface and that there is a lower amount of adhered material for the T1L2F100 condition, as seen in Figure 4c.

Regarding the influence of feed rate on the tools' wear and registered mechanisms, although there was an influence on the amount of wear that was sustained by the tools, the identified mechanisms were present in all the tested conditions, albeit in different levels of severity. In Figure 7, the tools' rake and clearance faces are depicted for the 75% and 125% feed conditions.

In Figure 7, it can clearly be observed that both the 75% feed rate condition and the 125% feed rate condition produced a more severe wear than that registered for the 100% feed rate (Figure 4); this was also observed in the previous section with the values of flank wear.

As previously mentioned, the main wear mechanisms that were detected during the analysis of this tool were abrasion, coating cracking, delamination and some adhesion. These are common wear mechanisms that are developed when machining DSS alloys [45,46]. Some evidence of abrasion can be observed in Figure 8.

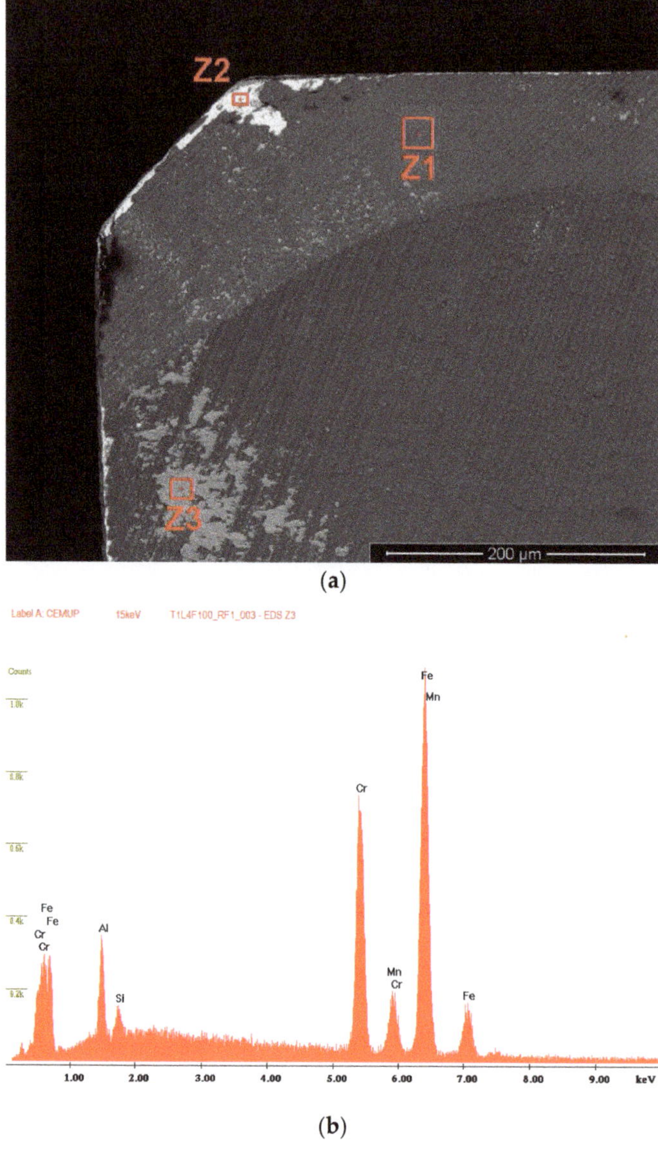

**Figure 6.** Rake face of the T1L4F100 tool with the various zones picked for analysis (**a**), and analysis of zone 3 (Z3), where there are signs of adhesion (**b**).

As seen in Figure 8, for a test conducted at a 2-m cutting length, there were already parts of the tool's rake face with exposed substrate. There were abrasion marks, as pointed to in the picture. This was concluded to be due to the thinning of the coating in the marked area. Furthermore, there was evidence of coating spalling in the area pointed to in Figure 8. Regarding the adhered material, material adhesion was promoted by an increase in cutting length, as Figure 8 shows that the material tended to adhere to the machining grooves present in the substrate, promoting abrasive wear in that area and eventually leading to coating delamination. Some cracking in the tool's coating was also registered, especially under the conditions of a lower feed rate, which can be observed in Figure 4. In Figure 9, this cracking process can be observed in a more detailed manner.

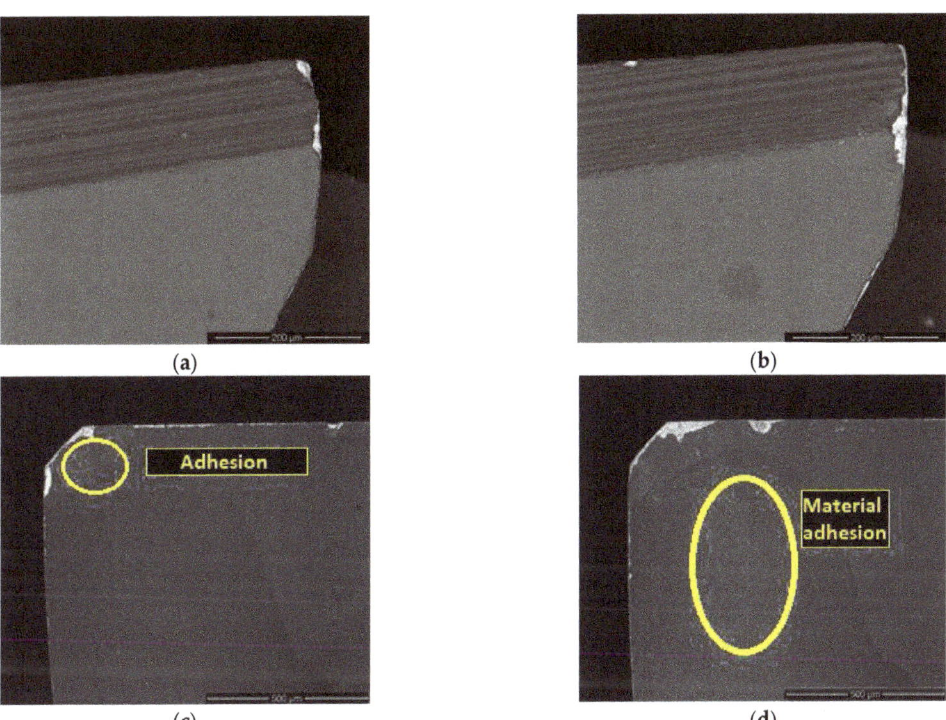



**Figure 7.** Clearance face, with a 500× magnification, of the T1L2F75 (**a**) and T1L2F125 (**b**) tools, and rake face, with a 500× magnification, of the T1L2F75 (**c**) and T1L2F125 (**d**) tools.

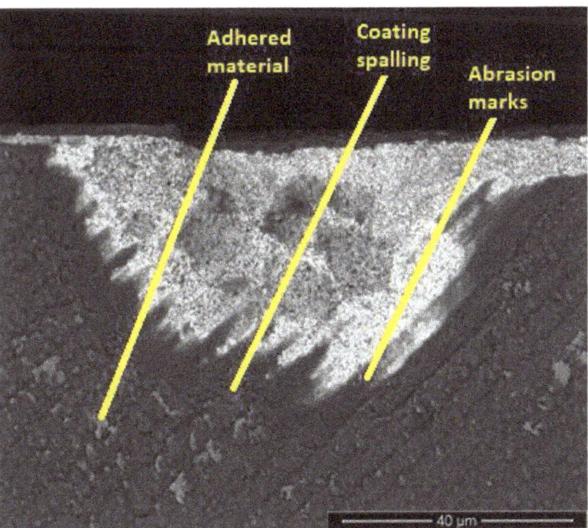



**Figure 8.** Wear mechanism analysis of the T1L2F125 rake face.

**Figure 9.** Coating cracking present on the tool's clearance face for the T1L2F75.

### 3.5.2. T2—Wear Mechanism Analysis

As observed in the case of T1, for this case the wear was also noticeable with the increase in cutting length. Regarding the wear mechanisms previously reported in the T1 tool, these were also present in T2; however, there were different severities. For example, in the case of the amount of adhered material, this was more severe in T2 than that observed in T1. The main wear mechanisms sustained by these tool types were abrasion, material adhesion, coating delamination and some coating cracking, as seen in T1, albeit at different levels.

As described in the section regarding the wear measurements performed on these tools, they exhibited more wear for longer cutting lengths; however, this wear was not as significant as recorded for the previous tools. Contrary to what was seen in T1, adhesion started to form at lower cutting lengths, as seen in Figure 10.

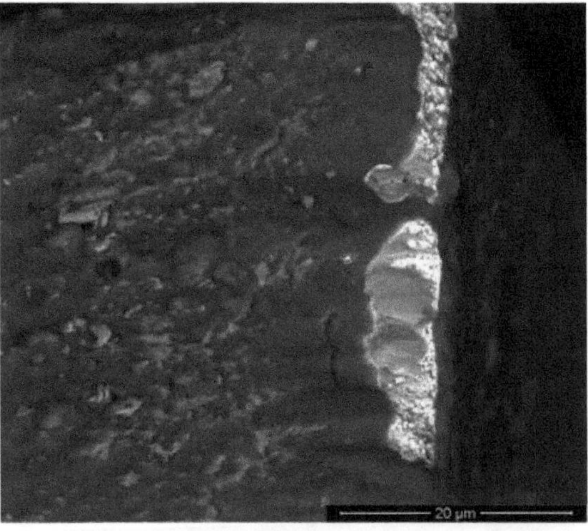

**Figure 10.** Adhered material in the T2 tool's flank, tested under the T2L2F100 condition.

This adhesion was registered in the coating's clearance and rake faces, forming primarily at lower feed rates. This adhesion to the tool's coated surface eventually leads to coating delamination by adhesive wear, as seen in Figure 11, where zones of adhered material are marked in yellow.

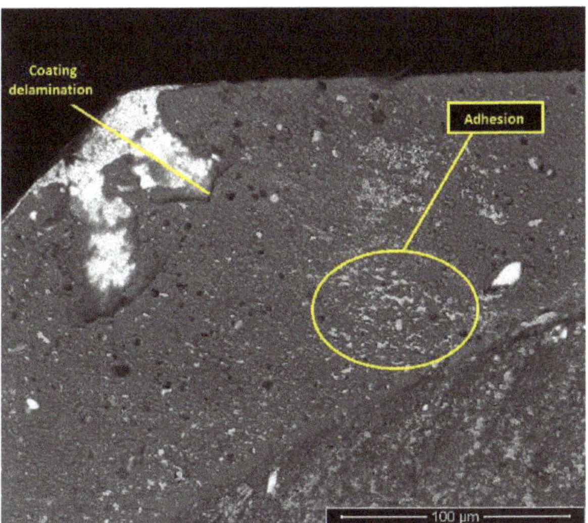


**Figure 11.** Coating delamination registered in the rake face of a T2L2F100 tool.

At higher feed rates, primarily at a 125% feed rate, the amount of adhesive wear was lower than that registered at lower feed rate values. However, the tools presented abrasion marks and cracking, exhibiting high levels of wear as well. In Figure 12, the abrasive wear can be observed in a T2 tool, tested at a 125% feed rate. The coating was abraded, exposing the substrate. However, there was some coating delamination, especially in the tool's rake face, possibly caused by adhesive wear, as shown in Figure 13.

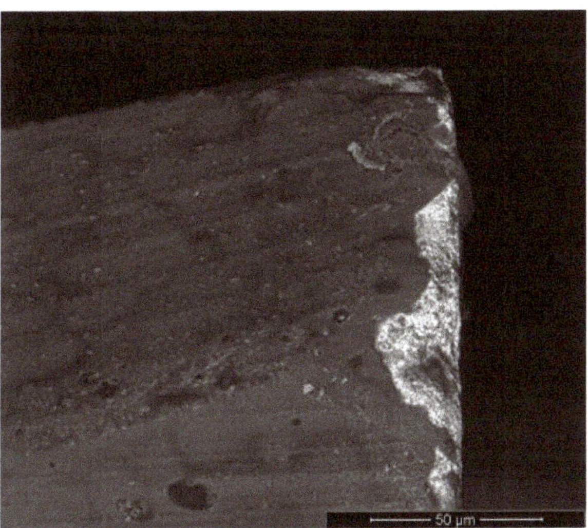


**Figure 12.** Abrasive wear on the clearance face of T2L4F125 tool.

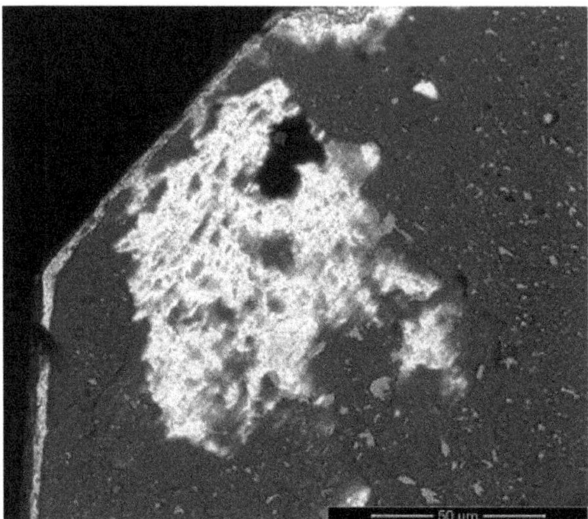

**Figure 13.** Rake face of a T2 tool tested at 4-m cutting length and 125% feed rate showing adhesive wear and coating delamination.

### 3.5.3. T3—Wear Mechanism Analysis

Regarding the wear mechanisms sustained by T3 tools, these exhibited high levels of material adhesion to the tools' surfaces, especially at high feed rates, and even originated a built-up edge. This can be observed in Figure 14.

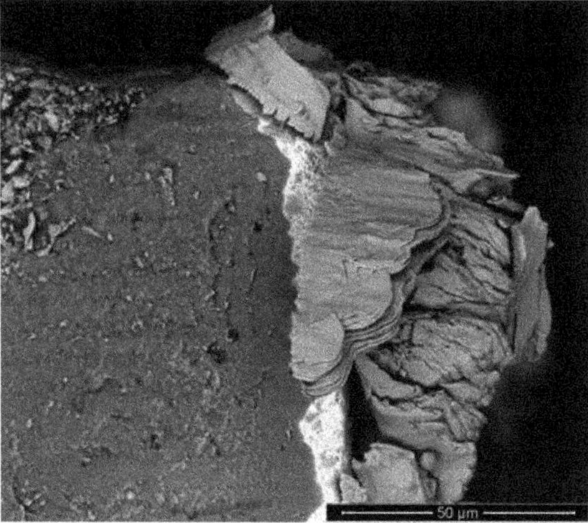

**Figure 14.** Built-up edge detected on the clearance face of a T3 tool, tested at 125% feed rate.

Other wear mechanisms were identified, being the same as the reported in the previously mentioned tools. There was evidence of abrasive wear and coating delamination, as seen in Figure 15. This delamination was promoted by the high amount of material adhesion and abrasion of the workpiece material on the tool's surface. The material adhesion promoted abrasive wear on these adhesion zones, eventually resulting in the spalling of the tool's coating (through adhesive wear).

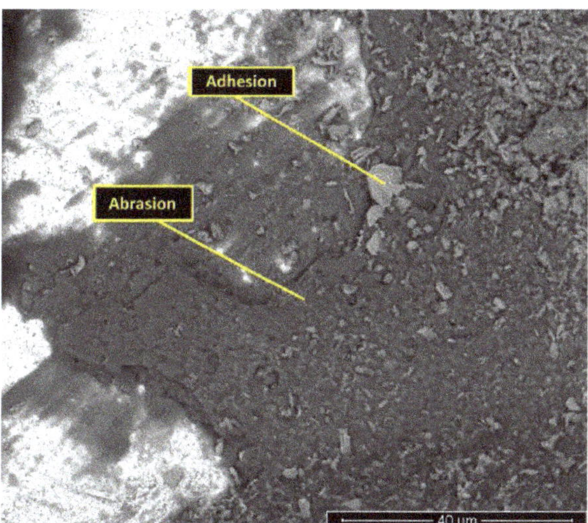



**Figure 15.** Material adhesion and abrasive wear detected on a T3 tool's rake face, tested at 125% feed rate.

Unlike in T1 and T2 tools, no major cracks were registered; however, some cracks in the nanometric scale were identified. These "nanocracks" can eventually lead to bigger ones, resulting in coating spalling and delamination. The propagation of these cracks is known to be slowed down by multi-layered coatings, such as the T3 coating in use [16,17]. This can be observed in Figure 16.

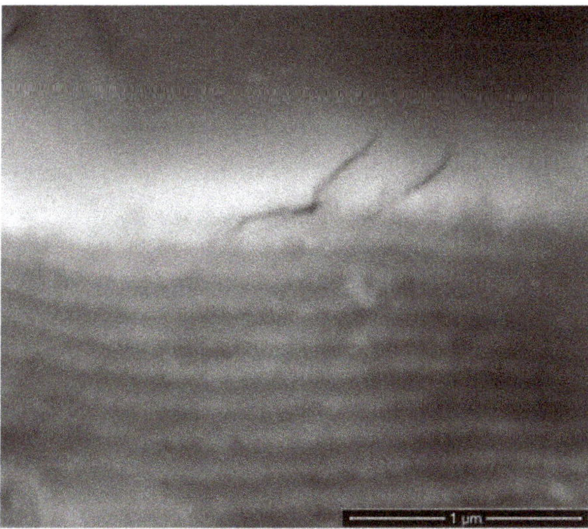



**Figure 16.** Cracks, detected in the coating of a T3 tool, tested at 75% feed rate and 4 m of cutting length.

As seen from figures presented in this subsection, this tool exhibited high amounts of adhesion and abrasion, exposing the substrate in some areas. There was, however, a low amount of flank wear reported, and this tool produced satisfactory results in terms of machined surface quality. Comparing this tool to T1, coated with the same AlCrN coating, it can be concluded that there is a clear benefit in using a tool with four flutes to machine these types of DSS alloys.

### 3.5.4. T4—Wear Mechanism Analysis

T4 tools exhibited the best wear performance of all the tools, exhibiting the lowest flank wear value, especially for high feed conditions. However, these tools suffered considerable damage to their rake faces. Regarding the wear mechanisms identified on these tools, all the previously mentioned were present, especially abrasive wear, coating delamination and material adhesion to the tools' surfaces. The process of coating delamination was the same as identified in the previously analyzed tools, as can be seen in Figure 17.

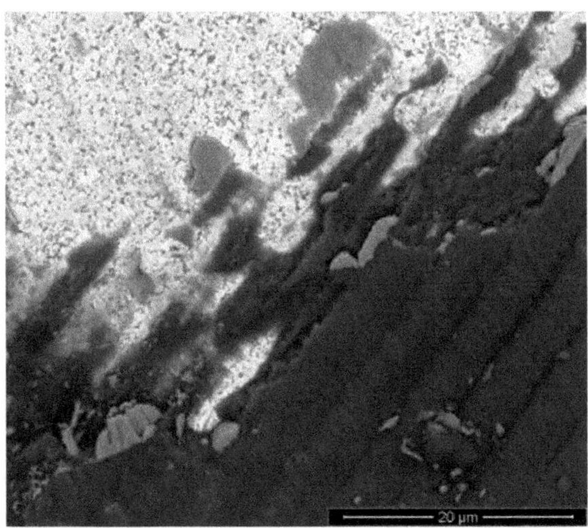

**Figure 17.** Adhesive wear detected on the T4 tool's rake face, tested at 4-m cutting length and at 100% feed rate.

The substrate's machining grooves, originated from the substrate's grinding process, promoted material adhesion on the tool's surface, thus promoting abrasion and coating delamination on these areas. This was a common trend throughout all the analyzed tools.

The wear sustained by the tools tested at 4 m of cutting length was more severe than that presented by the tools tested at 2 m, with this being a clear factor on wear mechanisms, such as adhesion, at lower feed rates. For higher feed rates, adhesion was identified, but in a smaller amount than that registered at lower feed rates. There was clear evidence of abrasion being the dominant wear mechanism for this tool at a 125% feed rate. Evidence of abrasion at a high feed rate can be observed in Figure 18.

There was no evidence of cracks detected in the analysis of this coating; also, it exhibited the least amount of flank wear for higher and lower feed rates. However, these tools sustained some wear on the rake faces, which can be attributed to chip formation and removal, and the material adhering to the rake faces being pushed upward, promoting adhesion to the tools' surfaces. As seen in all the previously presented tools, this adhesion promoted the delamination of the coating, resulting in the high wear of the rake faces.

### 3.6. Summary of the Analyses' Results

In the present subsection, the summary of all the conducted analyses is performed. A comparison of the results obtained for the produced surface roughness, flank wear and registered wear mechanisms is made.

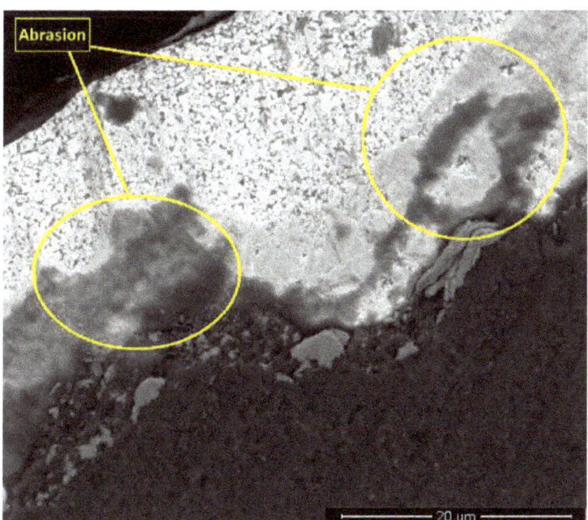



**Figure 18.** Abrasion marks detected on T4 tool's rake face, tested at 125% feed rate.

### 3.6.1. Surface Roughness Analysis Summary

In this section, the various measurements made using all the tool types in all the test conditions are presented. Both the Ra measurements taken in the radial and tangential directions are presented in Figures 19 and 20, and the test conditions are presented as seen in Section 2. The number after "L" indicates the cutting length used in that test, and the number that follows "F" indicates the percentage of feed rate that was used in the machining test.

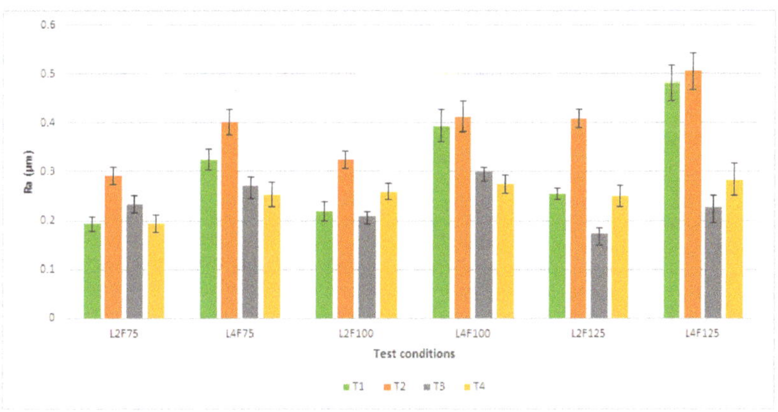



**Figure 19.** Average values of Ra, measured in the radial direction, produced by T1, T2, T3 and T4 tools for all the different test conditions.

As previously noticed in the subsections dedicated to the individual tool analysis, the surface roughness tended to increase with the increase in cutting length; this was expected, as the tool wear is more intense for higher values of cutting length. This was verified in all the tested tools, and there was also a variation in surface roughness with feed variation, as this parameter is known to have high influence on the surface roughness [14,15]. Regarding the variation in surface roughness with feed rate, the predominant trend was that lower feed rates will confer a better surface finish to the machined part, except for T3, where an increase in feed rate produced a better surface quality than at lower values of feed rate. It is worthy to note, however, that the surface roughness values produced by T3 were the best, being closely followed by T4. In fact, the surface quality produced at the 125% feed rate by the T3 and T4 tools was on par with the surface roughness quality produced by the T1 and T2 at the 75% feed rate, indicating that the four-fluted AlCrN-coated end-mill and the four-fluted TiAlSiN-coated end-mill are good choices for the conduction of finishing operations in this kind of material. The number of flutes has a great influence in the machining of this material, especially regarding the AlCrN-coated tools, as it can be observed that the two flutes' tools produced a worse surface quality than that of the four-fluted AlCrN-coated tool [44]. In Figure 21, the variation of machined surface quality for different feed rate values can be observed for the 2-m and 4-m cutting lengths (in the radial direction, as the identified trends were the same for both directions and the values were similar). It was observed that the trends were common for both 2 and 4 m of cutting length; however, the T4 tools seemed to be the least affected by the increase in cutting length, with the roughness values not increasing drastically for higher cutting length values, especially when compared to T1 and T2 tools. The T3 tools were also resistant to this variation. Furthermore, the results obtained for lower feed rate values were higher than those registered for T4.

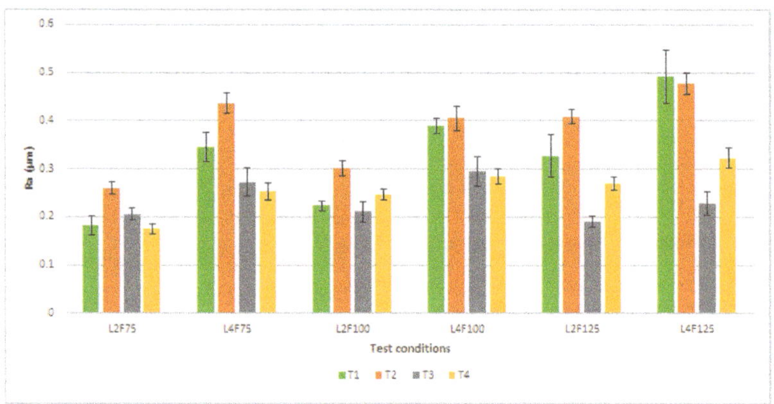

**Figure 20.** Average values of Ra, measured in the tangential direction, produced by T1, T2, T3 and T4 tools for all the different test conditions.

It is worthy to note that the amount of registered tool wear negatively impacts the machined surface quality, with the conditions that produced the highest amount of wear producing the worst machined surface quality [14,15]. Tools that exhibited higher levels of VB tended to produce a worse machined surface quality, primarily since the substrate was exposed (adhesive wear and coating delamination/spalling), and were subject to wear mechanisms such as abrasive wear. This altered the cutting edge's geometry, inducing this higher surface roughness value. However, as previously mentioned, the machining parameters highly impacted the machined surface quality, with the T3 tool being a prime example of this fact.

### 3.6.2. Flank Wear Measurement Summary

In the present subsection, all the calculated average values of flank wear, measured on the clearance faces of all the tested tools, are discussed. As seen in the section regarding surface roughness analysis, all the collected values are presented in graphical form in Figures 22 and 23.

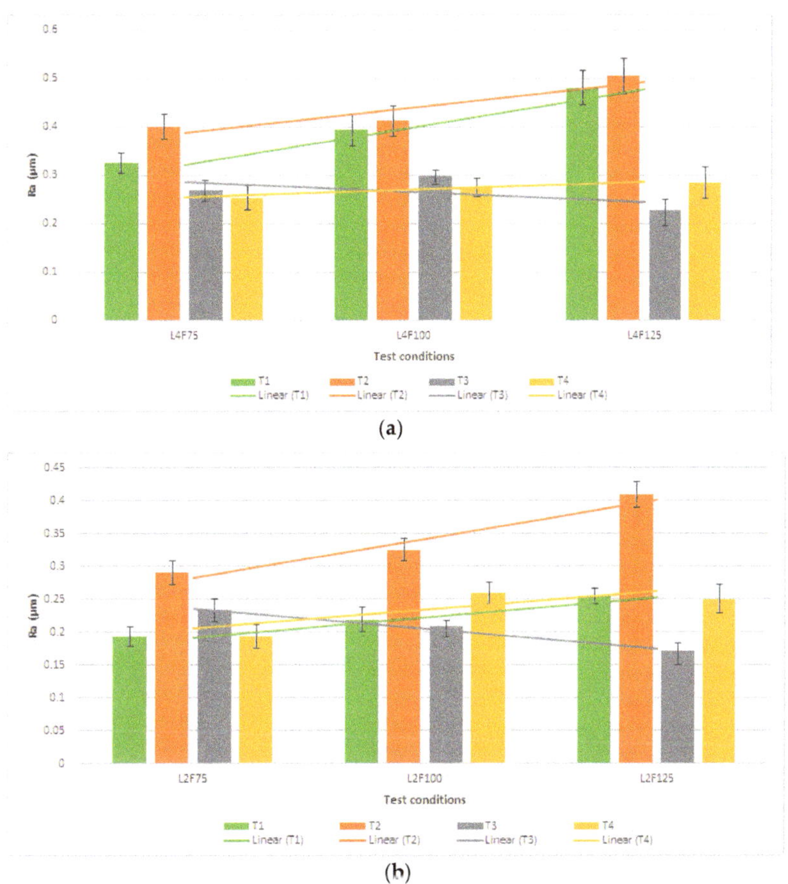

**(a)**

**(b)**

**Figure 21.** Average values of Ra, measured in the radial direction for T1, T2, T3 and T4 tools, for the test conditions conducted at 2-m cutting length (**a**); and 4-m cutting length (**b**).

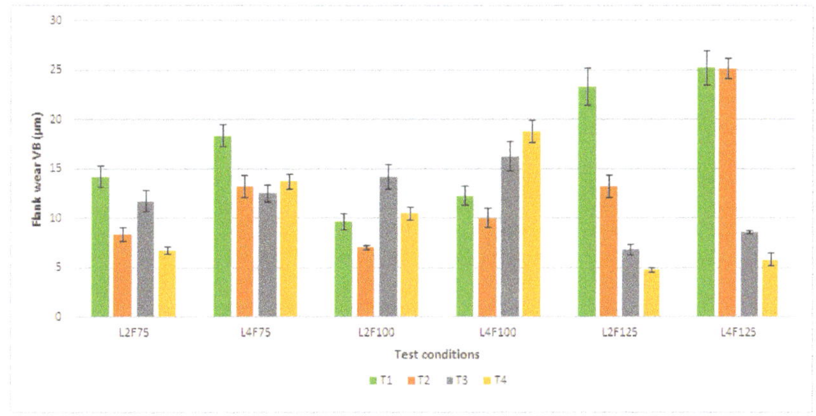

**Figure 22.** Average values of VB, measured on the clearance faces of the T1, T2, T3 and T4 tools, for all the different test conditions.

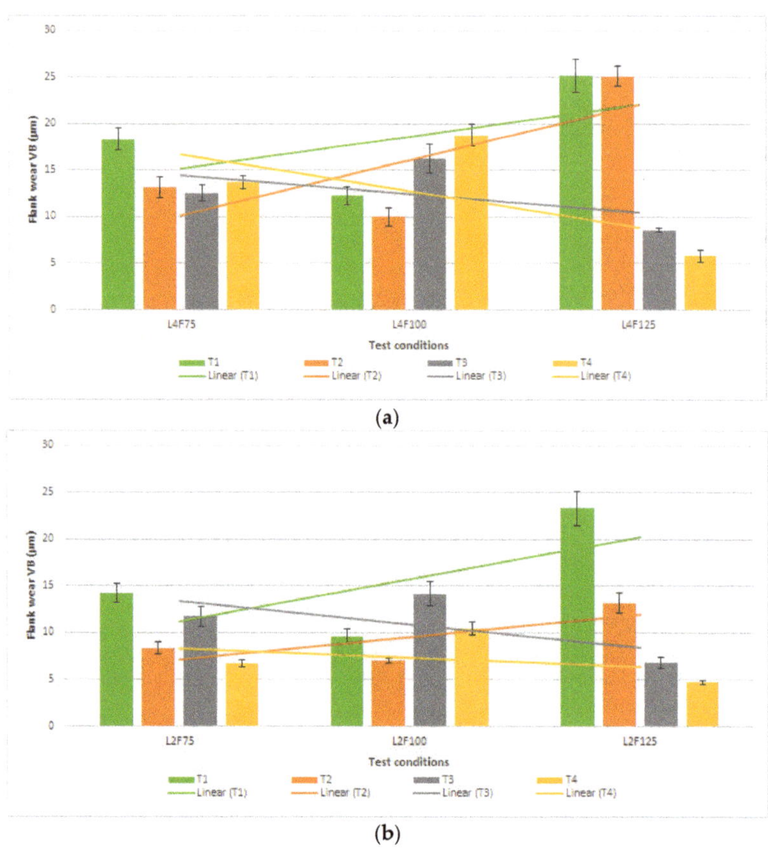

**Figure 23.** Average values of VB, measured on the clearance faces of the T1, T2, T3 and T4 tools, for the test conditions conducted at 2-m cutting length (**a**); and 4-m cutting length (**b**).

In Figures 22 and 23, a clear increase in wear with an increase in cutting length and feed rate was clearly observed for T1 and T2 tools. Furthermore, T1 tools seemed to be the most affected, in terms of wear, by feed rate variation (positive or negative), whereas T2 seemed to be the most affected by the variation in cutting length, especially at higher feed rate values. Regarding tools T3 and T4, these showed the most wear of all tools for the original feed rate conditions, with T3 being slightly affected by cutting length variation, andT4 being significantly more affected by the variation of this parameter at a 100% feed rate. However, T3 and T4 tools produced low wear for the lower feed rate values, producing, as well, the lowest amount of flank wear registered for the high feed rate test conditions. In fact, the wear presented by these tools was minimal for these test conditions, indicating that in terms of wear behavior these would be more suited for the machining of this material, especially at high feed values. Regarding once again tools T1 and T2, although these produced the best behavior at the original feed rate, and T2 even registered the second lowest values for flank wear tested under the condition of a 2-m cutting length and 75% feed rate, these tools exhibited high wear under the conditions of a 4-m cutting length and under the test conditions of a higher feed rate (125%), meaning that the original parameters were the most suited for machining with these tool types. It is shown that T1 was more sensitive to feed rate variations, whereas T2 was more sensitive to cutting length variations, in terms of flank wear.

### 3.6.3. Wear Mechanism Analysis Summary

The wear mechanisms that were observed in the tools, mainly abrasice and adhesive wear, are expected when machining DSS alloys [46,47]. It is worthy to note that the tool type that suffered the least amount of adhesive wear was T1, since this tool only has two flutes. This enables a better chip evacuation, as the lubricant can cool and clean a larger area. However, in this tool there was evidence of abrasion, coating delamination and cracking, despite the AlCrN coating's structure. This is due to, once again, the tool having only two flutes. Thus, a four-fluted or plus design is preferred to machine this type of material.

T3 tools exhibited high amounts of adhesion when compared to the other tools, especially at higher cutting lengths. This led to the formation of a built-up edge. However, the flank wear that was registered for these tools was not very severe. The tools coated with TiAlSiN showed the best wear performance. Indeed, this coating is well known to perform well in terms of wear, even in the machining of hard materials [56]. Although these tools exhibited the same wear patterns as the other tools, these were not so severe in the flank. However, these tools suffered damage on their rake faces, due to the buildup of adhered material in this area. In Table 16, the summary of the wear analysis performed on all tested tools is presented, showing the main mechanisms registered for these.

**Table 16.** Summary of the SEM wear analysis for each of the tested tools, mentioning the main wear type and mechanism identified.

| Tool | SEM Wear Analysis Summary |
| --- | --- |
| T1 | - Abrasive wear;<br>- Coating delamination;<br>- Coating cracking;<br>- Material adhesion. |
| T2 | - Material adhesion;<br>- Abrasive wear;<br>- Adhesive wear;<br>- Coating delamination;<br>- Coating cracking. |
| T3 | - Material adhesion;<br>- Adhesive wear;<br>- Built-up edge;<br>- Abrasive wear;<br>- Coating delamination. |
| T4 | - Abrasive wear;<br>- Material adhesion;<br>- Adhesive wear;<br>- Coating delamination. |

## 4. Conclusions

In the present work, an analysis of the wear behavior of four milling tools used in machining operations of DSS was presented. Four tools with different geometries and coatings were evaluated. One tool had two flutes (T1) and three tools had four flutes (T2, T3 and T4). The tested coatings were TiAlN (T2), AlCrN (T1 and T3) and TiAlSiN (T4). Additionally, the influence of cutting length and feed rate was evaluated based on the tools' wear and production quality (machined surface quality). Furthermore, the wear of the tested tools was analyzed, and the present wear mechanisms identified.

Regarding the tested tools' performance, it was found that T1 (two-flute, AlCrN) and T2 (TiAlN) tools were not the most suited to machine this DSS alloy, as they presented severe wear for the highest feed rate conditions. Their flank wear was heavily promoted by an increase in cutting length. This can be related to the coating's properties in the case of T2, as this coating did not present the best mechanical properties (of all evaluated coatings). However, in the case of T1, the increased wear and lack of performance can be attributed to the fact that this tool only had two cutting edges. The AlCrN coating's properties were quite good, presenting high values of H/E (0.11) and $H^3/E^2$ (0.395) (when compared to the other analyzed coatings) [53,54], which are representative of a good coating wear behavior. However, the AlCrN-coated tools presented a high $E_c/E_s$ ratio value (0.581, the highest of all analyzed coatings), indicating that it is quite prone to phenomena such as coating cracking and delamination [55].

As for the performances of T3 (four-fluted, AlCrN) and T4 (TiAlSiN), these tools exhibited the best performances, with the T3 tool producing the best results in terms of machined surface quality (for highest feed rate values). However, this tool presented more wear when compared to T4. Furthermore, for the rest of the test conditions (75% and 100% feed rate value), the T4 tool produced the best machined surface quality. As for flank wear, the T4 tool clearly outperformed the rest of the tested tools. This can be attributed to, not only the four-fluted geometry, but also to the mechanical properties of the TiAlSiN coating, which presented a high value of H/E (0.081, which is very close to 0.1) and a high $H^3/E^2$ value (0.142; these ratios' values were the second highest of all the analyzed coatings, being very close to the AlCrN coatings), and a lower $E_c/E_s$ ratio value (0.435), when compared to the AlCrN coating (in fact, this value was the lowest for all the analyzed substrate/coating systems).

Regarding the identified wear mechanisms, these were common among all evaluated tools, albeit at different levels of severity. These were: abrasive wear, adhesive wear and coating delamination/spalling. The coating failure method seemed to be common to all tools, as the workpiece material adheres to the machining grooves left by the substrate's grinding process. This adhesion promotes abrasive wear and further material buildup in these areas, eventually resulting in the spalling of the tool coating.

**Author Contributions:** V.F.C.S.: investigation, formal analysis and writing—original draft; F.J.G.S.: conceptualization, methodology, project administration, resources, supervision and writing—review and editing; R.A.: coating development, formal analysis, resources and visualization; G.P.: formal analysis, validation and writing—review and editing; A.B.: formal analysis, validation and writing— review and editing; J.S.F.: investigation and supervision. All authors have read and agreed to the published version of the manuscript.

**Funding:** The present work was conducted and funded under the scope of the project ON-SURF (ANI I P2020 I POCI-01-0247-FEDER-024521, co-funded by Portugal 2020 and FEDER, through the COMPETE 2020-Operational Programme for Competitiveness and Internationalisation.

**Data Availability Statement:** No data is made available regarding this work.

**Acknowledgments:** F.J.G. Silva thanks INEGI—Instituto de Ciência e Inovação em Engenharia Mecânica e Engenharia Indústria for its support. The authors would like to thank Rui Rocha from CEMUP (Porto, Portugal), due to his active collaboration in getting the best SEM pictures and helping with his critical analysis of some phenomena. The authors also would like to thank Ing. Ricardo Alexandre for his extremely important role in providing all the coatings through the TEandM company, and Eng. Nuno André for providing the substrate material and uncoated tools through the Inovatools company.

**Conflicts of Interest:** The authors declare no conflict of interest.

## References

1. Cheng, X.; Wang, Y.; Li, X.; Dong, C. Interaction between austein-ferrite phases on passive performance of 2205 duplex stainless steel. *J. Mater. Sci. Technol.* **2018**, *34*, 2140–2148. [CrossRef]
2. Jebaraj, A.V.; Ajaykumar, L.; Deepak, C.R.; Aditya, K.V.V. Weldability, machinability and surfacing of commercial duplex stainless steel AISI2205 for marine applications—A recent review. *J. Adv. Res.* **2017**, *8*, 183–199. [CrossRef] [PubMed]
3. Nomani, J.; Pramanik, A.; Hilditch, T.; Littlefair, G. Chip formation mechanism and machinability of wrought duplex stainless steel alloys. *Int. J. Adv. Technol.* **2015**, *80*, 1127–1135. [CrossRef]
4. Chail, G.; Kangas, P. Super and hyper duplex stainless steels: Structures, properties, and applications. *Procedia Struct. Integr.* **2016**, *2*, 1755–1762. [CrossRef]
5. Nomani, J.; Pramanik, A.; Hilditch, T.; Littlefair, G. Machinability study of first generation duplex (2205), second generation duplex (2507) and austenite stainless steel during drilling process. *Wear* **2013**, *304*, 20–28. [CrossRef]

6. Koyee, R.D.; Heisel, U.; Eisseler, R.; Schmauder, S. Modeling and optimization of turning duplex stainless steels. *J. Manuf. Processes* **2014**, *16*, 451–467. [CrossRef]
7. Gowthaman, P.S.; Jeyakumar, S.; Saravanan, B.A. Machinability and tool wear mechanism of Duplex stainless steel—A review. *Mater. Today Proc.* **2020**, *26*, 1423–1429. [CrossRef]
8. Sahithi, V.V.D.; Malayadrib, T.; Srilatha, N. Optimization of Turning Parameters on Surface Roughness Based on Taguchi Technique. *Mater. Today Proc.* **2019**, *18*, 3657–3666. [CrossRef]
9. Tlhabadira, I.; Daniyan, I.A.; Masu, L.; Van Staden, L.R. Process Design and Optimization of Surface Roughness during M200 TS Milling Process using the Taguchi Method. *Procedia CIRP* **2019**, *84*, 868–873. [CrossRef]
10. Vardhan, M.V.; Sankaraiah, G.; Yohan, M.; Rao, H.J. Optimization of Parameters in CNC milling of P20 steel using Response Surface methodology and Taguchi Method. *Mater. Today Proc.* **2017**, *4*, 9163–9169. [CrossRef]
11. Zhang, J.Z.; Chen, J.C.; Kirby, E.D. Surface roughness optimization in an end-milling operation using the Taguchi design method. *J. Mater. Process. Technol.* **2007**, *184*, 233–239. [CrossRef]
12. Kumar, S.; Saravanan, I.; Patnaik, L. Optimization of surface roughness and material removal rate in milling of AISI 1005 carbon steel using Taguchi approach. *Mater. Today Proc.* **2019**, *22*, 654–658. [CrossRef]
13. Selvaraj, D.P. Optimization of cutting force of duplex stainless steel in dry milling operation. *Mater. Today Proc.* **2017**, *4*, 11141–11147. [CrossRef]
14. Airao, J.; Chaudhary, B.; Bajpai, V.; Khanna, N. An Experimental Study of Surface Roughness Variation in End Milling of Super Duplex 2507 Stainless Steel. *Mater. Today Proc.* **2018**, *5*, 3682–3689. [CrossRef]
15. Policena, M.R.; Devitte, C.; Fronza, G.; Garcia, R.F.; Souza, A.J. Surface roughness analysis in finishing end-milling of duplex stainless steel UNS S32205. *Inter. J. Adv. Manuf. Technol.* **2018**, *98*, 1617–1625. [CrossRef]
16. Sousa, V.F.C.; Silva, F.J.G. Recent Advances in Turning Processes Using Coated Tools—A Comprehensive Review. *Metals* **2020**, *10*, 170. [CrossRef]
17. Sousa, V.F.C.; Silva, F.J.G. Recent Advances on Coated Milling Tool Technology—A Comprehensive Review. *Coatings* **2020**, *10*, 235. [CrossRef]
18. Martinho, R.P.; Silva, F.J.G.; Baptista, A.P.M. Cutting forces and wear analysis of $Si_3N_4$ diamond coated tools in high speed machining. *Vacuum* **2008**, *82*, 1415–1420. [CrossRef]
19. Paiva, J.M.F.; Amorim, F.L.; Soares, P.C.; Veldhuis, S.C.; Mendes, L.A.; Torres, R.D. Tribological behavior of superduplex stainless steels against PVD hard coatings on cemented carbide. *Inter. J. Adv. Manuf. Technol.* **2019**, *90*, 1649–1658. [CrossRef]
20. Silva, F.J.G.; Martinho, R.; Andrade, M.; Baptista, A.P.M.; Alexandre, R. Improving the Wear Resistance of Moulds for the Injection of Glass Fibre-Reinforced Plastics Using PVD Coatings: A Comparative Study. *Coatings* **2017**, *7*, 28. [CrossRef]
21. Silva, F.J.G.; Martinho, R.P.; Alexandre, R.J.D.; Baptista, A.P.M. Increasing the wear resistance of molds for injection of glass fiber reinforced plastics. *Wear* **2011**, *271*, 2494–2499. [CrossRef]
22. Silva, F.J.G.; Martinho, R.P.; Baptista, A.P.M. Characterization of laboratory and industrial CrN/CrCn/Diamond-like carbon coatings. *Thin Solid Films* **2014**, *550*, 278–284. [CrossRef]
23. Silva, F.J.G.; Fernandes, A.J.S.; Costa, F.M.; Teixeira, V.; Baptista, A.P.M.; Pereira, E. Tribological behaviour of CVD diamond films on steel substrates. *Wear* **2003**, *255*, 846–853. [CrossRef]
24. Silva, F.J.G.; Fernandes, A.J.S.; Costa, F.M.; Baptista, A.P.M.; Pereira, E. Unstressed PACVD diamond films on steel pre-coated with a composite multilayer. *Surf. Coat. Technol.* **2005**, *191*, 102–107. [CrossRef]
25. Baptista, A.; Silva, F.J.G.; Porteiro, J.; Míguez, J.L.; Pinto, G. Sputtering physical vapour deposition (PVD) coatings: A critical review on process improvement and market trend demands. *Coatings* **2018**, *8*, 402. [CrossRef]
26. Baptista, A.; Silva, F.J.G.; Porteiro, J.; Míguez, J.L.; Pinto, G.; Fernandes, L. On the Physical Vapour Deposition (PVD): Evolution of Magnetron Sputtering Processes for Industrial Applications. *Procedia Manuf.* **2018**, *17*, 746–757. [CrossRef]
27. Martinho, R.P.; Silva, F.J.G.; Martins, C.; Lopes, H. Comparative study of PVD and CVD cutting tools performance in milling of duplex stainless steel. *Int. J. Adv. Manuf. Technol.* **2019**, *102*, 2423–2439. [CrossRef]
28. Ginting, A.; Skein, R.; Cuaca, D.; Herdianto, P.; Masyithah, Z. The characteristics of CVD- and PVD-coated carbide tools in hard turning of AISI 4340. *Measurement* **2018**, *129*, 548–557. [CrossRef]
29. Koseki, S.; Inoue, K.; Morito, S.; Ohba, T.; Usuki, H. Comparison of TiN-coated tools using CVD and PVD processes during continuous cutting of Ni-based superalloys. *Surf. Coat. Technol.* **2015**, *283*, 353–363. [CrossRef]
30. Caliskan, H.; Panjan, P.; Kurbanoglu, C. 3.16 Hard coatings on cutting tools and surface finish. *Compr. Mater. Finish.* **2017**, 230–242.
31. Paiva, J.M.F.; Torres, R.D.; Amorim, F.L.; Covelli, D.; Tauhiduzzaman, M.; Veldhuis, S.; Dosbaeva, G.; Fox-Rabinovich, G. Frictional and wear performance of hard coatings during machining of superduplex stainless steel. *Int. J. Adv. Manuf. Technol.* **2017**, *92*, 423–432. [CrossRef]
32. Klocke, F.; Krieg, T. Coated tools for metal cutting—Features and applications. *CIRP Ann.* **1999**, *48*, 515–525. [CrossRef]
33. Fernández-Abia, A.I.; Barreiro, J.; Fernández-Larrinoa, J.; de Lacalle, L.N.L.; Fernández-Valdivielso, A.; Pereira, O.M. Behaviour of PVD Coatings in the Turning of Austenitic Stainless Steels. *Procedia Eng.* **2013**, *63*, 133–141. [CrossRef]
34. Vasu, M.; Nayaka, H.S. Investigation of Cutting Force Tool Tip Temperature and Surface Roughness during Dry Machining of Spring Steel. *Mater. Today Proc.* **2018**, *5*, 7141–7149. [CrossRef]
35. Phokobye, S.N.; Daniyan, I.A.; Tlhabadira, I.; Masu, L.; Van Staden, L.R. Model Design and Optimization of Carbide Milling Cutter for Milling Operation of M200 Tool Steel. *Procedia CIRP* **2019**, *84*, 954–959. [CrossRef]

36. Strafford, K.N.; Audy, J. Indirect monitoring of machinability in carbon steels by measurement of cutting forces. *J. Mater. Process. Technol.* **1997**, *67*, 150–156. [CrossRef]
37. Venkatesan, K.; Manivannan, K.; Devendiran, S.; Mathew, A.T.; Ghazaly, N.M.; Aadhavan; Benny, S.M.N. Study of Forces, Surface Finish and Chip Morphology on Machining of Inconel 825. *Procedia Manuf.* **2019**, *30*, 611–618. [CrossRef]
38. Caudill, J.; Schoop, J.; Jawahir, I.S. Numerical Modeling of Cutting Forces and Temperature Distribution in High Speed Cryogenic and Flood-cooled Milling of Ti-6Al-4V. *Procedia CIRP* **2019**, *82*, 83–88. [CrossRef]
39. Fernández-Abia, A.I.; Barreiro, J.; de Lacalle, L.N.L. Behavior of austenitic stainless steels at high speed turning using specific force coefficients. *Int. J. Adv. Manuf. Technol.* **2012**, *62*, 505–515. [CrossRef]
40. Batuev, V.A.; Batuev, V.V.; Ardashev, D.V.; Shipulin, L.V.; Degtyareva-Kashutina, A.S. Analytical Calculation of Cutting Forces and Analysis of their Change at 3-D Milling. *Procedia Manuf.* **2019**, *32*, 42–49. [CrossRef]
41. Davoudinejad, A.; Chiappini, E.; Tirelli, S.; Annoni, M.; Strano, M. Finite Element Simulation and Validation of Chip Formation and Cutting Forces in Dry and Cryogenic Cutting of Ti-6Al-4V. *Procedia Manuf.* **2015**, *1*, 728–739. [CrossRef]
42. Bhopale, S.; Jagatap, K.R.; Lamdhade, G.K.; Darade, P.D. Cutting Forces during Orthogonal Machining Process of AISI 1018 Steel: Numerical and Experimental Modeling. *Mater. Today Proc.* **2017**, *4*, 8454–8462. [CrossRef]
43. Mebrahitom, A.; Choon, W.; Azhari, A. Side Milling Machining Simulation Using Finite Element Analysis: Prediction of Cutting Forces. *Mater. Today Proc.* **2017**, *4*, 5215–8521. [CrossRef]
44. Gouveia, R.; Silva, F.J.G.; Reis, P.; Baptista, A.P.M. Machining duplex stainless steel: Comparative study regarding end mill coated tools. *Coatings* **2016**, *6*, 51. [CrossRef]
45. Ahmed, Y.S.; Paiva, J.; Covelli, D.; Veldhuis, S. Investigation of Coated Cutting Tool Performance during Machining of Super Duplex Stainless Steels through 3D Wear Evaluations. *Coatings* **2017**, *7*, 127. [CrossRef]
46. Dos Santos, A.G.; da Silva, M.B.; Jackson, M.J. Tungsten carbide micro-tool wear when micro milling UNS S32205 duplex stainless steel. *Wear* **2018**, *414–415*, 109–117. [CrossRef]
47. Diniz, A.E.; Machado, A.R.; Corrêa, J.G. Tool wear mechanisms in the machining of steels and stainless steels. *Int. J. Adv. Manuf. Technol.* **2016**, *87*, 3157–3168. [CrossRef]
48. Silva, F.; Martinho, R.; Martins, C.; Lopes, H.; Gouveia, R. Machining GX2CrNiMoN26-7-4 DSS Alloy: Wear Analysis of TiAlN and TiCN/Al2O3/TiN Coated Carbide Tools Behavior in Rough End Milling Operations. *Coatings* **2019**, *9*, 392. [CrossRef]
49. Krolczyk, G.M.; Nieslony, P.; Legutko, S. Determination of tool life and research wear during duplex stainless steel turning. *Arch. Civ. Mech. Eng.* **2015**, *15*, 347–354. [CrossRef]
50. Rajaguru, J.; Arunachalam, N. Coated tool Performance in Dry Turning of Super Duplex Stainless Steel. *Procedia Manuf.* **2017**, *10*, 601–611. [CrossRef]
51. Suárez, A.; de Lacalle, L.N.L.; Polvorosa, R.; Veiga, F.; Wretland, A. Effects of high-pressure cooling on the wear patterns on turning inserts used on alloy IN718. *Mater. Manuf. Process.* **2017**, *32*, 678–686. [CrossRef]
52. *ISO 8688-2:1986*; Tool Life Testing in Milling—Part 2: End Milling. International Organization for Standardization: Geneva, Switzerland, 1986.
53. Leyland, A.; Matthews, A. On the significance of the H/E ratio in wear control: A nanocomposite coating approach to optimised tribological behaviour. *Wear* **2000**, *246*, 1–11. [CrossRef]
54. Beake, B.D. The influence of the H/E ratio on wear resistance of coating systems—Insights from small-scale testing. *Surf. Coat. Technol.* **2022**, 128272. [CrossRef]
55. Beake, B.D.; Vishnyakov, V.M.; Harris, A.J. Nano-scratch testing of (Ti,Fe)Nx thin films on silicon. *Surf. Coat. Technol.* **2017**, *309*, 671–679. [CrossRef]
56. Sousa, V.F.C.; Silva, F.J.G.; Alexandre, R.; Fecheira, J.S.; Silva, F.P.N. Study of the wear behaviour of TiAlSiN and TiAlN PVD coated tools on milling operations of pre-hardened tool steel. *Wear* **2021**, *476*, 203695. [CrossRef]

 *metals*

*Article*

# Delamination of Fibre Metal Laminates Due to Drilling: Experimental Study and Fracture Mechanics-Based Modelling


Francisco Marques [1], Filipe G. A. Silva [2], Tiago E. F. Silva [2,*], Pedro A. R. Rosa [3], António T. Marques [2,4] and Abílio M. P. de Jesus [2,4]

[1]  Palbit S.A., Product Development Department, 3854-908 Branca, Portugal; fmarques@palbit.pt
[2]  INEGI, Faculty of Engineering, University of Porto, Rua Dr. Roberto Frias 400, 4200-465 Porto, Portugal; fsilva@inegi.up.pt (F.G.A.S.); marques@inegi.up.pt (A.T.M.); ajesus@fe.up.pt (A.M.P.d.J.)
[3]  IDMEC, Instituto Superior Tecnico, University of Lisbon, Avenida Rovisco Pais 1, 1049-001 Lisboa, Portugal; pedro.rosa@tecnico.ulisboa.pt
[4]  DEMEc, Faculty of Engineering, University of Porto, Rua Dr. Roberto Frias 400, 4200-465 Porto, Portugal
*   Correspondence: tesilva@inegi.up.pt



**Abstract:** Fibre metal laminates (FML) are significantly adopted in the aviation industry due to their convenient combination of specific strength, impact resistance and ductility. Drilling of such materials is a regular pre-requisite which enables assembly operations, typically through rivet joining. However, the hole-making operation is of increased complexity due to the dissimilarity of the involved materials, often resulting in defects (i.e., material interface delamination), which can significantly compromise the otherwise excellent fatigue strength. This work explores the potential of three different drill geometries, operating under variable cutting speeds and feeds on CFRP-AA laminates. In addition, the usage of sacrificial back support is investigated and cutting load, surface roughness and delamination extension are examined. In order to predict delamination occurrence, ADCB tests are performed, enabling the calculation of fracture energy threshold. Drill geometry presents a very significant influence on delamination occurrence. The usage of specific step-tools with secondary cutting edge showed superior performance. Despite its simplicity, the applied critical force threshold model was able to successfully predict interface delamination with good accuracy.

**Keywords:** fibre metal laminates; drilling; cutting tool; modelling



**Citation:** Marques, F.; Silva, F.G.A.; Silva, T.E.F.; Rosa, P.A.R.; Marques, A.T.; de Jesus, A.M.P. Delamination of Fibre Metal Laminates Due to Drilling: Experimental Study and Fracture Mechanics Based Modelling *Metals* **2022**, *12*, 1262. https://doi.org/10.3390/met12081262

Academic Editor: Francisco J. G. Silva

Received: 30 May 2022
Accepted: 23 July 2022
Published: 27 July 2022

**Publisher's Note:** MDPI stays neutral with regard to jurisdictional claims in published maps and institutional affiliations.




## 1. Introduction

Fibre metal laminates (FML) are hybrid materials comprised of alternating metal sheet and fibre composite layers, that can be bonded in distinct sequences. Unlike hybrid multi-material stacks, in which thicker composite and metal layers are simply stacked and fastened by means of rivets, adhesive joints or bolted connections, the FML layers' thickness is typically less than 1 mm and consolidated through hot-curing cycles [1,2]. The superior performance of the laminate combination when compared to the isolated composite and metallic materials is highly relevant in structural applications, such as the transportation sector, where high specific strength is required, while also maintaining good impact and bending resistance [3,4]. Moreover, when such materials are adopted in the aviation industry, energy savings of approximately 30% are achieved [5,6]. Effectively illustrating the FML expression in aircraft applications and its increasing adoption [7] is the excellent fatigue strength, damage tolerance and overall durability of these materials, due to fibres acting as a barriers, thus delaying metal crack propagation [8]. Furthermore, worth mentioning is the good thermal insulation, corrosion and flame resistance properties of such materials [7,9].

FML can be currently manufactured in near-net shape geometries, such as large fuselage panels and stringers in aeronautics and also complex-shaped floor assemblies in the automotive industry [10]. Nonetheless, the fastening of multiple components relies mostly on mechanical joints, such as rivets or bolts, which can amount up to 3 million in a commercial aircraft [11]. Hole-making is, therefore, intensively performed for parts assembly, enabling riveting of aircraft panels such as the fuselage, wings and stabilizers [12,13] and despite the existent non-conventional feasible alternatives, such as laser machining and water-jet cutting, drilling remains the most employed technique [14]. The heterogeneity of FML allied to the highly abrasive properties of fibre reinforcement make drilling operations a challenging task. Their success may be compromised by simultaneous occurrence of (i) the well-known entry (peel-up) and exit (push-out) delamination, matrix thermal damage, fibre pull out and formation of abrasive fibre particles in composite layers as well as (ii) strain-hardening, continuous chip formation and thermal softening in the metal layers [15–17].

The lack of research regarding drilling operations on FML is evidenced by Bonhin et al. [16], especially in what concerns aluminium alloy (AA) and carbon-fibre-reinforced polymers (CFRP) configurations with thermoset matrices. To the authors' knowledge, no studies have been performed in AA-CFRP laminates with thermoplastic matrices. The enhanced sustainability of the thermoplastic polymers (promoted by their improved recyclability) has encouraged the increasing usage of thermoplastic-based FML in relevant sectors, such as the aeronautical industry. Ekici et al. [18] experimentally analysed hole quality and delamination on AA-CFRP material samples, using Physical Vapour Deposition (PVD)-coated and uncoated drills. Despite the little number of holes each drill performed, the authors found that the uncoated condition (carbide) outperforms (PVD) coated drills in terms of entry delamination and hole nominal size. Sridhar et al. [19] systematically analysed the influence of operative conditions on drilling performance indicators such as thrust force and roughness, being able to identify ideal cutting parameters for AA-CFRP laminates using a conventional drill geometry. Despite the more intricate process kinematics of helical milling when compared to drilling, Bolar et al. [20] report advantages of the former, concerning cutting load, thermal impact, chip evacuation and hole nominal size. The implementation of analytic/numerical models capable of delamination prediction is of key relevance for both FML processing/assembly and drill tool manufacturers. Although some work can be found for CFRP materials, no data are available regarding these novel AA-CFRP material configurations. Feito et al. [21] compare the predictability of both complete simulation of a drilling operation with a simplistic model, in which the drill acts as a punch that pierces the laminate. The latter yields very reasonable cost-effective results with slight overestimation of delamination factor, setting upper limits that are highly valuable as support decision techniques.

## 2. Materials and Methods

This work focuses on the assessment of three distinct drill geometries with regard to their cutting performance of AA-CFRP hybrid laminates, through conventional single-step drilling operations. Roughness measurement and load monitoring were carried out for distinct sets of cutting parameters on each geometry. Delamination, which may account for 60% of the rejected parts [22] is thoroughly analysed and a critical load threshold was estimated for its occurrence, based on delamination modelling of AA-CFRP interface through asymmetric double cantilever beam experimental testing procedure. Previous knowledge of the critical loads associated with the drilling operation is highly convenient in the design and selection of appropriate tooling solutions for hole quality compliance.

Specimens were built from CFRP and AA, with a stacking configuration of three CFRP layers (two external, one internal) and two internal aluminium layers, as illustrated in Figure 1a. The CFRP layers consisted of four 0.13 mm thick prepreg plies for the internal CFRP layer and three of the same plies for the external CFRP layers. Each aluminium layer was composed of 0.2 mm thick AA 5754 sheet. The composite material was constituted by polyamide 6 (PA 6) thermoplastic matrix reinforced by uni-directional carbon fibre with a volume fraction of 48.5%. Layer adhesion was promoted using conventional pre-treatment techniques such as degreasing and laser texturing of the metallic sheets. The laminate was submitted to a hot plate press curing process at a temperature interval between 240 and 280 °C and a pressure of 2 to 6 bar. Fibre direction was kept the same in all composite layers (uni-directional). Rectangular-shaped plates (240 × 250 mm$^2$) were manufactured, with a resulting thickness (post-curing) of 1.2mm, which were posteriorly cut into 40 × 225 mm$^2$ strips. Table 1 exhibits the mechanical properties of each material, according to the respective datasheets [23,24]. The experimental tests were conducted in a DMG Mori DMU60eVo series machining centre (25 kW), equipped with a piezoeletric dynamometer (Kistler 9272) and a signal amplifier (Kistler 5070A), connected to a data aquisition system (Advantech USB4711). A clamping system was developed for fixation of the material strips to the load cell, enabling drilling operations with and without sacrificial back support. The laminate strip is secured in-between two circular plates (top and bottom, refer to Figure 1b,c) which have a centre hole ($\phi$ 36 mm), enabling the drilling operation and placement of a PTFE cylinder under the laminates for sacrificial back support (when used). A constant torque was applied on the bolts which hold the top plate against the material and lower plate. With regard to the cutting tools, diamond-coated (through Chemical Vapour Deposition, CVD) tungsten carbide drills with a diameter of 6 mm were employed. This drill material and coating configuration has been increasingly used in the hole-making of hybrid materials [25,26]. Moreover, three different drill geometries were tested: (i) a conventional drill geometry (herein referred to as CNV) with a 120° point angle, 30° helix angle, 20° rake angle and 10° clearance angle, refer to Figure 1d; (ii) a chip-breaking drill geometry (herein referred to as CBR) identical to CNV, with v-shaped grooves on the principal cutting edge periphery, refer to Figure 1e; (iii) a double-point angle tool (herein referred to as 2PA) also identical to the CNV with a 60° secondary point angle (2:1 ratio) and the same geometry as the previous tools, refer to Figure 1f.

**Table 1.** Mechanical properties of the AA and CFRP separate materials.

| Materials | AA 5754 | CFRP (PA 6) |
| --- | --- | --- |
| Density [g/cm$^3$] | 2.67 | 1.45 |
| Young's modulus [GPa] | 65-75 | 100 |
| Tensile strength [MPa] | 200-350 | 1910 |
| Strain at fracture [%] | <25 | 1.76 |

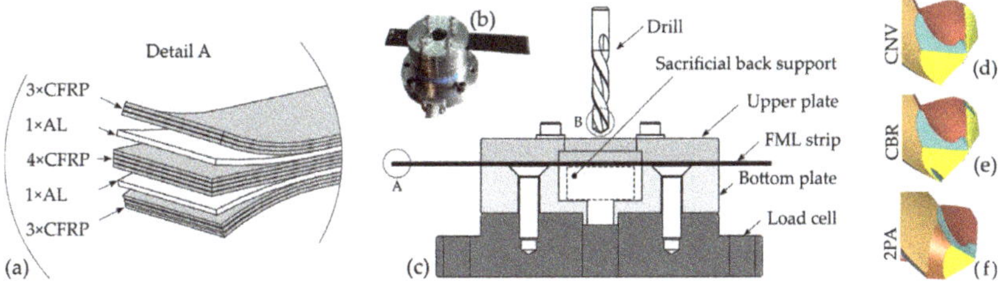


**Figure 1.** Experimental setup used in drilling operations: (**a**) Fibre metal laminate sequence and layer number of CFRP plies and AA sheets; (**b**) Clamping system and load cell assemblage with mounted FML strip; (**c**) Cross-sectional view scheme showing internal placement of back support; Drill tip geometry (detail B in **Figure 1c**) of conventional drill (**d**), chip-breaking drill (**e**) and double-point angle drill (**f**).

Despite the constant search for novel drill geometries capable of generating fewer defects, the conventional drill geometry (such as the CNV in the current study) still constitutes a widely employed solution, which in this work has been used for reference/control and comparison with other geometries. It is also relevant to note that their performance can sometimes match or exceed newer, more intricate geometries regarding drilled hole quality [27]. Diamond-coated double-point angle drills, such as the considered 2PA, can be effectively employed in hole drilling of CFRP materials given the consequent action of lower cutting forces on the drill step (secondary cutting edge), that is mainly responsible for the final surface condition of the drilled hole [28]. Their overall good performance has motivated its study in fibre metal laminated hole drilling. With regard to chip-breaking features on drills (such as the considered CBR drill) the goal is to promote more efficient chip evacuation by creating grooves on the drill geometry (typically on principal cutting edge) capable of chip segmentation and width dividing, thus minimizing load and torque. Such concern is particularly relevant when drilling materials with thermoplastic resin (such as PA 6) which unlike the often employed thermoset resins (i.e., epoxy) promotes long chip morphology rather than fragmented chips. By dividing the chip width, a more convenient scenario of chip removal could be attained (i.e., more fragile resultant split chips, clogging minimization at flute). Moreover, in order to avoid excessive friction of the chip with the newly generated hole surface, the groove has been positioned at the cutting edge margin in order to act as a chip relief that tentatively minimizes delamination due to smaller chip-hole contact. Cutting parameters testing range was selected based on the literature and tool manufacturer indications for laminate materials. Table 2 illustrates the tested levels of each considered variable. The full combination of parameters was tested using a random order generated by the Response Surface Methodology (RSM) Design Expert 13 software. Moreover, the operative conditions' influence on cutting load, roughness and delamination was investigated through analysis of variance (ANOVA). In order to mitigate the occurrence of wear mechanisms, each drill performed a maximum of 20 holes.

**Table 2.** Variables and respective levels used for cutting parameter assessment in experimental drilling operation tests, considering a full factorial testing plan.

| Variables | Levels | | |
|---|---|---|---|
| Cutting speed [m/min] | 80 | 100 | 120 |
| Feed [mm/rev] | 0.03 | 0.05 | 0.07 |
| Drill geometry | CNV | CBR | 2PA |
| Back support | with | without | - |

Delamination defects were observed through radiographic image analysis, using Satelec X-Mind X-Ray generator and a Kodak RVG 5100 digital sensor. For this, the samples were submitted to a diidomethane bath for a period of 30 min, which enables contrast creation between delaminated and non-delaminated zones. A fixed exposure time of 0.16 s and a radiographic contrast of 70 kVp were selected. The obtained X-ray images were post-processed (converted into binary maps) allowing for delamination assessment and quantification, using the the criteria shown in Equations (1)–(3), where: $D_{max}$ and $A_{max}$ correspond to the maximum diameter of the delamination area and its area, respectively; $A_d$ to the actual delamination area; $D_{nom}$ and $A_{nom}$ to the nominal hole diameter and area, respectively. For the calculation of the delamination factors, a Matlab script capable of measuring $A_{max}$ and $D_{max}$ from the previously generated binary maps was used. This method ensures control process repeatability and minimization of data analysis effort.

$$F_d = D_{max}/D_{nom} \tag{1}$$

$$F_a = A_d/A_{nom} \tag{2}$$

Although the diameter-based ($F_d$) and area-based ($F_a$) delamination factors are the most employed criteria, they do not fully portray the drilled hole quality [11]. The diameter-based delamination factor ($F_d$) may account for the same delamination values (same maximum diameter around hole), for instance, in two very different scenarios: (i) whole delamination of a full annular section area or (ii) crack delamination of a crack (very small area). Similar interpretation errors can occur when considering an area-based delamination factor, given that (i) uniform damage and (ii) uniform damage with small cracks may result in the same area. In sum, whereas $F_d$ accounts only for the delamination maximum extent in the radial direction, $F_a$ cannot account for crack delamination, prone to occur in CFRP, as only the area is used for its calculation. For this reason, an adjusted delamination factor $F_{da}$, proposed by Davim et al. [29] has been used. It tends to $F_d^2$ values with uniformly distributed delamination and to $F_d$ values when it is strongly directional, allowing for a more accurate estimate of delamination shape and its extension.

$$F_{da} = F_d + \frac{A_d}{(A_{max} - A_{nom})}(F_d^2 - F_d) \tag{3}$$

With regard to roughness analysis, it was optically estimated using the 3D measurement system (Alicona Infinite Focus SL). A three-measurement average was calculated for each drilled hole.

*Delamination Modelling*

Delamination has long been viewed from a fracture mechanics perspective as a crack propagation phenomenon and the critical force at its onset ($C_F$) can be calculated according to Equation (4), proposed by Cheng et al. [15], assuming a point load applied on an isotropic-circular-clamped plate, where $G_{IC}$ corresponds to the mode I fracture energy associated to the material interface delamination, $E$ is the Young's modulus, $v$ is the Poisson coefficient and $h$ is the depth of uncut material under the drill tool.

$$C_F = \pi \left[ \frac{8G_{IC}Eh^3}{3(1 - v^2)} \right]^{1/2} \tag{4}$$

In order to estimate the fracture energy of the CFRP-AA interface, asymmetric double cantilever beam (ADCB) tests were performed. In this type of test, a traction load is applied to the specimen arms, inducing the propagation of an existent pre-crack at a specified specimen plane, with a length and thickness of $a_0$ and $t$ (refer to Figure 2a). A specially built testing machine coupled with a 50N capacity load cell (Tedea-Huntleigh Model 1042), intended for fracture characterization, was employed (refer to Figure 2b). The specimens arms were bonded (Araldite 2052-1 structural adhesive) to aluminium blocks with a 6 mm hole to allow for ADCB specimen gripping in the testing machine. The challenging real-time monitoring of crack propagation can be avoided using an equivalent crack length ($a_e$) procedure [30,31]. A relationship between $a_e$ and specimen compliance (defined as the ratio between the applied displacement, $\delta$, and load, $P$), can be obtained considering the strain energy ($U$) of the specimen due to bending and shear effects (Timoshenko beam theory) and applying the Castigliano theorem ($\delta = dU/dP$). In this context, specimen current compliance can be defined as shown in Equation (5), where $B$, $h_u$ and $h_l$ correspond to specimen dimensions, $D_u$ and $D_l$ are the bending stiffness of upper and lower arms. Although critical load estimation relies exclusively on mode I fracture, in this work the ADCB specimens were selected given the relative difficulty in inducing a pre-crack in middle layer when compared to layer interfaces. Moreover, the fracture mechanism in drilling is consistent with the representation of mixed mode fracture with predominant mode I [32].

$$C = \frac{a_e^3}{3}\left(\frac{1}{D_u} + \frac{1}{D_l}\right) + \frac{6a_e}{5BG_{13}}\left(\frac{1}{h_u} + \frac{1}{h_l}\right) \tag{5}$$

Combining Equation (5) with the Irwin–Kies relation, Equation (6) can be derived, providing the total strain energy release under mixed mode I + II (with predominant mode I) as a function of $a_e$. Other approaches could be applied, namely within the scope of the linear-elastic fracture mechanics, such as that described in [33,34].

$$G_T = \frac{P^2}{2B}\left[a_e^2\left(\frac{1}{D_u} + \frac{1}{D_l}\right) + \frac{6}{5BG_{13}}\left(\frac{1}{h_u} + \frac{1}{h_l}\right)\right] \tag{6}$$

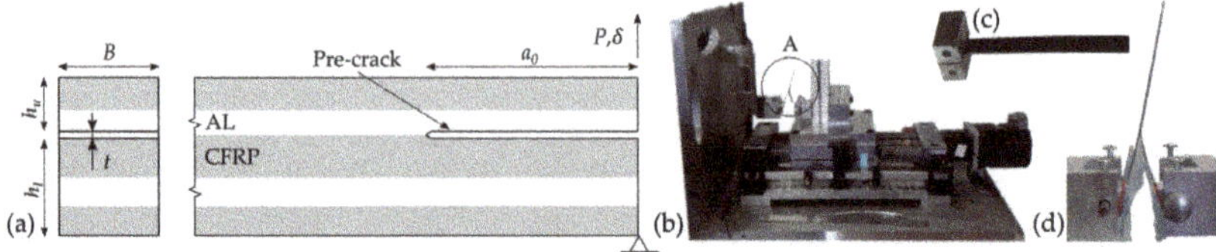

**Figure 2.** Experimental details of delamination fracture energy evaluation procedure: (**a**) Scheme of ADCB specimens loading, showing pre-crack location. (**b**) Fracture testing machine equipped with load cell. (**c**) ADCB specimen with bonded aluminium blocks on both upper and lower arms. (**d**) Detail A of **Figure 2b**, showing ADCB specimen with aluminium blocks assembled to testing machine.

## 3. Results and Discussion

The thrust force ($F_t$) evolution of each tested tool geometry and its association with current drill point position is exhibited in Figure 3. A specific load signature can be identified and significant correspondence can be made with each layer of the laminate material. Common to all drill geometries, a rise in $F_t$ is observed due to the contact increase (between the drill's primary cutting edge and the laminate material), from the start of the drilling operation up to instant B. A steeper increase of the force is noticed from instants B to C, corresponding to the CFRP material layer, evidencing the higher cutting resistance of this material as compared to the AA. This effect is further highlighted by the subsequent $F_t$ decrease in the C to D path. Load curve tends to another maximum as the drill exits the laminate material (E). Up to this instant, $F_t$ signature is rather indistinct of drill geometry, which is coherent with the identical point and helix angles of the three tested drills.

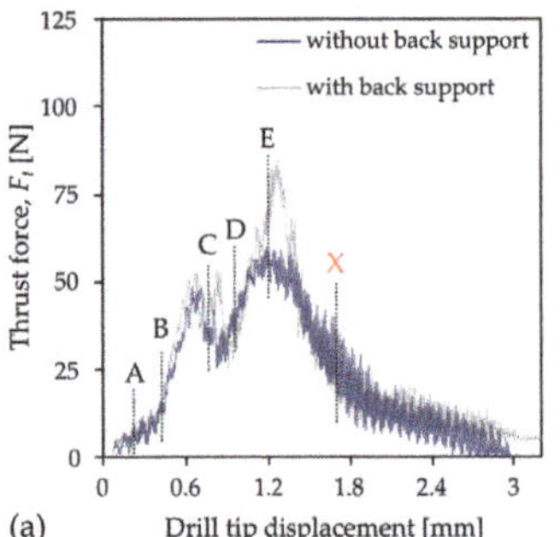

(a)

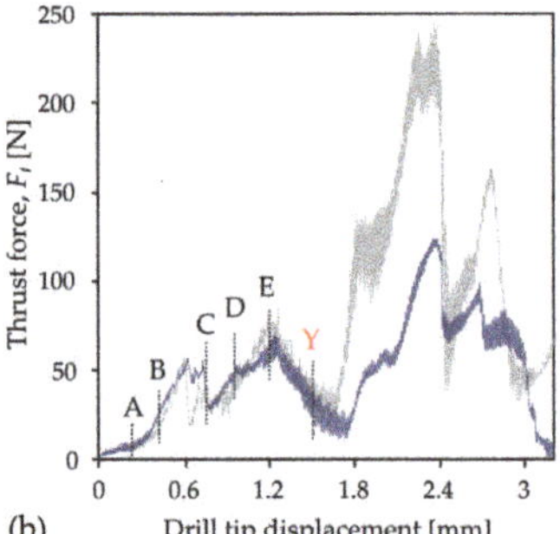

(b)

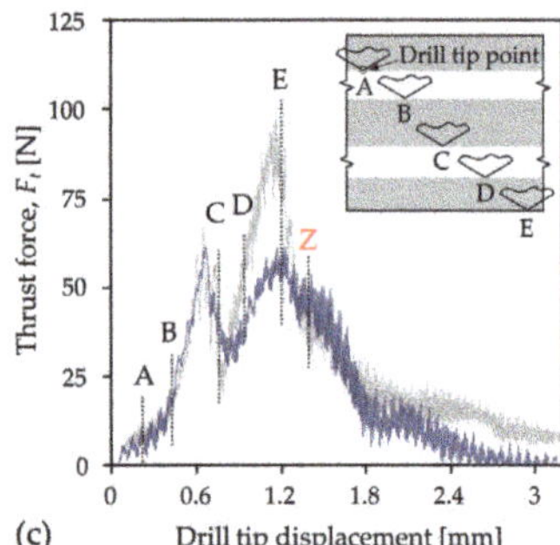

(c)

**Figure 3.** Maximum thrust force evolution in function of drill displacement. (**a**) CNV drill geometry. (**b**) CBR drill geometry. (**c**) 2PA drill geometry.

Instant X, in Figure 3a, indicates the moment the CNV drill's secondary cutting edge engages the first layer of the laminate material. A slight transition (increase) in the curve's slope is noticed. Similarly, and referring to Figure 3c, the instant Z corresponds to the moment the second point angle of the 2PA drill engages the laminate material. It is interesting to note that with the addition of a second (smaller) point angle, $F_t$ conveniently decreases more rapidly. In contrast, as the chip-breaking groove of CBR drill engages the laminate material (instant Y of Figure 3b) a very significant load increase is noticed, peaking at approximate double $F_t$ values of the A-Y drill tip path. Given the assumption that higher loads may contribute to higher delamination, the CBR drill might be inadequate. The load signature differences between drilling operations with and without back support are illustrated in Figure 3. It is possible to note that the overall thrust force signature does not significantly change. Still, the maximum values (peak of the $F_t$ curves) are consistently higher for all drill geometries, promoted by the stiffness increase of the clamping system.

Moreover, at the end of the drilling operation, a load plateau is maintained, corresponding to the cutting of the PTFE disk (back support provider).

The analysis of variance (ANOVA) conducted for the maximum thrust force revealed significant impact of drill geometry, feed as well as the usage of back support. Such is illustrated by the <0.05 $p$-values respective of those variables, in Table 3. Moreover, despite the slight increase of maximum $F_t$ with cutting speed, it did not present a relevant influence, especially when compared with the other considered variables, as shown in Figure 4. Drill geometry CBR seems to develop much higher thrust forces (approximately two-times higher), when compared with the other two drill configurations, which show identical results (slightly lower axial force with the CNV), as illustrated in Figure 4c. Therefore, as thrust force magnitude may be an indicator of delamination severity, the CBR drill may not perform adequately.

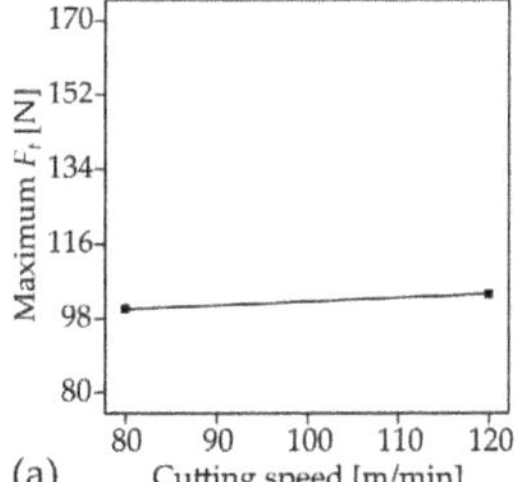
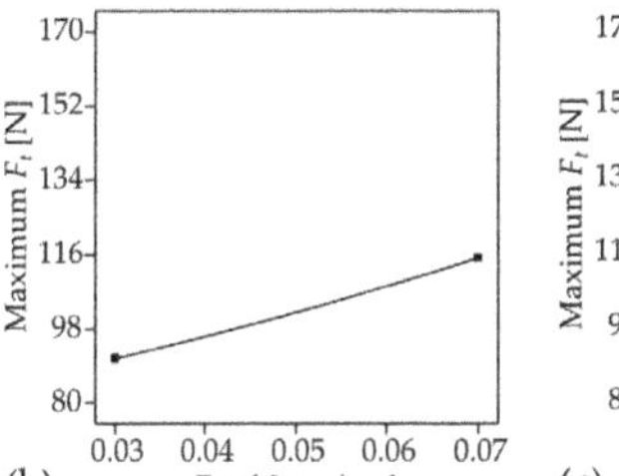
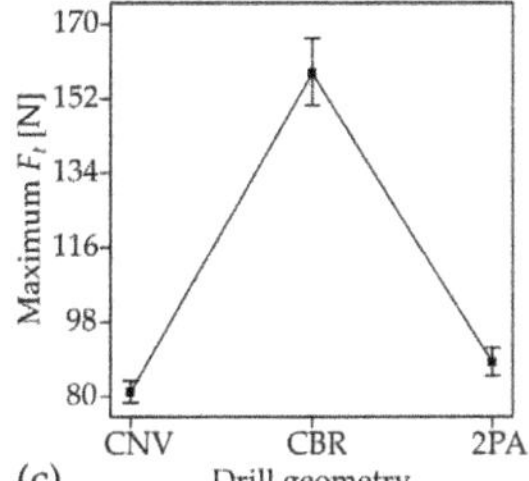
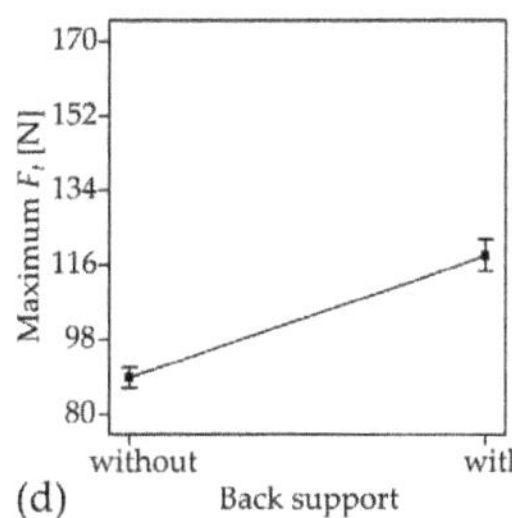

**Figure 4.** ANOVA results for axial force variability in function of: (**a**) Cutting speed. (**b**) Feed. (**c**) Drill geometry. (**d**) Back support.

**Table 3.** ANOVA results on maximum axial force of the conducted experimental campaign.

| Source | Sum of Squares | df | Mean Square | $F$-Value | $p$-Value |
| --- | --- | --- | --- | --- | --- |
| Model | 0.0192 | 5 | 0.0038 | 89.98 | <0.0001 |
| Cut. speed | 0.0000 | 1 | 0.0000 | 0.851 | 0.3539 |
| Feed | 0.0018 | 1 | 0.0018 | 41.72 | <0.0001 |
| Drill geom. | 0.0126 | 2 | 0.0063 | 147.7 | <0.0001 |
| Back support | 0.0038 | 1 | 0.0038 | 88.96 | <0.0001 |
| Residual | 0.0030 | 70 | - | - | - |

Figure 5a–c illustrate some representative examples of delamination occurrence on the machined holes using each drill geometry. The X-ray analysis has enabled the observation of the otherwise indiscernible defects. Figure 5a shows the delamination type mostly associated with the usage of CNV drill. The high directionality of damage occurrence (aligned with fibre orientation) is coherent with the push-out delamination mechanism resultant from AA-CFRP material de-bonding caused by the drill thrust. Since this interface de-bonding is predominantly mode I fracture, the developed modelling towards delamination prediction using ADCB is in accordance with the obtained results.

Alternatively, uniformly distributed delamination (as illustrated in Figure 5b) was more prone to occur with the CBR drill. The damage around the hole contour may be associated with the chip-breaking v-grooves on the principal cutting edge of the drill. These structures have seemingly failed to control chip morphology, which was identical regardless of the employed tool as well as operative conditions: continuous (ribbon) chips constituted of an aluminium core and with discontinuous bonded CFRP, as illustrated in Figure 5d. Unable to improve chip segmentation or breakage (comparatively to CNV and 2PA drills), the v-shaped grooves on the CBR drill seem to have caused internal delamination due to chip imprisonment. Repositioning of the groove towards a more central position of the drill's cutting edge or increasing the number of grooves along the cutting edge may promote better chip splitting. Further research on the identification of suitable drill morphologies towards effective chip partition in fibre metal laminates is required, which may be supported using more advanced numerical methods. Constitutive and damage modelling may be convenient towards accurate portrayal of chip flow. Apart from the delamination effects, an example of a delamination-free drilled hole (using 2PA drill) is shown in Figure 5c.

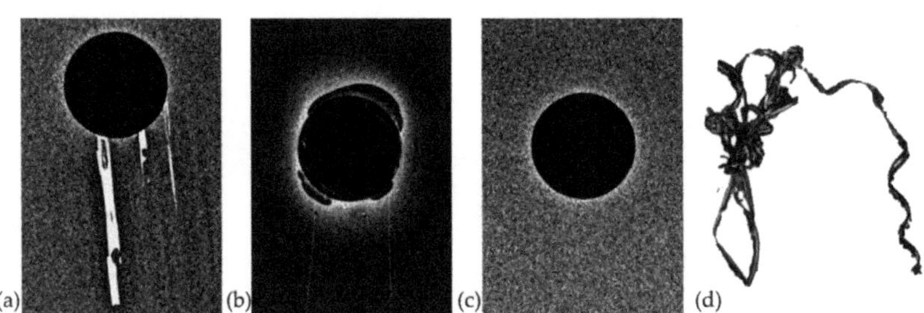

**Figure 5.** X-ray images of representative delamination defects on drilled holes using: (**a**) CNV drill geometry and (**b**) CBR drill geometry. (**c**) Example of delamination-free drilled hole using 2PA drill geometry. (**d**) Typical chip morphology obtained from drilling operations of FML, regardless of the employed drill.

The influence of the tested variables on the considered delamination factors results is shown in Figure 6. In addition, Tables 4–6 present the analysis of variance details. Despite having a negative impact on maximum thrust force (refer to Figure 4d), back support is commonly employed with the goal of increasing the fixture stiffness and minimizing delamination (preventing displacement of FML layers up to fracture initiation and propagation). Figure 6 shows the influence of the tested variables on the calculated delamination factors ($F_a$, $F_d$, $F_{ad}$) for each used drill tool. Although it is not expressive for the $F_d$ and $F_{da}$ delamination factors, a significant correlation between back support employment and delamination factor minimization ($F_a$) is observed in Figure 6d, illustrating its decreasing tendency with back support usage. Cutting speed and feed did not show accountable statistical impact (*p*-value higher than 0.05), as illustrated in Figure 6a,b. Drill geometry is the most influential variable on delamination results. The CBR drill yields the worst case scenario regarding delamination values for all calculated factors (up to three-times higher than CNV). In addition, 2PA seems to slightly outperform the CNV drill geometry. It is important to note the consistency of delamination results with the previous maximum load measurements, illustrating the importance of load prediction in metal cutting operations.

**Table 4.** ANOVA results on $F_a$ delamination factor of the conducted experimental campaign.

| Source | Sum of Squares | df | Mean Square | *F*-Value | *p*-Value |
|---|---|---|---|---|---|
| Model | 30,963 | 5 | 6192 | 17.74 | <0.0001 |
| Cut. speed | 46 | 1 | 46 | 0.133 | 0.715 |
| Feed | 241 | 1 | 241 | 0.691 | 0.408 |
| Drill geom. | 28,572 | 2 | 14286 | 40.92 | <0.0001 |
| Back support | 1675 | 1 | 1675 | 4.80 | 0.0318 |
| Residual | 24,438 | 70 | 349 | - | - |

**Table 5.** ANOVA results on $F_d$ delamination factor of the conducted experimental campaign.

| Source | Sum of Squares | df | Mean Square | F-Value | p-Value |
|---|---|---|---|---|---|
| Model | 1.61 | 5 | 0.3220 | 5.68 | 0.0002 |
| Cut. speed | 0.0064 | 1 | 0.0064 | 0.1126 | 0.7382 |
| Feed | 0.0112 | 1 | 0.0112 | 0.1973 | 0.6583 |
| Drill geom. | 1.59 | 2 | 0.7947 | 14.02 | <0.0001 |
| Back support | 0.0048 | 1 | 0.0048 | 0.0850 | 0.7715 |
| Residual | 3.97 | 70 | 0.0567 | - | - |

**Table 6.** ANOVA results on $F_{da}$ delamination factor of the conducted experimental campaign.

| Source | Sum of Squares | df | Mean Square | F-Value | p-Value |
|---|---|---|---|---|---|
| Model | 1.74 | 5 | 0.3479 | 7.56 | <0.0001 |
| Cut. speed | 0.0045 | 1 | 0.0045 | 0.0970 | 0.7564 |
| Feed | 0.0041 | 1 | 0.0041 | 0.0883 | 0.7672 |
| Drill geom. | 1.73 | 2 | 0.8645 | 18.79 | <0.0001 |
| Back support | 0.0030 | 1 | 0.0030 | 0.0659 | 0.7982 |
| Residual | 3.22 | 70 | 0.0460 | - | - |

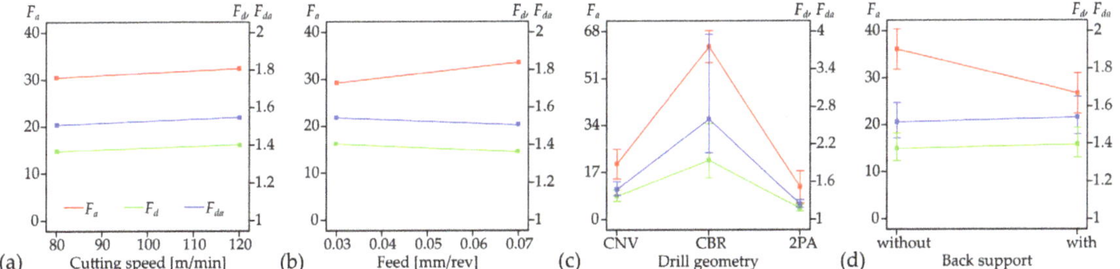


**Figure 6.** ANOVA results on delamination factors in function of: (**a**) Cutting speed. (**b**) Feed. (**c**) Drill geometry. (**d**) Back support.

The arithmetical mean height roughness (*Ra*) has been estimated on a 5 mm length profile of the generated hole surface. Three measurement repetitions were performed for each hole and the average values were taken into consideration for ANOVA. The ANOVA statistical results show that drill geometry is the only relevant variable with regard to roughness (*Ra*) values (refer to Figure 7). Still, both cutting speed and feed p-values range relatively close to the 0.05 limit, from which significant impact can be inferred, thus showing slight tendencies for smaller roughness values when higher cutting speed and smaller feed operative conditions are applied. With regard to drill geometry, an identical trend to the tested variables has been identified, meaning that lower surface quality holes have resulted from hole making with the CBR drill. Moreover, from all machined holes, only 20% were above the 3.2 μm surface roughness limit (*Ra*). The majority of those were performed using the CBR drill (87%) with the remainder using the CNV drill. Only the 2PA drill was capable of attaining *Ra* < 3.2 μm in all machined holes. This criterion has been a useful indicator of the *Ra* quality in industrial conditions, with special relevance to the aeronautics sector [35].

The arithmetical mean height roughness (*Ra*) has been estimated on a 5 mm length profile of the generated hole surface. Three measurement repetitions were performed for each hole and the average values were taken into consideration for ANOVA. The ANOVA statistical results show that drill geometry is the only relevant variable with regard to roughness (*Ra*) values (refer to Figure 7). Still, as can be seen from 7, both cutting speed and feed p-values range relatively close to the 0.05 limit, from which significant impact can be inferred, thus showing slight tendencies for smaller roughness values when higher cutting speed and smaller feed operative conditions are applied. With regard to drill geometry, an identical trend to the tested variables has been identified, meaning that lower surface quality holes have resulted from hole making with the CBR drill. Moreover, from all machined holes, only 20% were above the 3.2 μm surface roughness limit (*Ra*). The majority of those were performed using the CBR drill (87%) with the remainder using the CNV drill. Only the 2PA drill was capable of attaining *Ra* < 3.2 μm in all machined holes. This criterion has been a useful indicator of the *Ra* quality in industrial conditions, with special relevance to the aeronautics sector [35].

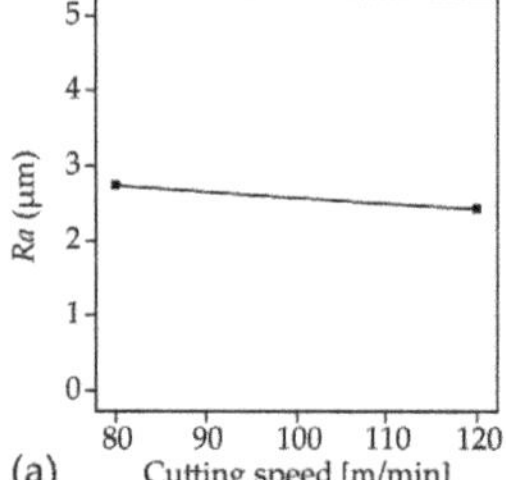
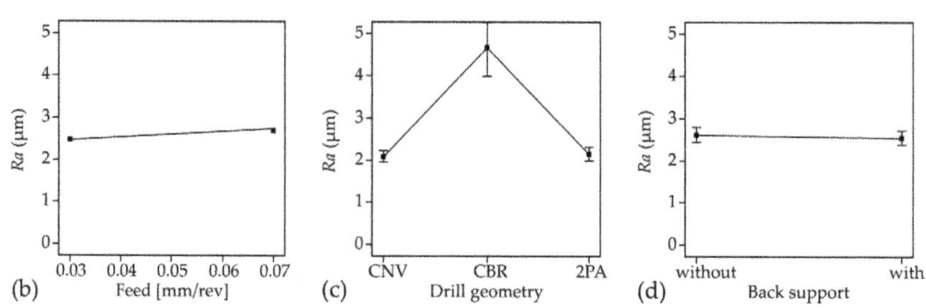

**Figure 7.** ANOVA results on measured surface roughness (*Ra*) in function of: (**a**) Cutting speed. (**b**) Feed. (**c**) Drill geometry. (**d**) Back support.

**Table 7.** ANOVA results on *Ra* surface roughness conducted during the experimental campaign.

| Source | Sum of Squares | df | Mean Square | *F*-Value | *p*-Value |
|---|---|---|---|---|---|
| Model | 0.8707 | 5 | 0.1741 | 15.23 | <0.0001 |
| Cut. speed | 0.0249 | 1 | 0.0249 | 2.18 | 0.1447 |
| Feed | 0.0126 | 1 | 0.0126 | 1.10 | 0.2979 |
| Drill geom. | 0.8245 | 2 | 0.4122 | 36.06 | <0.0001 |
| Back support | 0.0016 | 1 | 0.0016 | 0.1408 | 0.7086 |
| Residual | 0.8002 | 70 | 0.0114 | - | - |

The load–displacement results of the ADCB fracture tests are shown in Figure 8a. The resistance curves (refer to Figure 8b) were obtained to determine the energy release rate, needed to estimate a delamination critical force ($C_F$) through Equation (5). From the analysis of Figure 8b, an energy value plateau of approximately 0.249 N/mm with upper and lower boundaries of 0.29 and 0.20, respectively, is identified. Although delamination is more likely to occur as the drill approaches material exit (commonly known as exit delamination and promoted by the lack of subsequent material layers), the critical force, $C_F$ (or delamination onset load) has been calculated for three distinct interfaces as the tool advances on the laminate. These are labelled and highlighted in Figure 8c. Since the ADCB fracture tests were conducted at an AA-CFRP interface, delamination prediction is limited to those interfaces within the considered laminate. It is important to note that the critical force value obtained by Equation (5) refers to a single point at the drill tool path. This point is defined by the distance between the drill point and the bottom surface of the laminate (also called depth of uncut material, $h$).

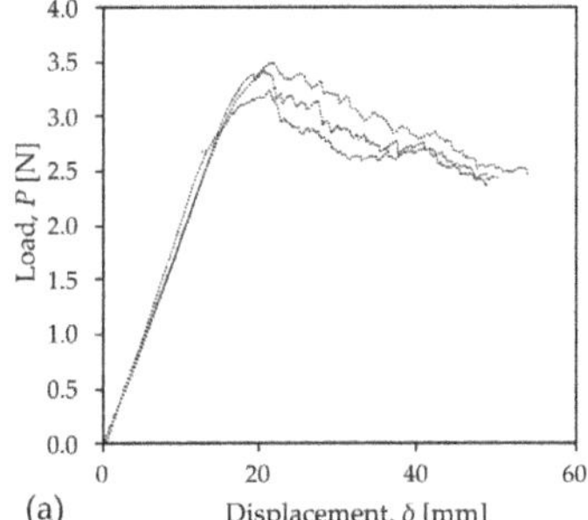
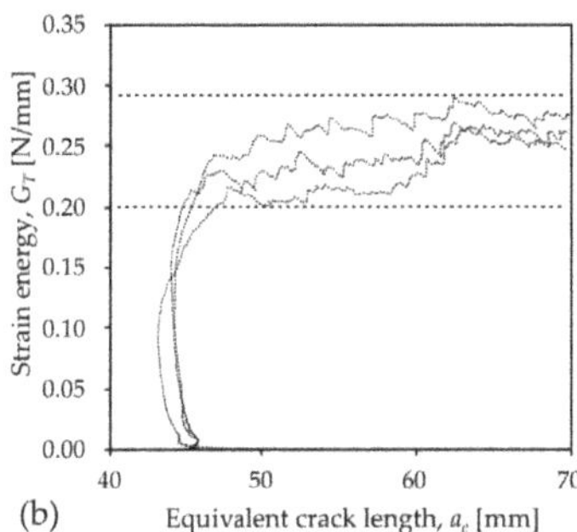
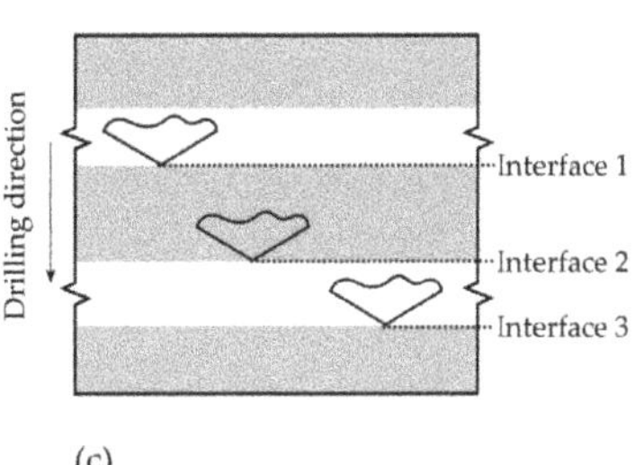



**Figure 8.** Load–displacement and energy release rate results obtained from ADCB fracture tests: (**a**) Load–displacement curves. (**b**) Resistance curves. (**c**) Interfaces where delamination occurrence has been analysed.

Table 8 presents the range of critical forces for each identified interface of Figure 8c, based on the upper and lower boundaries energy values. Elastic modulus has been calculated based on weighted average of metal volume fraction (MVF) of the uncut material, which in the case of interface 3 corresponds exclusively to CFRP material. A Poisson ratio of 0.4 was considered for the $C_F$ estimation.

**Table 8.** Estimation of critical force range for each respective interface, based on ADCB fracture tests critical energies.

| Interface Number | Drill Point Distance [mm] | MVF | Elastic Modulus [MPa] | $C_F$ Range [N] |
|---|---|---|---|---|
| 1 | 0.76 | 0.26 | 92,200 | 503.6–585.1 |
| 2 | 0.44 | 0.45 | 86,500 | 214.9–249.7 |
| 3 | 0.24 | 0 | 100,000 | 93.1–108.1 |

The estimated critical force $C_F$ range can be seen as a threshold of values from which delamination is likely to occur. This range has been compared with several drilling operation thrust force signatures and the X-ray images of drilled holes showing delamination occurrence (or the absence thereof). It is important to note that the usage of back support hinders delamination by bending prevention (and thus interface de-bonding) of the laminate layers. For this reason, the tests conducted with sacrificial back support were not considered in this part of the study. In addition, the model proposed by Cheng et al. [15], is not valid for such support conditions.

Figure 9a shows the critical force range thresholds comparison with the thrust force loading signature of the CNV drill using a cutting speed of 120 m/min and 0.03 mm/rev feed. Since maximum thrust force was consistently below the drill point path corresponding threshold, delamination is not predicted, which is coherent with its absence in the X-ray image of the corresponding hole. Figure 9b shows a similar example for a CNV drill with a cutting speed of 100 m/min and 0.07 mm/rev of feed, where maximum thrust force surpasses the minimum delamination threshold at "interface 3" (refer to Figure 8c for interface relative position) resulting in delamination occurrence, as verified in the corresponding X-ray image of the drilled hole.

One of the shortcomings of the presented methodology is illustrated by Figure 9c. Although the critical force limit is not attained within the considered tool path, it may have been surpassed by the action of the drill's chip grooves. Such a possibility is consistent with the delamination morphology and its occurrence at an internal interface (contrary to the exit delamination of the previous examples in Figure 9a,b). The estimated critical force of Equation (4) relies on the assumption of a point load (associated with the cutting phenomenon) and thus, in the current case, its applicability is compromised since it is only valid to drill features that are effectively cutting (principal cutting edge). Still, when considering a critical load value that is independent of drill position, the developed model would correctly predict delamination occurrence.

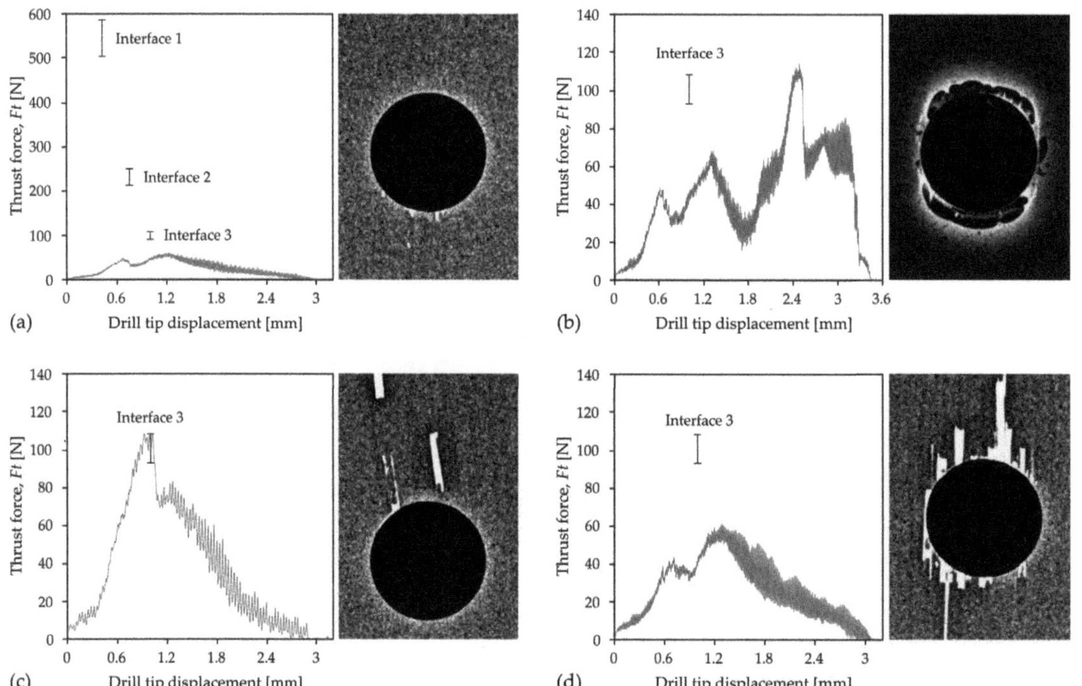

**Figure 9.** Comparison between load–tool displacement curves, calculated critical load and X-ray images of drilled hole for: (**a**) CNV drill at 120 m/min and 0.03 mm/rev; (**b**) CNV drill at 100 m/min and 0.07 mm/rev; (**c**) CBR drill at 120 m/min and 0.03 mm/rev; (**d**) CNV drill at 80 m/min and 0.03 mm/rev.

Given that the load has not surpassed the critical force threshold, Figure 9d shows an example of unexpected defects. It is, however, noticeable that the pattern of damage occurrence (indicated by red arrows) is compatible with the delamination type of chip formation in fibre-reinforced polymers [32,36], suggesting fibre-matrix interface failure (and crack propagation) within the composite material. In drilling, the relative position constantly changes with each rotation and when the cutting direction and fibre direction are the same (occurring in two distinct instants) mixed mode fracture occurs. Thus, predominant type I or II, depending on rake angle, develops within fibre-composite material, promoting crack initiation and its propagation along the fibre-reinforcement interface. This observation explains the non-compliance of the developed criterion (only valid for AA-CFRP interfaces).

## 4. Conclusions

In this study, the impact of tool geometry and operative conditions on the hole drilling of thermoplastic-based fibre metal laminates was investigated. Cutting load, delamination and internal roughness were assessed. Drill geometry shows very significant influence on all measured machinability indicators. The inclusion of grooves on the primary cutting edges of the drill revealed neither an appropriate technique toward efficient chip control nor delamination mitigation. In contrast, the usage of a double-point angle drill stands out as an effective method toward improved hole-making in fibre metal laminates, with minimal delamination. Back support was shown to be an adequate alternative method for minimizing delamination, especially when in conjunction with smaller feed rates. Fracture energy was successfully estimated using asymmetric double cantilever beam tests, enabling the calculation of critical force thresholds for delamination occurrence, which adequately predicts the delamination with some exceptions duly justified from known model limitations.

**Author Contributions:** Conceptualization, A.M.P.d.J. and A.T.M.; methodology, F.M. and F.G.A.S.; software, F.M. and F.G.A.S.; validation, F.M. and F.G.A.S.; formal analysis, P.A.R.R., A.M.P.d.J. and A.T.M.; investigation, F.M., F.G.A.S. and T.E.F.S.; resources, A.M.P.d.J. and A.T.M.; data curation, F.M. and T.E.F.S.; writing—original draft preparation, F.M. and T.E.F.S.; writing—review and editing, T.E.F.S., P.A.R.R. and A.M.P.d.J.; visualization, F.G.A.S. and A.T.M.; supervision, F.G.A.S. and A.M.P.d.J.; project administration, A.M.P.d.J. and A.T.M.; funding acquisition, A.M.P.d.J. and A.T.M. All authors have read and agreed to the published version of the manuscript.

**Funding:** This work has been conducted under the scope of MAMTool (PTDC/EME-EME/31895/2017) and AddStrength (PTDC/EME-EME/31307/2017) projects, funded by Programa Operacional Competitividade e Internacionalização, and Programa Operacional Regional de Lisboa funded by FEDER and National Funds (FCT). This work was supported by Add.Additive project (POCI-01-0247-FEDER-024533) funded by ANI under the scope of Programa Operacional Competitividade e Internacionalização and funded by FEDER and National Funds (FCT).

**Institutional Review Board Statement:** Not applicable

**Informed Consent Statement:** Not applicable

**Data Availability Statement:** Not applicable.

**Acknowledgments:** Palbit S.A. is acknowledged for providing necessary tools for experimental cutting tests. INEGI is also acknowledged for providing the materials used in this study.

**Conflicts of Interest:** The authors declare no conflict of interest.

## Abbreviations

The following abbreviations are used in this manuscript:

| | |
|---|---|
| 2PA | Two Point Angle Drill Geometry |
| AA | Aluminium Alloy |
| ADCB | Asymmetric Double Cantilever Beam |
| ANOVA | Analysis Of Variance |
| CBR | Chip-Breaking Drill Geometry |
| CFRP | Carbon-Fibre Reinforced Polymer |
| CNV | Conventional Drill Geometry |
| CVD | Chemical Vapour Deposition |
| FML | Fibre Metal Laminate |
| MVF | Metal Volume Fraction |
| PA | Polyamide |
| PVD | Physical Vapour Deposition |
| RSM | Response Surface Methodology |

## References

1. Sinmazçelik, T.; Avcu, E.; Bora, M.Ö.; Çoban, O. A review: Fibre metal laminates, background, bonding types and applied test methods. *Mater. Des.* **2011**, *32*, 3671–3685. [CrossRef]
2. He, W.; Wang, L.; Liu, H.; Wang, C.; Yao, L.; Li, Q.; Sun, G. On impact behavior of fiber metal laminate (FML) structures: A state-of-the-art review. *Thin-Walled Struct.* **2021**, *167*, 108026. [CrossRef]
3. Chandrasekar, M.; Ishak, M.R.; Jawaid, M.; Leman, Z.; Sapuan, S.M. An experimental review on the mechanical properties and hygrothermal behaviour of fibre metal laminates. *J. Reinf. Plast. Compos.* **2017**, *36*, 72–82. [CrossRef]
4. Alderliesten, R. *Fatigue and Fracture of Fibre Metal Laminates*; Springer: New York, NY, USA, 2017.
5. Hamill, L.; Hofmann, D.C.; Nutt, S. Galvanic corrosion and mechanical behavior of fiber metal laminates of metallic glass and carbon fiber composites. *Adv. Eng. Mater.* **2018**, *20*, 1700711. [CrossRef]

6.   Tyczyński, P.; Lemańczyk, J.; Ostrowski, R. Drilling of CFRP, GFRP, glare type composites. *Aircr. Eng. Aerosp. Technol. Int. J.* **2014**, *86*, 312–322. [CrossRef]

7.   Vlot, A.; Gunnink, J.W. (Eds.)  *Fibre Metal Laminates: An Introduction*; Springer Science & Business Media: New York, NY, USA, 2011.

8.   Baker, A.A. *Composite Materials for Aircraft Structures*; American Institute of Aeronautics and Astronautics (AIAA): Reston, VA, USA, 2004.

9.   Botelho, E.C.; Silva, R.A.; Pardini, L.C.; Rezende, M.C. A review on the development and properties of continuous fiber/epoxy/aluminum hybrid composites for aircraft structures. *Mater. Res.* **2006**, *9*, 247–256. [CrossRef]

10.  Ding, Z.; Wang, H.; Luo, J.; Li, N. A review on forming technologies of fibre metal laminates. *Int. J. Lightweight Mater. Manuf.* **2021**, *4*, 110–126. [CrossRef]

11.  Aamir, M.; Tolouei-Rad, M.; Giasin, K.; Nosrati, A. Recent advances in drilling of carbon fiber–reinforced polymers for aerospace applications: A review. *Int. J. Adv. Manuf. Technol.* **2019**, *105*, 2289–2308. [CrossRef]

12.  Soutis, C. Fibre reinforced composites in aircraft construction. *Prog. Aerosp. Sci.* **2005**, *41*, 143–151. [CrossRef]

13.  Brinksmeier, E.; Fangmann, S.; Rentsch, R. Drilling of composites and resulting surface integrity. *CIRP Ann.* **2011**, *60*, 57–60. [CrossRef]

14.  Liu, D.; Tang, Y.; Cong, W. L. A review of mechanical drilling for composite laminates. *Compos. Struct.* **2012**, *94*, 1265–1279. [CrossRef]

15.  Ho-Cheng, H.; Dharan, C.K.H. Delamination during drilling in composite laminates. *J. Eng. Ind.* **1990**, *112*, 236–239. [CrossRef]

16.  Bonhin, E.P.; David-Müzel, S.; de Sampaio Alves, M.C.; Botelho, E.C.; Ribeiro, M.V. A review of mechanical drilling on fiber metal laminates. *J. Compos. Mater.* **2021**, *55*, 843–869. [CrossRef]

17.  Giasin, K.; Gorey, G.; Byrne, C.; Sinke, J.; Brousseau, E. Effect of machining parameters and cutting tool coating on hole quality in dry drilling of fibre metal laminates. *Compos. Struct.* **2019**, *212*, 159–174. [CrossRef]

18.  Ekici, E.; Motorcu, A.R.; Yıldırım, E. An experimental study on hole quality and different delamination approaches in the drilling of CARALL, a new FML composite. *FME Trans.* **2021**, *49*, 950–961. [CrossRef]

19.  Sridhar, A.K.; Bolar, G.; Padmaraj, N.H. Comprehensive experimental investigation on drilling multi-material carbon fiber reinforced aluminum laminates. *J. King Saud Univ.-Eng. Sci.* **2021**. [CrossRef]

20.  Bolar, G.; Sridhar, A.K.; Ranjan, A. Drilling and helical milling for hole making in multi-material carbon reinforced aluminum laminates. *Int. J. Lightweight Mater. Manuf.* **2022**, *5*, 113–125. [CrossRef]

21.  Feito, N.; López-Puente, J.; Santiuste, C.; Miguélez, M.H. Numerical prediction of delamination in CFRP drilling. *Compos. Struct.* **2014**, *108*, 677–683. [CrossRef]

22.  Phadnis, V.A.; Makhdum, F.; Roy, A.; Silberschmidt, V.V. Drilling in carbon/epoxy composites: Experimental investigations and finite element implementation. *Compos. Part A Appl. Sci. Manuf.* **2013**, *47*, 41–51. [CrossRef]

23.  CELSTRAN CFR-TP PA6 CF60-03 Datasheet Celanese Corporation. Available online: https://tools.celanese.com/products/datasheet/SI/CELSTRAN%C2%AE%20CFR-TP%20PA6%20CF60-03 (accessed on 18 January 2022).

24.  Aluminum/Magnesium Foil Al97/Mg 3 Material Properties, Goodfellow GmbH. Available online: https://www.goodfellow.com/de/en-us/displayitemdetails/p/al01-fl-000150/aluminum-magnesium-foil (accessed on 26 January 2022).

25.  Kuo, C.L.; Soo, S.L.; Aspinwall, D.K.; Bradley, S.; Thomas, W.; M'Saoubi, R.; Pearson, D.; Leahy, W. Tool wear and hole quality when single-shot drilling of metallic-composite stacks with diamond-coated tools. *Proc. Inst. Mech. Eng. Part J. Eng. Manuf.* **2014**, *228*, 1314–1322. [CrossRef]

26.  Zhang, L.; Liu, Z.; Tian, W.; Liao, W. Experimental studies on the performance of different structure tools in drilling CFRP/Al alloy stacks. *Int. J. Adv. Manuf. Technol.* **2015**, *81*, 241–251. [CrossRef]

27.  Marques, A.T.; Durão, L.M.; Magalhães, A.G.; Silva, J.F.; Tavares, J.M.R. Delamination analysis of carbon fibre reinforced laminates: Evaluation of a special step drill. *Compos. Sci. Technol.* **2009**, *69*, 2376–2382. [CrossRef]

28.  Heisel, U.; Pfeifroth, T. Influence of point angle on drill hole quality and machining forces when drilling CFRP. *Procedia Cirp* **2012**, *1*, 471–476. [CrossRef]

29.  Davim, J.P.; Rubio, J.C.; Abrao, A.M. A novel approach based on digital image analysis to evaluate the delamination factor after drilling composite laminates. *Compos. Sci. Technol.* **2007**, *67*, 1939–1945. [CrossRef]

30.  Ramírez, F.M.; de Moura, M.F.; Moreira, R.D.; Silva, F.G. Experimental and numerical mixed-mode I+ II fracture characterization of carbon fibre reinforced polymer laminates using a novel strategy. *Compos. Struct.* **2011**, *263*, 113683. [CrossRef]

31.  Moreira, R.D.F.; de Moura, M.F.S.F.; Silva, F.G.A.; Ramírez, F.M.G.; Rodrigues, J.S. Mixed-mode I+ II fracture characterisation of composite bonded joints. *J. Adhes. Sci. Technol.* **2020**, *34*, 1385–1398. [CrossRef]

32.  Sheikh-Ahmad, J.Y. Mechanics of chip formation. In *Machining of Polymer Composites*; Springer: Boston, MA, USA, 2009; pp. 63–110.

33.  Valvo, P.S. On the calculation of energy release rate and mode mixity in delaminated laminated beams. *Eng. Fract. Mech.* **2016**, *165*, 114–139. [CrossRef]

34.  Altenbach, H.; Meenen, J. Single Layer Modelling and Effective Stiffness Estimations of Laminated Plates. In *Modern Trends in Composite Laminates Mechanics*; Springer: Vienna, Austria, 2003; pp. 1–68.

35. Coromant, S. Machining carbon fibre materials. In *Sandvik Coromant User's Guide-Composite Solutions*; Sandvik: Stockholm, Sweden, 2010.
36. Geier, N.; Davim, J.P.; Szalay, T. Advanced cutting tools and technologies for drilling carbon fibre reinforced polymer (CFRP) composites: A review. *Compos. Part A Appl. Sci. Manuf.* **2019**, *125*, 105552. [CrossRef]

173

*Review*

# A Review of $CO_2$ Coolants for Sustainable Machining

Leon Proud [1,2,*], Nikolaos Tapoglou [1,3,*] and Tom Slatter [2,*]

[1] Advanced Manufacturing Research Centre (AMRC), University of Sheffield, Catcliffe Way, Rotherham S60 5TZ, UK

[2] Department of Mechanical Engineering, University of Sheffield, Mappin Street, Sheffield S1 3JD, UK

[3] Industrial Engineering and Management Department, International Hellenic University, 57001 Thessaloniki, Greece

* Correspondence: l.proud@amrc.co.uk (L.P.); ntapoglou@iem.ihu.gr (N.T.); tom.slatter@sheffield.ac.uk (T.S.)

**Abstract:** In many machining operations, metalworking fluids (MWFs) play an invaluable role. Often, proper application of an intelligent MWF strategy allows manufacturing processes to benefit from a multitude of operational incentives, not least of which are increased tool life, improved surface integrity and optimised chip handling. Despite these clearly positive implications, current MWF strategies are often unable to accommodate the environmental, economic and social conscience of industrial environments. In response to these challenges, $CO_2$ coolants are postulated as an operationally viable, environmentally benign MWF solution. Given the strong mechanistic rationale and historical evidence in support of cryogenic coolants, this review considers the technological chronology of cryogenic MWF's in addition to the current state-of-the-art approaches. The review also focuses on the use of $CO_2$ coolants in the context of the machining of a multitude of material types in various machining conditions. In doing so, cryogenic assisted machining is shown to offer a litany of performance benefits for both conventional emulsion (flood) cooling and near dry strategies, i.e., minimum quantity lubrication (MQL), as well as aerosol dry lubrication (ADL).

**Keywords:** cryogenic machining; coolant; metalworking fluids; carbon dioxide; liquid nitrogen; minimum quantity lubrication; tool wear

**Citation:** Proud, L.; Tapoglou, N.; Slatter, T. A Review of $CO_2$ Coolants for Sustainable Machining. *Metals* **2022**, *12*, 283. https://doi.org/10.3390/met12020283

Academic Editor: Francisco J. G. Silva

Received: 22 December 2021
Accepted: 25 January 2022
Published: 5 February 2022

**Publisher's Note:** MDPI stays neutral with regard to jurisdictional claims in published maps and institutional affiliations.

## 1. Introduction

In order to accommodate the demands of the current technological landscape, engineers have developed novel materials that meet an increasingly expansive range of lofty performance indices. One pertinent example is that of an aero-engine disk. In such an application, the subject material must exhibit high flexural rigidity, fatigue strength and creep resistance, all whilst being exposed to operating temperatures in excess of 800 °C [1]. In this case, it crucial that the material chosen exhibits both high shear strength and low thermal conductivity. Whilst these properties serve a clear purpose in an aeroengine, they generate a challenging machining environment, which must be met with a robust metalworking fluid (MWF) strategy. Unfortunately, however, whilst they remain an operational necessity, the use of conventional MWF strategies is accompanied by a series of negative environmental, social and ergonomic implications. To elaborate, their disposal is not only unsustainable, but both extremely costly and time consuming; they pose a health risk to machine operators and their maintenance requires a significant time commitment that could be allocated to more productive tasks. In contrast, whilst cryogenic and $CO_2$ cutting fluids have their own limitations, they offer markedly improved sustainability (Section 4); they have been shown (situationally) to produce significantly improved machinability outcomes (Section 5), and by virtue of their return to the atmosphere post machining cycle, they remove the burden of fluid maintenance and disposal.

Despite the prospective benefits of cryogenic and $CO_2$ MWF strategies, the use of $CO_2$ as an MWF remains confined to niche applications. This is due to the fact that whilst $LN_2$

is, at this point, extensively researched, the available literature on $CO_2$ is (comparatively) very much in its infancy. As such, if $CO_2$ MWF strategies are to become commonplace, it is crucial that the body of literature is compiled in a means that is both accessible and categorised according to research interest. With this in mind, the proceeding review looks to outline the scenarios in which $CO_2$ MWF strategies have, at this stage, been employed, and further, serves to identify avenues for future research.

## 2. Review Structure

Given the extensive research that focuses upon the use of liquid nitrogen ($LN_2$) MWFs, the proceeding text is primarily focused upon the emerging use of $CO_2$ coolants, with only an ancillary consideration of $LN_2$. The review first provides a background (Section 3) on the subject matter of cryogenic and $CO_2$ MWFs, focusing foremost upon the historical use of cryogenic MWF's (Section 3.1) followed by their motivation for use (Section 3.2) and the mechanism of action for $scCO_2$ coolants specifically (Section 3.3). Thereafter, the document considers the sustainability of $CO_2$ coolants in lieu of conventional MWFs (Section 4), giving primary consideration to the environmental, social and ergonomic implications of the widespread disruptive use of $CO_2$ coolants. The review next focuses upon the current state-of-the-art by way of outlining the available literature on $CO_2$ MWFs, categorizing said literature foremost according to the workpiece material considered (Section 5) and then according to a range of other operational variables (Section 6), i.e., feeds and speeds, tool design, etc. Finally, the document is brought to a conclusion (Section 7). When locating research for this review article, the following search terms were used: "$CO_2$", "Machining", "Cryogenic", "MQL", "Milling" and "Turning".

## 3. Background

### 3.1. History of Cryogenic MWFs

In many regards, formal research on cryogenic cutting fluids began in 1967, when Okoshi [2] observed, and later went on to present the chip handling benefits of $LN_2$ at the annual general meeting of the Japan Society of Precision Engineers (JSPE). Despite sparking interest, it was not until 1969 that the first journal articles were published in the field. In the original paper on cryogenic machining, Uehara and Kumagai [3] examined the variability of cutting temperature, chip formation and tool force with workpiece temperature for plain carbon steel, stainless steel and commercially pure titanium. In their research, Uehara and Kumagai employed a simple turning model where $LN_2$ was applied in situ to the three workpiece materials. Thereafter, two separate thermocouples were employed to measure temperature (at the workpiece and chip-tool interface), and a dynamometer utilised to measure force components. The ultimate findings of the paper noted that whilst the cryogenic machining strategy led to an improved surface finish in the carbon steel workpiece, in the stainless steel and titanium samples, surface roughness was either comparable or worsened by the application of $LN_2$. Moreover, the paper found that during the machining of titanium sample, workpiece temperature and cutting force followed a negative correlation. The authors went on to conclude that the variability with which the different materials responded to reduced workpiece temperature was largely due to the extent to which a given material exhibited low temperature brittleness. As the amount of shear flow stress implicit during low temperature machining increases, surface roughness is worsened and cutting force is generally increased (if not offset by increased brittleness).

In their following paper, Uehara and examined the tool life implications of a cryogenic, workpiece cooling strategy [4]. In this regard, the first truly promising tool life rationale for the use of cryogenic coolants was made, where tool wear was noted to decrease at reduced temperatures in both the carbon steel and titanium test pieces. Moreover, this paper went on to define the conditionality of cryogenic cooling strategies, noting that the extent of flank wear in the stainless steel workpiece followed the reverse trend to that of both carbon steel and titanium. As was the case in their original paper, this article again concludes by defining three variables responsible for the observed phenomenon: the temperature dependency of tool wear, low temperature brittleness of the workpiece and the protective effect of built-up edge (BUE) against tool wear (the latter being heavily cited in the authors reasoning for the conditionality of recorded results). Interestingly, whilst both titanium and stainless steel failed to exhibit low temperature brittleness, the impact of workpiece temperature on tool life differed dramatically between the two materials. The authors went on to state that this discrepancy is likely a consequence of the protective effect of a BUE having a much more significant impact on tool life in the machining of stainless steel (then titanium), and as such, the extinguishment of the BUE at cryogenic temperatures was a more relevant phenomenon in the machining of stainless steel. Additionally, it can be argued that in the machining of titanium, temperature dependency of tool wear is higher based upon the low thermal conductivity of titanium and the known susceptibility of titanium and titanium alloys to diffusion dominated wear mechanisms such as crater wear.

Given the early research of Uehara and Kumagai, many authors have since followed suit in the study of cryogenic MWF's. Generally, much of this research has historically focused upon the use of $LN_2$ coolants. Whilst these articles are the subject of a great deal of academic interest, much of their contents have already been previously considered in earlier review papers [5,6]. Additionally, it is the rationale of the author that the significant incentives offered by $CO_2$ MWFs (Sections 3.2 and 4) make further $LN_2$ focused reviews inconsistent with the current research trend of the cryogenic machining community. Given this statement, a case is thus made in support of the proceeding work's focus upon $CO_2$ (rather than $LN_2$) MWF's.

### 3.2. CO₂ Mechanism of Action

In a conventional $LN_2$ cryogenic machining strategy, the tool is cooled by proximity to the cryogenic medium. In such a scenario, the extremely cold fluid serves to function as a heat sink, cooling the tool primarily by convection, driven by temperature gradient (between the tool and cryogen). $CO_2$ coolants on the other hand are delivered at ambient temperature (Figure 1), and as such, their function as a coolant is not a consequence of their cryogenic temperature at delivery, but rather, the Joule–Thomson effect. In a Joule–Thomson expansion, the fluid is adiabatically throttled through the exit valve of the coolant system. As the gas is throttled, it undergoes a sudden drop in pressure. This reduction in pressure is accompanied by an increase in the potential energy of the fluid owing to the increasing Van der Waals attraction at a larger atomic separation. As such, in order for enthalpy to remain constant across the throttle (which is observed in a Joule–Thomson expansion), the thermal kinetic energy of the fluid post throttle must reduce to accommodate this increase in potential energy, ultimately generating a cooling effect in the system. The throttling process also corresponds a phase-transformation where, in the case of $CO_2$, $LCO_2$ becomes a combination of 60% solid (generally in the form of $CO_2$ "snow") and 40% gas [7], which equally has a novel impact upon the mode of cooling, and the machining process in general.

### 3.3. Motivation for CO₂ Coolants

Whilst $LN_2$ can be shown to outperform conventional MWF's across a broad range of performance indices (i.e., demonstrated improvements to tool life, surface integrity, etc.), the current research landscape has largely made strides to transition away from $LN_2$, and instead focus upon $CO_2$; this is likely a consequence of the following factors:

- By using $CO_2$ coolants, there is a reduced risk of cold burns to the operator due to their delivery at ambient temperature.
- $CO_2$ coolant strategies do not create the same work-holding complexities as $LN_2$ coolant strategies owing to their higher temperature and more localised application.
- $LN_2$ coolants can impede the normal operation of the machining centre, i.e., by freezing the lubricating grease on the spindle; $CO_2$ is not associated with the same risks.

- $CO_2$ storage does not require the use of an insulated pressure vessel and pipe network, making the process of designing and retrofitting machining centres with $CO_2$ more convenient than is the case with $LN_2$.
- Due to the extreme cold, there is a risk of near surface microstructural transformation when using $LN_2$ coolants, this may be mitigated with $CO_2$.

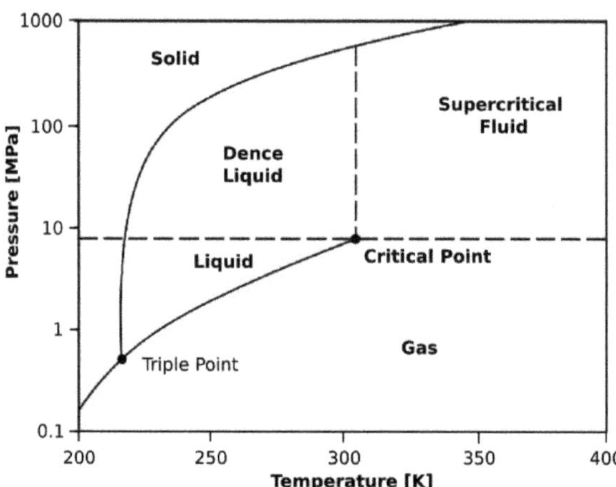


**Figure 1.** Adapted $CO_2$ pressure–temperature phase diagram. Reprinted from reference [8].

## 4. Sustainability of $CO_2$ Coolants

Although the performance benefits of cryogenic cooling strategies are highly contingent upon a range of a factors, from a sustainability standpoint, their benefits are much more widespread. Whilst emulsion cooling has been able to meet a range of performance indices (i.e., long tool life, low surface roughness and cutting forces), it remains a highly imperfect solution from an ergonomic and environmental perspective. So much so that a large portion of the research into cryogenic cutting fluids is primarily motivated by the ultimate unsustainability of conventional MWF's rather than a commercial desire to reduce operating costs (although, this is clearly a significant auxiliary benefit). As such, the following section examines a range of cited sustainability issues associated with the use of conventional MWF's, and thereafter explores the ways in which these issues are entirely, or partially, mitigated by cryogenic MWF strategies. Moreover, the section will additionally outline the processing route by which $CO_2$ is obtained, as well as address any concerns associated with the abundant release of $CO_2$ into the atmosphere (as is associated with its use as an MWF).

### 4.1. Environmental Implications of $CO_2$ Coolants

Given that $CO_2$ is a greenhouse gas, it is reasonable to assume that should it displace conventional emulsion coolant strategies, the net release of $CO_2$ into the atmosphere would invariably increase in concert. This is a particularly concerning proposition given that the UN's Intergovernmental Panel on Climate Change (IPCC) has stated that a 45% reduction in global net $CO_2$ emission is necessary (by 2030) to meet global warming targets [9]. Clearly, on the surface, these concerns make intuitive sense. Unrestricted $CO_2$ usage would seem to be a significant backwards step in meeting the UN's lofty climate change goals; however, it is the argument of the author that these concerns are a consequence of a fundamental misunderstanding of the $CO_2$ processing route. If $CO_2$ was indeed purposefully generated in order to meet coolant needs, this would be a pertinent point; however, this is not the case. In reality, $CO_2$ is generated in a range of essential industrial processes (i.e., in the production of ammonia and alcohol), and is ordinarily released into the atmosphere at the point of production. By instead reclaiming this $CO_2$ for use in MWF's (amongst other things), the rate of $CO_2$ production is unchanged relative to its baseline industrial production, and as such, the otherwise wasted $CO_2$ now serves a purpose prior to being released into the atmosphere (as it ordinarily would be). Ultimately, the only additional energy expenditure associated with $CO_2$ as an MWF comes from that which is consumed in the recycling process; otherwise, its use corresponds to a net zero change in emissions relative to the industrial baseline. The use of $CO_2$ coolants could also lead to an overall lower carbon footprint of the machining process, since the targeted application of such coolants could lead to increased throughput of the machining process. Furthermore, carbon capture techniques could be used to recycle and reuse the $CO_2$ used in the machining process.

In contrast with the sentiments of the previous paragraph, the use of emulsion and oil-based coolants can be extremely problematic from an environmental sustainability standpoint. Whilst mineral oils were once regarded as an inconvenient by-product of petroleum production, this is no longer the case. In fact, it can be reasonably suggested that the vast demand for mineral oils serves as an ancillary driving force for the crude oil industry. Of course, many of the concerns associated with the crude oil industry can be subverted by formulating MWF's that are comprised of oils which are non-derivative of mineral oil, i.e., vegetable oils and synthetic formulations. Indeed, whilst synthetic and vegetable oil-based MWF's are generally more expensive, this statement is largely correct, and in fact, many high performance MWF's are formulated without the use of mineral oil. Despite the improved solution offered by synthetic oil formulations, there nonetheless persists a range of problems that apply to both synthetic, mineral and vegetable oil-based MWF's, not least of which the issue of their ultimate disposal.

Whilst cryogenic MWF's re-enter the atmosphere after a given machining cycle, conventional MWF's are repeatedly recirculated in the machining centre. Although this is a comparatively environmentally frugal system of coolant delivery relative to single use, oil-based coolant strategies, it fails to address the issue of the ultimate disposal of the MWF. This disposal becomes necessary as the coolant is recycled throughout the machine owing to a degradation in quality by means of microbial, and processing by-product contamination [10]. In fact, in a 1988 article, this degradation is explained as an intuitive consequence of the underlying MWF chemistry, whereby "mineral oil base stocks, glycols, fatty acid soaps, amines and other metalworking fluid concentrates" are said to "provide all of the essential nutrients required for growth" of bacterial and fungal organisms [11]. Whilst this ultimate microbial spoilage is necessary to facilitate the ultimate biodegradation of the fluid, it equally accompanies a range of adverse implications for machine operation, some of which include:

- Reduction in emulsion stability.
- Increased propensity to contribute to workpiece, or machine corrosion.
- Decrease in MWF pH (and thereby, increased alkalinity of the fluid).
- Clogging of coolant delivery system including MWF lines and screens.
- Reductions in tool life.
- Development of an unpleasant odour.

In addition to these consequences, microbial spoilage corresponds to an increased incidence of dermatitis and skin irritation; however, this is explored in greater detail in Section 4.2.

Evidently there exists a range of problems associated with the prolonged use of spent coolant. In order to subvert these concerns, machine coolant reservoirs must frequently be drained, cleaned and replenished with fresh MWF, whilst the spent coolant should be disposed of, or recycled in an environmentally cognisant manner. The process of cleaning an MWF system is, however, not without its concerns. A common cleaning protocol involves the use of compressed airlines and/or the administration of synthetic cleaning products, which are often harmful to life [12]. Whilst this use of compressed air is effective in displacing MWF biofilm and clogged particulates, it equally leads to the mobilising of a rancid MWF mist, corresponding to the generally poor ergonomics of the cleaning process. This phenomenon is equally concerning given the often small, or confined working areas associated with machine shop floors, whereby fresh, uncontaminated air is often unable to easily recirculate. Moreover, this cleaning protocol is generally iterative and may have to be repeated several times until the extent of bacterial contamination is satisfactorily reduced (as tested by a dip slide). In addition to the challenging cleaning protocol, the recycling process by which MWFs are treated is typically extremely convoluted and, in general, requires the adoption of a robust separation procedure. Whilst there are currently a range of developments being explored in the treatment of spent MWF (i.e., nanofiltration, ion exchange resins, etc.), the general processing route by which spent MWF's are treated still largely fails to generate a reusable MWF, but rather, a less hazardous waste substance which can be disposed of in a more financially, or environmentally sustainable manner [13]. For these reasons, the use of cryogenic MWFs, which return to the atmosphere post use, is of clear benefit in terms of the relative reduction in environmental, financial and time cost to the manufacturing sector.

### 4.2. Social Implications of $CO_2$ Coolants

As an aside to the environmental limitations of conventional MWFs, their usage is equally accompanied by a range of negative health repercussions, persisting during handling, use and ultimate disposal. Whilst it is true that $CO_2$ usage as an MWF (and elsewhere) creates a non-negligible risk of asphyxiation, this health concern can largely be mitigated by reasonable precaution such as the use of proper ventilation and local $O_2/CO_2$ monitoring systems. By contrast, conventional MWFs present a much broader encompassing range of health concerns, elevated in part by their capacity for contact exposure via a range of media (primarily dermal, ocular and via inhalation). As dermal contact has been shown to account for as much as 80% of the occupational risk of MWFs [14], it is unsurprising that research into the negative health implications of cutting fluids has, historically, been heavily focused upon the deleterious effect of MWFs on skin health. One 2014 systematic review [15] notes currently established causal links between cutting fluids and the following skin conditions: irritant contact dermatitis, allergic contact dermatitis, folliculitis, oil acne, keratosis and carcinomas. Despite these concerns, much of the risk to skin health can effectively be mitigated by the combined protocol of wearing suitable gloves and adopting a thorough hand cleanliness routine. Whilst this strategy is simple, skin conditions such as those outlined previously are often extremely prevalent in the manufacturing sector. This is likely a consequence of poor adherence to the PPE guidelines, an assertion supported by a 2012 survey that found that 82% of safety professionals had observed a failure to use the requisite PPE in their workplace [16]. Clearly it is unrealistic to expect universal adherence to even the simplest of PPE guidelines, and as such, the negative dermal implications of conventional MWF's persist where they could otherwise be avoided by employing cryogenic cooling strategies.

In addition to the development of skin conditions, MWFs are linked to a range of ocular and respiratory conditions of varying severity. In general, the respiratory conditions associated with MWFs exposure are primarily a consequence of the inhalation of a fine particulate mist, which is generated as the MWF evaporates. Thereafter, either the MWF suspension or a fine atomised aerosol of non-aqueous MWF constituents (as water and water-soluble constituents are vaporised) is inhaled by individuals proximal to the machine. The consequences of which manifest as a range of negative health implications largely, but not entirely, focused upon the respiratory tract, including: airway irritation, chronic bronchitis, asthma, laryngeal cancer, alveolitis, and more abstractly, cancers of the pancreas, rectum and prostate [15]. Although these concerns could, in part, be addressed by the implementation of respirators, or even by assuring proper machine shop ventilation, this remains far from the industrial norm, with respirator use in machining typically being confined to grinding scenarios, wherein the goal is simply to filter out machined chips/dust, rather than MWF mist. Although these issues are perhaps the most severe of the negative health implications associated with MWF exposure, eye irritation, though non-permanent, is equally common amongst machinists. This ocular exposure to MWF can be extremely painful, particularly when contact results from direct splashing of the eye (as opposed to airborne particulates) [17], and necessitates the use of safety glasses. Whilst the use of safety glasses should equally be encouraged during cryogenic machining (owing to the risk of swarf projection), it is reasonable to assume that both the respiratory and ocular risks of cryogenic MWFs are invariably reduced.

*4.3. Economic Implications of $CO_2$ Coolants*

Whilst the use of cryogenic MWFs is incentivised by a multiplicity of environmental, operational and social factors, their widespread adoption in industry will of course be subject to their financial viability. Indeed, if a company is to convert to, or supplement their existing MWF strategies with, $CO_2$, there are a range of set-up and running costs that have to be weighed against the relative benefits of the strategy. With regard to set-up, at the moment, each machine must be retrofitted with a $CO_2$ coolant delivery system and both volumetric and Coriolis flow meters (i.e., Fusion Coolant Systems Pure Cut ©); of course, the cost of which is, at this point, relatively significant. With that being said, it is almost always the case that the early adoption and implementation of an emerging technology is costly; however this is not likely to be restrictive in the long term for two primary reasons. Foremost, it seems likely that the initial capital outlay will reduce as machining centres begin to be built from stock with integrated $CO_2$ coolant systems, or equivalently, as the option to incorporate a $CO_2$ coolant system as a manufacturer add-on becomes available (in the same way that high pressure coolant is often available as an add-on). Secondly, and perhaps most importantly, it is clear that if marginal improvements can be made to the productivity of the manufacturing sector, i.e., by reducing lead time via the use of more aggressive feeds and speeds or, equally by reducing machine down time via fewer tool changes, the cost saving incurred by adopters of the technology would invariably far surpass this initial capital outlay.

In addition to the initial capital outlay associated with the purchase of the coolant system, it is also important to consider the operating costs associated with large-scale $CO_2$ consumption and, additionally, the practicalities associated with the supply of suitable tooling. On the topic of $CO_2$ consumption, the cost of small-scale $CO_2$ usage is generally relatively high compared to the low cost of conventional MWFs. As such, it would likely be the case that the machine shop would be supported by a large cryogenic tanker retrofitted to the externalities of the premises. The tanker would thereafter be filled on site by a $CO_2$ supplier. In adopting this strategy, the machine shop would be able to benefit from significant economies of scale relative to small scale purchase of $CO_2$ pallets, wherein pallet rental, refill and delivery costs become significantly cost restrictive. With regard to the availability of tooling, in much of the cryogenic machining research, bespoke tool holders are employed to accommodate the $CO_2$ nozzle. Whilst this is suitable for small scale research, it is unsurprisingly not cost effective. This, however, is again a consequence of the infancy of the technology and the limited demand placed upon tooling manufacturers to produce tool holders capable of $CO_2$ cooling. It is thus likely that, as $CO_2$ cooling strategies become more commonplace, new 'off the shelf' tooling will be developed in concert: a hypothesis supported by patents filed by both SECO [18] and Kennametal [19] to protect their prospective tool holders and the novel $CO_2$ nozzle configurations that they will employ.

Although the initial capital outlay associated with the use of $CO_2$ cooling strategies is non-trivial, it is worthwhile to consider the multitude of ways in which these costs can be recouped. As such, it is important to outline the ways in which $CO_2$ MWF strategies can avoid the operational shortcomings of conventional MWFs and subsequently benefit from improved economic sustainability. With this in mind, it is noteworthy that one of the primary driving forces of cryogenic MWF research is an intent to find operationally superior MWFs, so that tool life can be prolonged. It thus follows that more aggressive feeds and speeds may be employed, and subsequently cycle time significantly reduced. In doing so, it is likely that less energy would be consumed in the machining process, and by proxy, both expenditure and carbon footprint should be reduced in kind.

Another benefit of $CO_2$ MWF strategies is the avoidance of post-process cleaning. This has the potential to incur a significant cost saving as, in many safety critical applications, parts are unable to enter service with a cutting fluid residue. There are a range of reasons for this, primarily focused upon the altered component tribology as well as general cleanliness and ease of handling. These cleaning procedures generally add a significant amount of lead time to the (often) already sedate process of manufacturing safety critical components, and as such, have a clear downstream impact upon the supply chain, and subsequently the cost per component (and ultimate cost of the overall system). This is a factor that can be entirely subverted by the preferential use of cryogenic cooling strategies (in lieu of conventional MWFs), wherein no residue remains post machining, but rather the coolant dissolves and returns to the atmosphere. Moreover, this benefit of cryogenic machining strategies may persist even in cases of machining with $CO_2$ + MQL, where, by design, MQL leaves little to no residue on the final machined component.

Evidently, the range of problems discussed makes for a compelling argument for the deracination of conventional MWFs (in favour of cryogenic fluids). Although the current widespread use of conventional cutting fluids is born of necessity, it is clear that such strategies are not only unsustainable, but equally, are potentially operationally sub-optimal across a range of performance indices (Section 5).

## 5. Applications of $CO_2$ MWF Strategies

The following work has been categorised according to the workpiece materials that are considered in a given article, such that the first section focuses upon the use of $CO_2$ MWFs for the machining of steels, the second section focuses upon titanium alloys and the third considers any other material species that have been the subject of research. The order is of materials is chosen to be coincident with the volume of research that is available on a given material, wherein steels and titanium alloys are subject to the most research in the field. In addition to the research outlined in Sections 5.1–5.3, there are a small number of articles that consider multiple materials [20–22]. As these articles serve as important case studies for use in determining the relative cryogenic machinability of a given material (and do not easily fit into the proceeding sections), they are retained for Section 6.

### 5.1. $CO_2$ Machining of Steels

For almost two centuries, steels have found a broad range of applications owing to their high strength and durability. For these reasons, it is intuitive that a significant portion of the cryogenic research landscape would be focused upon the machining of steel, and indeed this has historically proven to be the case. In recent times, these articles have generally focused upon the use of $CO_2$ cooling to improve the machinability otherwise generated by conventional MWFs. Such research is often motivated by the early success of $LN_2$ as an MWF for the machining of steel. Some of these trials are discussed in the following section, where they have been selected to provide a variety of steel species and to cover a range of research interests within the field of cryogenic machining.

In order to understand the fundamental physics of $CO_2$ as an MWF, it is useful to adopt a reductionist strategy towards the employed machining model. This approach was recently undertaken by Rahim et al. [23], who examined the impact of a $scCO_2$ MWF in the context of the orthogonal cutting of AISI 1045 medium carbon steel. The performance of said cryogenic coolant was thereafter measured across four primary performance indices (cutting force, chip thickness, tool-chip contact length and cutting temperature) and compared to an equivalent machining trial undertaken with MQL (where MQL was chosen as a competitive sustainable machining strategy). The paper notes that the $scCO_2$ machining strategy leads to 5–14% lower cutting forces, 0.5–2.5% thinner chips, 1–10% shorter tool-chip contact length and 15–30% lower cutting temperature than the MQL condition. Each of these variables is heavily correlated to machinability. Foremost, lower cutting forces are beneficial for a range of reasons including improving system dynamics and reducing tool holder deflection (thereby improving dimensional accuracy); thinner chips tend to be more breakable and, as such, pose a significant chip handling benefit. A shorter tool-chip contact length infers lower friction (and thus, superior lubricity). Finally, lower cutting temperature corresponds to the reduced severity of diffusion dominated wear mechanisms and allows the retention of a strong cutting edge. These findings make for a strong mechanistic rationale for the use of $CO_2$ as a cutting fluid, and further, leave a strong foundation upon which research with a higher degree of specificity to real, practical machining set-ups can be built.

A preceding paper by the same authors [24] utilised a similar set-up to trial the effectiveness of $scCO_2$ and $scCO_2$ + MQL as an MWF strategy for the turning of, again, AISI 1045 medium carbon steel. In this trial, the performance of the coolant was quantified according to the cutting temperature, cutting force and the surface roughness. Again, performance was compared to MQL in isolation. The paper noted that the use of a $scCO_2$ coolant was able to generate a reduction in cutting temperature of 6–30%, dependent upon presence (or lack thereof) of MQL, nozzle configuration and proximity to the cut, where the tool cooled with $scCO_2$ in isolation generated a lower cutting temperature. Despite $scCO_2$ in isolation generating the lowest temperature of the three coolant configurations trialled, the $scCO_2$ + MQL strategy corresponded to the lowest cutting force and surface roughness of the three strategies trialled. These findings further build upon the prospective positive impact of the usage of $scCO_2$ as an MWF, or in supplement to the use of MQL.

Whilst AISI 1045 steel is extremely well researched, other steel species are increasingly being considered, one such novel species being that of high thermal conductivity steel (HTCS). As part of their 2017 paper, Mulyana et al. [25] compared the performance of $scCO_2$ to MQL and dry machining in the milling of HTCS. In accordance with the previous research undertaken by Rahim et al. [23], the authors considered the impact of cutting fluid on temperature, cutting force and tool wear in addition to considering the underlying wear mechanisms associated with the machining process (by the assigned MWF). The author noted that across all major performance metrics, MQL, $scCO_2$ and $scCO_2$ + MQL outperformed dry machining and, further, each cryogenic strategy outperformed MQL in isolation. With regard to both cutting force and tool life, the $scCO_2$ + MQL strategy led to the best performance, generating 28–64% lower cutting forces and 60–85% longer tool life relative to MQL in isolation (Figure 2). Moreover, in each of the three performance indices considered in the paper (cutting force, tool life and cutting temperature), elevated $scCO_2$ input chamber pressure corresponded to superior performance. The author reasons that this phenomenon is a consequence of the elevated density of $scCO_2$ and higher volumetric flow rate of MQL.

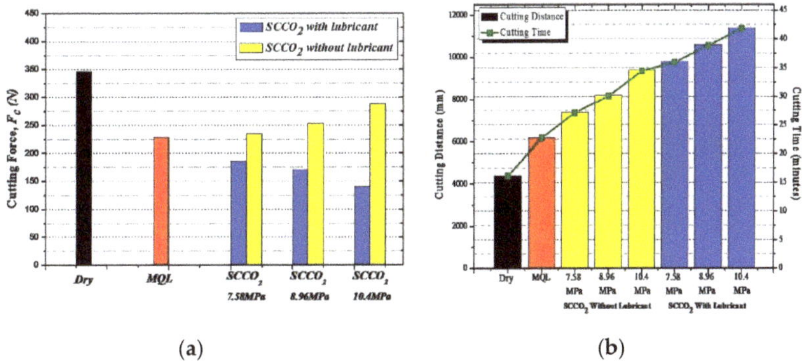


**Figure 2.** Graphs to show the variation in: (**a**) cutting force, (**b**) cutting distance with coolant and $scCO_2$ input chamber pressure. Reprinted with permission from reference [25], Copyright 2000 Elsevier.

In a 2014 conference proceeding, Cordes et al. [26] outlined a novel strategy for the internal, selective, multi-channel supply of $LCO_2$, MQL or a combination of the two. As part of the proceedings, they trialled their tool holder configuration in the machining of 1.4962 stainless steel. The article employed a "radial multiaxial turn milling strategy" (as is common in the roughing of aerofoils), where the extent of tool wear was measured at the end of the roughing cycle of one blade and thereafter compared to dry machining. Ultimately the obtained tool life was inferred to be vastly increased by the application of a $CO_2$ coolant. This inference came as a consequence of the cryogenic cooling strategy corresponding to a 63% lower flank wear (after one pass). Interestingly, the authors then went on to increase the cutting speed and feed rate of the cryogenic machining strategy to the point of which flank wear (after one pass) was equivalent to that which was observed in dry machining. They note that a 37.5% increase in feed rate and a 25% increase in cutting speed were required to elevate flank wear to that of the dry machining trial. In such a scenario, material removal rate (MRR) was increased by 72%, outlining the potential of $CO_2$ cooling in increasing the productivity of the manufacturing sector.

Similarly, in 2019, Fernández et al. [20] published a paper considering the application of $CO_2$ as an MWF in the cryogenic face milling of hard-to-cut materials. As part of their research, they undertook tool life testing on grade EA1N steel when machined with gaseous $CO_2$, indexing its performance against the tool life obtained by a Hysol XF (6–8%) emulsion. The authors noted that whilst the underlying wear mechanism, which led to tool failure, was, in both cases, abrasion, the rate at which the tool became abrasively worn was significantly reduced in the cryogenic trial. Moreover, in accordance with ISO 8688-1:1989, the authors utilised a tool life criterion wherein tool failure was inferred by insert fracture. As such, the article notes that $CO_2$ cooling is able to generate a 175% longer tool life than that obtained by a conventional emulsion strategy. This finding is of particular interest owing to the authors observing a favourable comparison between the performance of $CO_2$ and emulsion (rather than dry machining), which is generally best practice for the machining of most materials.

As is outlined in the proceeding work, much of the research that has thus far been conducted has focused upon the machining of steels, and although there remains a great deal of novelty yet to be explored, the field is becoming increasingly well understood. Whilst this is of great utility in the sense that steel is an extremely popular engineering material, it fails to address the demands of many performance-driven engineering sectors, namely aerospace, motorsport and biomedical. In many of these scenarios, weight restrictions are implicated into the design process and, as such, specific, rather than generic, mechanical properties are of great importance. As these applications are (in general) driven by factors such as performance and lead time (as opposed to cost and durability), they are generally better positioned to become early adopters of technological advancements in the manufacturing sector or elsewhere. For this reason, it is intuitive that early research efforts focus more heavily upon high-performance materials such as titanium alloys, nickel-based superalloys, composites, etc. Consequently, the proceeding work examines $CO_2$ as an MWF in the context of these aforementioned materials.

### 5.2. $CO_2$ Machining of Titanium Alloys

As a consequence of their remarkable mechanical properties, corrosion resistance and ability to function effectively at elevated temperatures, titanium alloys have become commonplace at the high-performance end of the engineering sector. Although these exceptional properties allow titanium alloys to serve a broad range of demanding applications, they equally correspond to poor machinability. For this reason, it is intuitive that titanium alloys would be the focus of a great deal of cryogenic machining research, and indeed this has proven to be the case. Whilst the cryogenic machining of titanium is undoubtedly an area of significant research, many of the articles published focus specifically on Ti-6Al-4V [27–30]. Although this research direction is intuitive based upon Ti-6Al-4V being the most popular titanium alloy (accounting for 50% of titanium production worldwide as of

April 2020) [31], there remains a great degree of novelty to be explored in the machining of more exotic titanium alloys. The proceeding section is dedicated to the current portion of the research landscape focused upon the cryogenic machining of titanium alloys with $CO_2$ MWF strategies.

In much of the cryogenic machining research currently available, there is significant variance in the relative performance of the coolant strategies employed. To give an example, in 2016, Sadik et al. [32] published an article examining the use of $LCO_2$ as an MWF for the cryogenic face milling of Ti-6Al-4V. In the article, the authors note a tool life increase of 250–350% when $LCO_2$ was applied in lieu of an emulsion flood cooling strategy. In contrast, to the strong performance of $CO_2$ in the research of Sadik et al., the later work of Tapoglou and colleagues [7] noted that $CO_2$ failed to compare favourably to conventional (emulsion based) MWF strategies. In their research, Tapoglou and colleagues undertook shoulder end-milling trials on a Starrag LX051, 5-axis horizontal milling centre employing a $LCO_2$ cooling strategy. In contrast with the significant tool life improvements experienced by Sadik et al., Tapoglou observed markedly inferior tool life when utilizing both $CO_2$ in isolation and $CO_2$ + MQL strategies (in comparison to a both a flood and medium pressure through tool emulsion cooling strategy). Tapoglou did, however, note that the use of a $CO_2$ + MQL MWF strategy corresponded to an increase in tool life relative to both MQL and $CO_2$ in isolation.

Further to the variability of the data, An et al. [33] undertook side-milling trials on Ti-64 with three $CO_2$ MWF strategies: $scCO_2$ in isolation, $scCO_2$ with water based MQL and $scCO_2$ with vegetable oil-based MQL. As part of the trial, An and colleagues measured (and simulated) flank wear, cutting torque and surface morphology, indexing performance against dry cutting conditions. The authors noted that the $scCO_2$ strategy generated the most flank wear, followed by dry, $scCO_2$ with water-based MQL and, finally, $scCO_2$ with vegetable oil-based MQL. In the paper, An et al. hypothesises that the elevated wear state experienced during $scCO_2$-only machining strategy is likely a consequence of increased friction at the tool-workpiece interface, thermal cyclic fatigue and adhered chips accumulating on the tool. Interestingly, the extent to which flank wear was reduced by the $scCO_2$ + MQL strategies, relative to dry machining, was extremely marginal, wherein the difference in flank wear between the three strategies fell within the variance of the data set.

It is clear that, whilst there are non-trivial differences in the experimental design of the three papers in question (differing cutting speeds, $CO_2$ rather than $LCO_2$, end-milling/side-milling rather than face milling, etc.), their nonetheless exists a clear variance in the findings of Sadik, Tapoglou and An et al. It is thus the opinion of the author that this is illustrative of the conditionality of the machinability outcomes that are achieved when employing $CO_2$ MWF strategies. Further to this point, it is not immediately obvious as to which variables are of the most importance in manipulating performance indices such as tool life. For this reason, a compelling case is made to work towards establishing an appropriate processing window for $CO_2$ MWF strategies in a range of different machining context. In doing so, the processing parameters that allow the cryogenic media to offer comparative (or superior) tool life to conventional means would be outlined, thereby allowing manufacturers to employ the most appropriate MWF strategy for a given application.

One of the few articles to consider non Ti-6Al-4V titanium alloy ($CO_2$) machining was published in 2018 by Kaynak and Gharibi [34]. The paper was published with the outlook of examining the impact of $LN_2$ and $CO_2$ as MWF's for the turning of Ti-5553 and as such, undertook trials with the following performance indices in mind: cutting temperature, tool wear and dimensional accuracy. Thereafter, the performance of each of the two media was compared against each other, in addition to a dry machining baseline. The authors observed that in both the $CO_2$ and $LN_2$ trials, the measured maximum temperature was shown to be significantly decreased relative to the dry machining trials, whilst $LN_2$ cooling generally corresponded to a marginally lower maximum temperature than was observed in $CO_2$ machining. Moreover, the authors observed a general trend in both cryogenic media reducing the extent of flank wear, noting that this effect was particularly exaggerated at elevated cutting speeds, where, at the maximum trialled cutting speed, $CO_2$ cooling lead to a 22% reduction in flank wear (relative to dry machining) and $LN_2$ a 59% reduction in flank wear. In addition, although both of the cryogenic media trialled led to improvements in dimensional accuracy, $LN_2$ cooling corresponded to a lower maximum deviation from the nominal diameter than the $CO_2$ strategy. Whilst this article generally points to the superior performance of $LN_2$ (relative to $CO_2$) in this context, it is important to note that both cutting force and feed force were most optimally reduced by the $CO_2$ cooling strategy, whereby a noteworthy reduction was observed across the majority of cutting speeds. This reduction in feed force provides further evidence of the excellent lubricity, which can be obtained by employing $CO_2$ cooling strategy in a machining context.

In addition to the work of Kaynak and Gharibi, Machai and Biermann [35] undertook cryogenic OD turning trails on Ti-10V-2Fe-3Al. The authors compared the tool life performance of $CO_2$ snow (as a coolant) to an emulsion, flood coolant strategy. Moreover, Machai and Biermann went on to outline the transient wear progression of the cutting insert when subject to each media, as well as discuss the inherent mechanisms associated with said wear progression. The authors observed that the $CO_2$ cooling strategy corresponded to increased tool life at each of the cutting speeds trialled (Figure 3); moreover, whilst the emulsion-cooled tool was subject to significant notch wear, the $CO_2$ cooled tool was entirely devoid of notching. Moreover, $CO_2$ cooling additionally contributed to reduced feed force during later machining passes, despite initially generating elevated radial forces, findings that have been observed elsewhere in the literature. In addition to the positive tool wear implications of $CO_2$ cooling, Machai and Biermann observed that, during the later passes with emulsion cooling, burrs formed at the tools' exit from the cutting zone; in contrast, no such burrs were observed during the $CO_2$ machining trials (Figure 3). The authors went on to suggest that the presence of burrs (or lack thereof) was a consequence of the periodic impact of a worn, notched tool, and in this sense, the lack of burr formation in the $CO_2$ trials is an intuitive finding.

In conclusion, the current landscape of the literature around titanium alloy machining remains, at this stage, inconclusive. Although some of the variability between the current research is undoubtedly a result of the differences in employed processing parameters (Section 6), the significant contrast between otherwise experimentally similar remains a challenge that must be addressed by the research community. Clearly, it is possible to establish a range of operating conditions with which $CO_2$, or $CO_2$ + MQL could be rendered the optimal MWF strategy for a given material; however, in the available research, the efficacy of $CO_2$ cooling remains subjective. Further work should thus focus upon establishing a suitable operating range (for multiple titanium alloys), whereby $CO_2$ is able to function as a suitable coolant. Moreover, should future research become more aligned with the findings of Tapoglou, it would undoubtedly be beneficial for researchers, who have an interest in the adoption of cryogenic machining technologies to further develop the consortium of literature studying the auxiliary benefits of $CO_2$ as an MWF, rather than simply focusing upon tool life. Alternatively, it may also be beneficial to consider directing further research towards the cryogenic machining of alternative titanium alloys or even entirely new material species. With this recommendation in mind, the following section will consider current examples of $CO_2$ usage as an MWF for the machining of various other material species.

### 5.3. Additional Applications of $CO_2$ MWFs

Although steel and titanium alloys are the focus of a large portion of the current cryogenic machining literature, $CO_2$ is speculated to function as a potentially efficacious MWF strategy for a range of other materials. As an example, in nickel-based superalloys, the combination of high hardness and low thermal conductivity leads to excessive heat forming in the cutting zone, and by proxy, the tool. This, in turn, leads to a litany of problems, not least of which are reduced tool life and thermally induced geometric distortion. Despite presenting a clear cause for concern, these challenges are certainly not unique to nickel-based superalloys, and in fact are commonplace in titanium alloy machining. Given this realisation, it is rational to extrapolate the potential benefits of $CO_2$ in the machining of titanium alloys to a prospective application in the machining of nickel-based superalloys. Whilst this reasoning is undoubtedly useful in deciding upon the most likely avenues for future research, prior machining trials on mechanistically similar materials of course cannot be regarded as a direct citation of the efficacy of $CO_2$ for new, unproven material species. For this reason, the use of $CO_2$ as an MWF for the machining of nickel-based superalloys, whilst currently in its infancy, is proving to be a burgeoning field of cryogenic machining research.

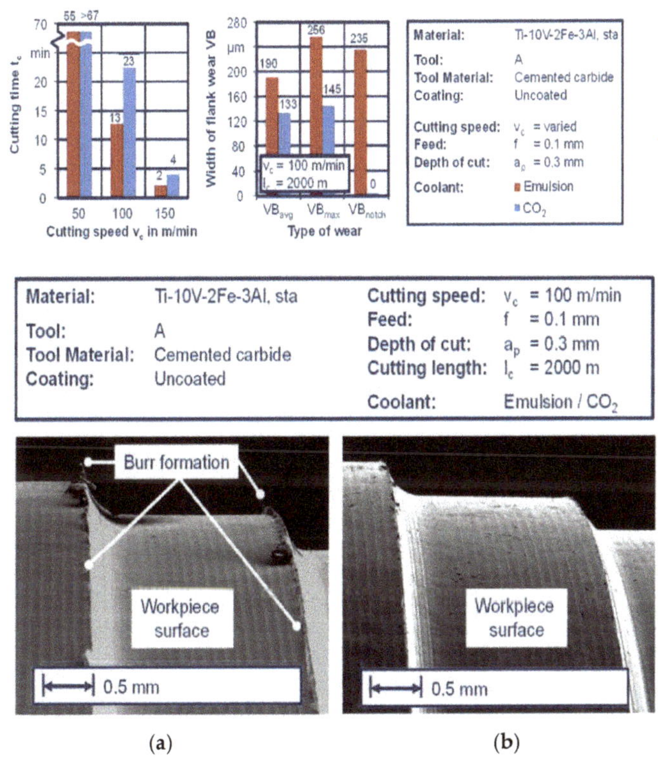


**Figure 3.** Tool wear and burr formation during the cryogenic machining of Ti-10V-2Fe-3Al. (a) Emulsion; (b) $CO_2$. Reprinted with permission from reference [35], Copyright 2011 Elsevier.

Whilst superalloys have, in recent times, been the subject of an increasing number of promising cryogenic machining trials (with $LN_2$) [36], $CO_2$ trials are, despite being supported by a strong mechanistic rationale, relatively unexplored. Of the limited research that is currently available, results are generally mixed. In 2016, Busch et al. [21] undertook turning trials with coated CNMG 120,408 cemented carbide inserts on Inconel 718. The authors went on to compare the performance of both $CO_2$ in isolation, and ADL + $CO_2$ to a high-pressure emulsion coolant strategy, as is regarded as the current best practice for the machining of this material. The performance of the three coolant strategies were thereafter assessed primarily according to tool life, specific energy consumption and general machined surface quality. Whilst the authors noted a generally invariable energy consumption across each of the trialled media, both the $CO_2$ and ADL + $CO_2$ strategies generated markedly lower tool life relative to the use of high-pressure coolant. In addition to the poor tool life, the authors observed no clear chip handling benefit of ADL + $CO_2$ and further went on to note the presence of adhered material at the machined surface when $CO_2$ cooling was applied in isolation. Whilst the article makes note of the operational convenience of retrofitting a $CO_2$ coolant delivery system, and equally the benefits of oil residue free machining, the paper states that any scope for the future efficacy of these $CO_2$ cooling strategies will be reliant upon future optimisation to the delivery of the cryogenic media (and the ADL).

In addition to these findings, Patil et al. [37] also conducted $CO_2$-assisted machining trials on Inconel 718. In accordance with the work of Bush et al. [21], the authors employed an OD turning model utilising TiAlN-coated SNMG120408-cemented carbide inserts. Thereafter, the authors examined both cutting forces and surface phenomena over a range of feeds and speeds, ultimately comparing the performance of $CO_2$ cooling to a dry machining condition. Whilst the authors observed an increase in both cutting and feed forces when $CO_2$ assistance was applied, this effect was observed to be of a much smaller magnitude than the impact of cutting speed and feed rate on tool force. Moreover, when compared to dry cutting, the presence of a $CO_2$ MWF led to a reduced surface roughness. This effect is visible across all feeds and speeds; however, it was most pronounced at 0.1 mm/rev, 100 m/min. When surface integrity was further analysed (by way of taking microhardness measurements of the machined surfaces), the $CO_2$ assisted trials corresponded to an increase in surface hardness (relative to bulk), whilst the dry machined surfaces were of equal or lesser hardness to the as-supplied bulk. The authors note that this finding is likely a consequence of the cold work hardening of the workpiece in the presence of $CO_2$, presumably in lieu of the elevated microstructural recovery, which would otherwise be observed at elevated temperatures (as in dry machining). This increase in hardness is regarded, by both the authors of the study and others [6], to be a desirable outcome for improved surface integrity. Further, where increased surface and subsurface hardness has previously been observed (in the $LN_2$ machining of Inconel 718) [38], it has been accompanied by higher compressive residual stress, which makes positive implications for fatigue life (via the inhibition of crack propagation).

Given the research of Busch and Patil [21,37], the holistic impact of $CO_2$ coolants in the machining of Inconel 718 remains uncertain. Clearly, future research is necessary to optimise the operational parameters used in the $CO_2$ machining of each workpiece material. This will likely involve optimising variables such as nozzle position, coolant flow rate, feed rate, cutting speed, tool material, tool geometry, etc. over a range of machining operations (Section 5). Moreover, whilst material classifications are helpful when analysing the density of research in a given domain, there remains no guarantee that similar species of material will behave similarly in an equivalent cryogenic machining context. In the case of superalloys specifically, although iron-, nickel- and cobalt-based superalloys are often employed in similar roles, their machinability is likely to vary in accordance with their metallurgical differences. Whilst this variability is likely to be most noticeable amongst dissimilarly based alloys, it will undoubtedly persist within each species. As an example, nickel-based superalloys used in aero-engine disks (i.e., RR1000) may, and likely will, exhibit markedly different machinability to alloys used in aero-engine turbine blades (i.e., Inconel 738LC) owing to their variation in mechanical, and thermal properties. For this reason, whilst a $CO_2$ MWF may not be regarded as efficacious for one alloy, it may be wholly suitable for another.

Whilst the cryogenic machining of superalloys is a burgeoning field of research, there remains a mechanistic rationale for many other materials. In fact, many of the most thought-provoking avenues for future cryogenic machining research will invariably focus upon materials that otherwise are ineffectively machined by way of their material properties; one pertinent example being that of viscoelastic polymers such as polydimethylsiloxane (PDMS) or ultra-high molecular weight polyethylene (UHMWPE). In general, the high elasticity and susceptibility to adhesion make for a material that is challenging to machine by conventional means. These properties correspond to a litany of adverse machining outcomes including geometric deformation of the workpiece material and undesirable chip formation. Moreover, the propensity of polymeric material to adhere to the cutting tool often corresponds to an increase in effective edge radius, which in turn corresponds to a ploughed, rough surface (owing to the size effect). Despite this, with the rise of the burgeoning field of microfluidics, small scale, geometrically accurate polymeric chips are becoming increasingly necessitated, not least of which within the pharmaceutical industry, wherein they serve a range of roles including drug screening and metabolic research [39]. Currently, microfluidic chips are manufactured via micromolding techniques, whereby the micromolds are photolithographically formed in the polymeric substrate. Whilst this technique allows for accurate fabrication, the process is not easily customisable. As such, future iterations of a chip, or equally new chip geometries, generally necessitate the development of a new, bespoke mould. Given this reality, and with the rise of micromachining technology, cryogenic machining may prove to become a cost-effective strategy for the manufacture of viscoelastic polymers.

Where the cryogenic machining of polymers has been undertaken, the strategy generally involves either a workpiece pre-cooling strategy or, further, the submersion of the polymer during the machining process. This strategy takes advantage of the changed chip-formation mechanisms that occur below the glass transition temperature ($T_g$) of the polymer. In a 2015 paper by Aldwell et al. [40], UHMWPE was submerged in a vat of liquid nitrogen for a period of 24 h prior to turning. Thereafter, cutting force, surface roughness and chip morphology were measured in both the pre-cooled and room temperature billets. As a consequence of the elevated elastic modulus of the polymer (sub-$T_g$), cutting force was elevated when cryogenic-assisted machining was applied. Ordinarily, increased cutting force has negative implications for productivity; however, as tool life is of limited relevance in the small batch machining of polymers, surface quality is a much more vital parameter. In this regard, cryogenic pre-cooling was shown to be efficacious from a surface finish perspective, whereby mean surface roughness was reduced by more than 20% in the pre-cooled material. In addition, the authors generally observed a greater propensity to produce chips (rather than dust) when $LN_2$ pre-cooling was employed. Although this effect was more apparent when worn tools were used, the finding is of potential utility to the manufacturing sector as, an inability to form chips is, in general, accompanied by an inability to dispel heat, which, particularly in the machining of polymers, can become extremely problematic.

In addition to the research of Aldwell and colleagues, Kakinuma et al. published an article employing the micro-end milling of cryogenically cooled PDMS [41]. The authors utilised a complete emersion strategy wherein the both the (single crystal diamond) milling cutter and PDMS workpiece were submerged in a vat of liquid nitrogen; thereafter, a ductile mode milling strategy was employed with a varying depth of cut. By employing such a strategy, the PDMS workpiece is held below the glass transition temperature such that it retains properties of low adhesion and elasticity. In abstract, this allows the thermal deformation (which is implicit during a machining operation) to be suppressed, and thus form inaccuracies to be minimised. When machining trials were conducted, the authors observed that the large quantity of heat generated in the cutting operation was sufficient to vaporise the proximal $LN_2$ and, subsequently, that the generated surface transparency was unsatisfactory. In response to this realisation, Kakinuma and colleagues went on to additionally apply cryogen directly to the cutting zone by way of $LN_2$ jet. In doing so, heat flux into the workpiece was minimised, and subsequently a high-quality surface with lower opacity and surface roughness was generated (Figure 4). Given this success, the authors went on to machine a series of microfluidic chips with the outlined approach, noting benefits over micromolding processes such as a comparative ease in forming stepped channels and a generally far reduced lead time.

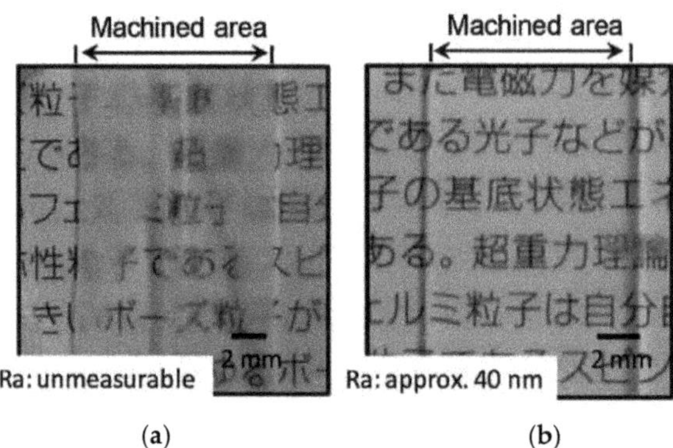



(a)         (b)

**Figure 4.** A demonstration of machined surface quality in micro-milled PDMS. (**a**) $LN_2$ immersion method; (**b**) $LN_2$ immersion + external supply cooling method. Reprinted with permission from reference [41], Copyright 2012 Elsevier.

Ultimately, although cryogenic MWFs are mostly well researched in the context of steels and titanium alloys, it is apparent that their early success has birthed widespread interest. In addition to the strides that have been made in researching the cryogenic machining of superalloys and polymers, cryogenic methods have already been applied to magnesium alloys [42], cobalt-chromium alloys [43] and composites [44], garnering generally positive results. Moreover, with the current pace of research, it is likely that trials will expand further into various other niche materials and manufacturing processes. Fortunately, in this regard, various avenues of research still exist. As an example, one material that may be suitable for future machining trials is that of oxygen free copper (OFC). Whilst OFC parts are in demand as high-conductivity electronic components, they are often extremely difficult to machine owing to poor chip handling and subsequent surface integrity. Given the prior use of cryogenic-assisted machining for chip-breaking purposes in other materials, it is reasonable to foresee future machining trials on OFC. In addition to examining novel materials, it may be equally worthwhile to consider alternative manufacturing process (to those commonly employed in a cryogenic machining context) as an example, rather than turning and milling, broaching has the scope to be significantly augmented by the addition of cryogenic cooling. By way of an explanation, in many articles researching the effects of $CO_2$ as an MWF, benefits (over flood coolant) are often relatively incremental. This incremental improvement in operational efficiency may be insufficient to justify the purchase of a $CO_2$ delivery system to an SME manufacturing company with a turning or milling focus. In contrast, broaching operations generally feature extremely expensive tooling, and create a great deal of added value. As such, if incremental process improvements can be made (i.e., to tool life) by cryogenic assisted machining, ample financial motivation may be available for early adoption of the technology. To summarise, cryogenic-assisted machining is an extremely interesting field of research, which, given its current promise, has scope to become a disruptive technology in the near future.

## 6. Practical Factors Impacting the Performance of Cryogenic and $CO_2$ MWFs

### 6.1. Workpiece and Tool Material

When reviewing cryogenic machining research, one of the most superficially apparent variables that dictates trial performance is the choice of workpiece material. In fact, it is reasonable to assume that the perception of success of a given research paper is, in absence of other compelling evidence, generally thought of as a consequence of the suitability, or lack thereof, of said workpiece material. Whilst this perception is somewhat short-sighted (given the extent to which cryogenic assisted machining trials can vary), the choice of machined material invariably plays a significant role in determining which MWF strategy is preferential for a given trial. Given this fact, it is unsurprising that this most fundamental dependency has been observed since the earliest iterations of cryogenic machining research. To be specific, the aforementioned work of Uehara and Kumagai [3,4] notes that a strategy of cryogenic pre and intra workpiece cooling lead to tool life improvements in both carbon steel and commercially pure titanium workpieces despite presenting no clear performance benefit in the context of stainless steel.

Needless to say, the trend of differing machining trial performance by workpiece material persists into the modern day and is in fact a topic of discussion in most, if not all, journal articles published in the field (not least of which this one). Despite this, it is also true that conclusions of workpiece suitability should not be equitably drawn from each cryogenic machining article on the basis of a lack of experimental control. In order to illustrate this point, take the example of two machining trials that both employ jet cooling, utilise equivalent feed rates, cutting speeds and depths of cut and consider two different workpiece materials (one per paper). In this example, it may seem appropriate to draw like for like comparisons between the two papers, particularly given their stark similarity. Whilst the results of these two papers would likely share some commonality, the problem with a direct comparison in this context is the invariable fact that across some parameters (i.e., tool nozzle diameter, or means/standards of measuring tool life), the employed experimental practice of the two papers will differ. Taking this into consideration, the most academically rigorous way to analyse the relative cryogenic machinability of different materials would involve the analysis of research papers that examine various material species in the one paper, or equivalently, those papers published as part of a research chronology by one set of authors (where DOE is unaltered). For this reason, Section 5, whilst citing papers utilising various materials, does not effectively serve to illustrate which materials are suited to cryogenic machining strategies. With this in mind, the following section employs such an approach with the outlook of understanding the conditionality with which different materials can be cryogenically machined.

One paper that considered various materials was the aforementioned work of Fernández et al. [20]. In their article, Fernández and colleagues undertook the cryogenic face milling of three challenging-to-machine materials: Grade EA1N steel, gamma Ti-Al and Inconel 718. Interestingly, the authors observed markedly different (tool life) performance across each trialled material. In the case of Inconel 718, both the $CO_2$ and emulsion strategies lead to equivalent tool life, such that the tools in both trials failed in each case after two machining passes (and a tool life of 5.3 min). Moreover, in both trials, the tools failed via a similar cracking mechanism said to be a consequence to the combined effects of abrasion adhesion and microchipping (Figure 5). In contrast, during the gamma Ti-Al trials, the authors observed that conventional emulsion cooling led to tool failure after the first machining pass, whilst the $CO_2$ cooled tools generally had not failed prior to undertaking the second machining pass and thus achieved a projected 100% increase in tool life. This performance was shown to extend to the EA1N steel trials, wherein the cryogenic cooling strategy led to a 175% increase in tool life relative to emulsion flood cooling. Moreover, despite this significantly reduced rate of tool wear, wear mechanisms were, in both cases, found to be predominantly abrasive despite being markedly more aggressive in the emulsion cooled trials. Ultimately the results of this paper would seem to suggest that both EA1N steel and gamma Ti-Al may be more suitable for $CO_2$-assisted machining than Inconel 718; however, this should be developed upon with further research.

In addition to the work of Fernández and colleagues, an earlier paper by Wang and Rajurka [22] was published examining the cryogenic machinability of Ti-6Al-4V, Inconel 718, Reaction Bonded Silicon-Nitride (RBSN) and (commercially pure) Tantalum. The paper employed a simple turning model with two different tool materials, Cubic Boron Nitride and Cemented Tungsten Carbide (WC-8 wt%Co). In contrast to the more recent research of Fernández, the authors employed a $LN_2$ remote-cooling strategy wherein $LN_2$ was recirculated in close proximity to the cutting edge via a network of copper tubes in a proximal tool cover. As part of their research, the authors primarily focused upon tool life, in addition to cutting temperature (via a mounted thermocouple), surface roughness and cutting force.

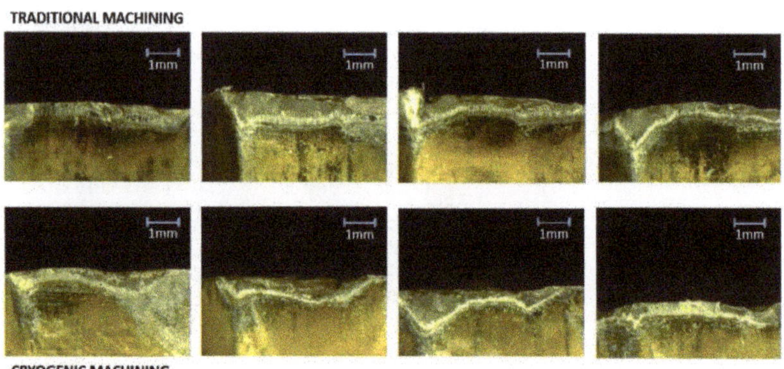



**Figure 5.** Worn inserts (at tool failure) during the machining of Inconel 718. Reprinted from reference [20].

During the machining of RBSN, three different varieties of CBN tools were employed, in all cases showing prolonged tool life during $LN_2$ remote cooling (relative to dry machining). Despite this commonality, not all tools were equally positively implicated by the $LN_2$ cooling strategy (Figure 6). Specifically, despite wearing most aggressively during dry machining, the CBN50 tool (produced by Sandvik) was most drastically impacted by $LN_2$ cooling, wherein after completing the 158 mm cutting length (which was allocated to each tool), a sub 0.5 mm flank wear was observed. These promising results equally persisted during the turning of Ti-6Al-4V (with H13A WC-8 wt%Co tools), whereby an over five times increased cutting length was required of the $LN_2$ cooled inserts to generate the equivalent flank wear to an 'oil cooled' equivalent.

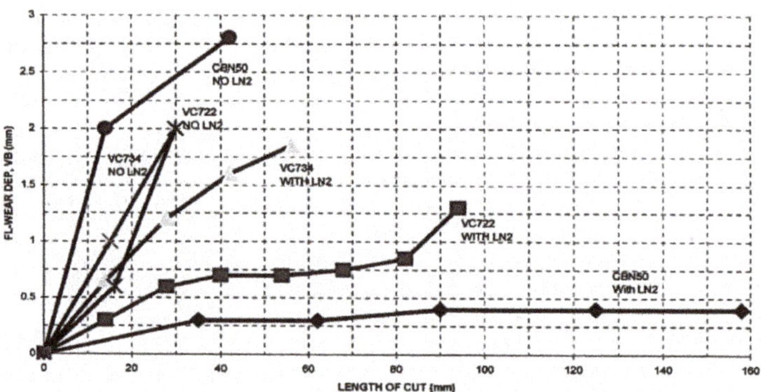



**Figure 6.** Wear progress curves during the machining of RBSN with three different CBN tools. Reprinted with permission from reference [22], Copyright 2000 Elsevier.

In addition to the trials conducted on RBSN and Ti-6Al-4V the author also considered the cryogenic machinability of Inconel 718 and Tantalum (with a H13A tool), again observing a markedly reduced rate of tool wear when $LN_2$ cooling was employed (in lieu of dry machining). Despite this clear trend, it is difficult to accurately extrapolate tool life from the figures present in the report. In the case of Tantalum, at the points where the tools were equivalently worn, the magnitude of flank wear was extremely low, which of course raises questions as to the efficacy of any conclusions which are drawn. With this limitation in mind, it was nonetheless observed that the $LN_2$-cooled tool required an approximately 90 mm cutting length to reach a flank wear of approximately 0.15 mm, whilst the uncooled tool reached such a wear state after only 20 mm (with $LN_2$ cooling). Unfortunately, however, when Inconel 718 is considered, the scope to predict the impact of cutting fluid on tool life is further hampered, wherein the available data are not at any point coincident and thus insufficient to make a sober estimate.

Whilst there are a limited number of papers to consider multiple workpiece materials, there are, of course, a few other examples of similarly inclined research, one such paper being that of Busch et al. [21], who conducted $CO_2$-assisted OD turning of Inconel 718 and Ti-6Al-4V. In contrast to some of the earlier promising research, Busch and colleagues observed generally poor performance markers for $CO_2$ cooling. The key observation made in this article (from a cryogenic machining perspective) was one of reduced tool life when $CO_2$ cooling was used in lieu of high-pressure (HP) coolant. This phenomenon was somewhat more pronounced during their Inconel 718 trials. Specifically, the tool life obtained during the $CO_2$ machining of Inconel 718 was almost 70% reduced relative to HP coolant, whilst the equivalent figure was closer to 60% in the Ti-6Al-4V trials. In this regard, the findings of Busch and colleagues are largely in keeping with the work of Wang and Rajurka, as both papers contribute to a future lack of candidacy of Inconel 718 for cryogenic machining research.

In summation, current research implies that Inconel 718 has, so far, not shown itself to be a particularly compelling candidate for future cryogenic machining research owing to generally poor susceptibility to cryogenic cooling, at least from a tool life perspective. Mechanistically, this may be a consequence of the markedly low thermal conductivity of Ni-based superalloys impeding the success with which $CO_2$ is able to effectively cool the cutting zone; however, this of course must be supported by more substantive data. Importantly however, although Inconel 718 is not shown to be particularly promising in any of the previously considered research, the suitability or lack thereof of superalloys remains to be seen. In contrast, cryogenic machining strategies are, at this stage, shown to have some scope to improve the machinability of both titanium alloys and steels.

*6.2. Feeds and Speeds*

One of the most critical aspects in machining is the appropriate selection of cutting parameters in order to optimise the overall efficiency of a production process. This selection must take into consideration not only the material that is being processed, but also other factors including the cutting tool characteristics and more importantly the lubri-cooling technique employed and the delivery method. Different cooling techniques would result in different cutting temperatures and loads on the cutting tool; therefore, the cutting parameters would need to be adjusted to allow for optimal cutting conditions. Given this complexity, it is often desirable to encompass various feed rates and cutting speeds centred around estimated best practice data established in the literature, or via practical machining guidelines.

It is undeniable that the success (or lack thereof) a machining trial is highly contingent upon the operational parameters employed during said trial. It follows that the way in which cryogenic coolants are perceived in a trial could, and likely would, equally be dependent upon the use of appropriate feeds and speeds. It is also apparent that any feed rate and cutting speed data will be incredibly specific to the material being machined. As way of an example, one online machining data base [45] recommends a cutting speed of 175 m/min for the turning of austenitic stainless steel, whilst equivalently recommending speeds as high as 860 m/min for wrought aluminium alloys. Although this effect is quite clearly exaggerated by the choice of materials, similar implications would undoubtedly persist across more (mechanically) similar species, i.e., titanium and nickel alloys. Given this, it is thus crucial that an extensive dataset of cryogenic machining performance is developed for a range of feeds, speeds and materials. Unfortunately, such data are not currently available in the public domain, and as such, investment into $CO_2$ cooling systems remains speculative until the research landscape is able to catch up.

### 6.3. Tool/Nozzle Design

When designing an experiment within the domain of machining science, one of the most important variables to optimise is the cutting tool. Often this takes the form of optimisation across parameters such as cutting insert material, tool geometry, tool holder design or nozzle configuration. Practically however, some of these variables are more easily established during experimentation. For example, if an experiment is not being run with the express intent of establishing novel tool materials (for use in a given context), it is likely that the chemical composition of the tool material is decided upon by external actors. To give a specific case, it is often taken as a given that the current best practice for the machining of challenging materials such as titanium and nickel alloys are cemented carbide tools. Of course, there remains a significant scope to optimise the specifics of the tool chemistry (e.g., WC-(Ti,Ta,Nb)C-(Co,Ni)); however, the nuance of tool composition is often removed from the experimental process by instead following the recommendations of a tool supplier (i.e., Sandvik, Seco or Kennametal), who generally offer tool compositions that have been optimised via extensive R&D to meet a range of similar machining challenges. With this in mind, and given both the relative complexity and financial (and informative) barriers to entry associated with tool composition optimisation, this section considers composition only in passing, instead focusing upon variables such as nozzle, tool and tool holder geometry.

Given this focus, one such article that considers the problem of tool design was recently published by Shokrani and Newman [46]. As part of their research, the authors focused upon the relationship between cutting insert geometry and tool life in the context of the cryogenic end milling of Ti-6Al-4V. Specifically, Shokrani and Newman mapped the relationship between both the rake and primary clearance angle on tool life, tool wear and surface roughness. The authors utilised 12 mm diameter solid WC end mills coated with TiSN-TiN to a thickness of approximately 3 μm. They employed a constant $LN_2$ nozzle position and flow rate (20 kg/h) throughout the trials. Within this experimental design, Shokrani and Newman noted the positive implications of both primary clearance angle and rake angle (Figure 7) on tool life. Moreover, they observed that, although the impact of increased clearance angle persisted across each rake angle, the effect was most pronounced at the highest rake angle, and likewise, the correlation between rake angle and tool life was most exaggerated at the higher clearance angle. Although these observations are of clear value, the authors do however note that (despite not being manifested in their trial) excessively increasing either rake, or clearance angle, can lead to weakening of the cutting edge and ultimately premature tool failure.

In addition to the tool life implications of manipulating clearance and rake angle, the relationship between surface roughness and insert geometry was also considered. In order to realise these goals Shokrani and Newman took readings at the start of experimentation in order to remove the impact of tool wear on average surface roughness (Ra). In doing so, the authors again noted the positive implications of both increased primary clearance and rake angles such that a tool with a sharper cutting edge corresponded to an equivalent or reduced surface roughness across the entirety of the data set (Figure 8a). Whilst it is difficult to qualitatively assess the relative effect of rake and primary clearance angle on this phenomenon, it is true that the lowest surface roughness was generated at a rake angle of 14° regardless of the primary clearance angle. Given this observation, it is also apparent that the impact of primary clearance angle is most significant at a rake angle of 12°. Whilst these observations make a compelling case for the use of steeper rake and clearance angles, it is important to note that the true impact of cutting edge geometry (on the surface mechanics of the machined workpiece) is not fully captured via Ra measurement. Rather, it is often true that two surfaces with an equivalent centre line average will exhibit varying degrees of asperity, and thus vastly different tribological performance [47] (Figure 8b).

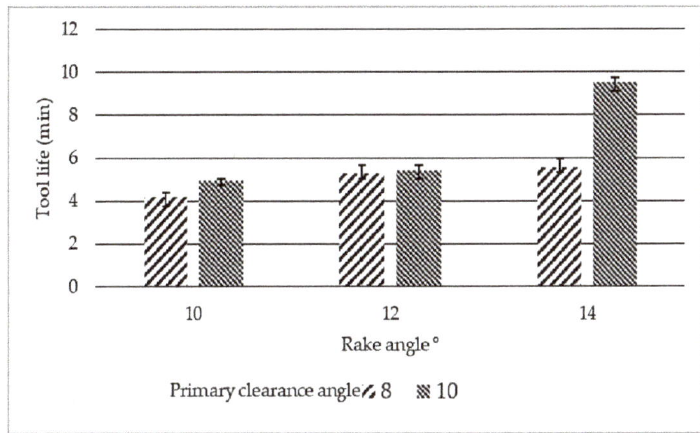

**Figure 7.** A graph to show the relationship between tool life and tool geometries during the cryogenic end milling of Ti-6Al-4V. Reprinted from reference [46].

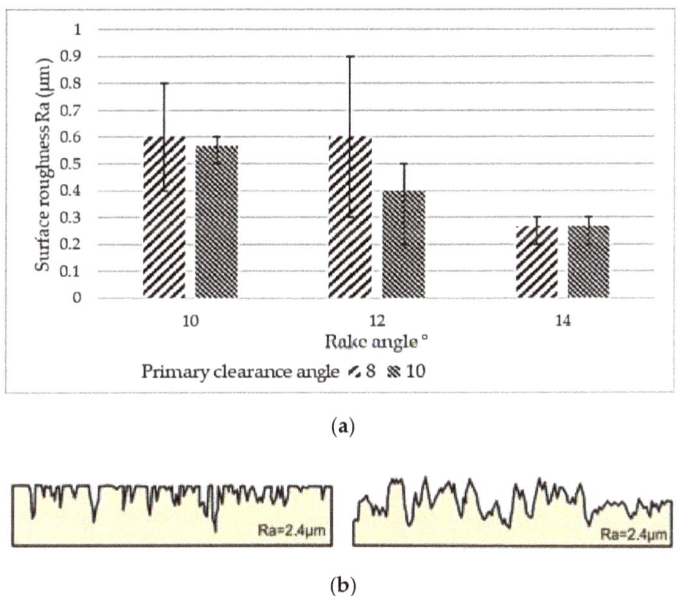

(a)



(b)

**Figure 8.** (a) A graph to show Ra and tool geometry during the cryogenic end milling of Ti-6Al-4V. Reprinted from reference [46]. (b) Schematic illustration of machined surfaces with equivalent values of Ra.

Furthermore, although the work of Shokrani and Newman provides an extremely useful case study to illustrate the importance of tool geometry in cryogenic machining, the research is not without limitation. Foremost, the authors chose to prematurely terminate machining trials (by failing to test rake angles in excess of 14° and clearance angles above 10°) at a point such that the negative implications of excessive clearance and rake angle were not realised. In this sense, their research may wrongly create the inference that a sharper cutting edge invariably improves performance, or equally, it may encourage future research to more aggressively select an insert than can be reasonably supported by their data set. Equally, although the variance in the tool life data was reasonable, the magnitude of the error bars in the surface roughness plot is somewhat excessive relative to the magnitude of the recorded data. For this reason, conclusions as to the most appropriate tool geometries (to optimise topology) should be made with a degree of reservation.

In addition to the efforts made in optimising tool geometry, Pereira and colleagues [48] also undertook research into the nozzle design for cryogenic machining. In contrast to the work of Shokrani and Newman, Pereira et al. instead focused upon the development and optimisation of CFD (ANSYS Fluent) simulation-derived nozzle outlets for $CO_2$ + MQL cooling, focusing primarily upon nozzle diameter. Thereafter, the authors validated their models with experimental trials, and ultimately developed a prototype nozzle in accordance with the data generated in the paper (taking as a given that a $CO_2$ velocity of 325 m/s is necessitated to aid in the cutting process). Given these research parameters, the authors inputted three nozzle diameters into their CFD model of 0.5, 1 and 1.5 mm. Pereira and colleagues thereafter qualified the relative performance of each nozzle geometry according to normal average velocity at a distance of 20 mm from the outlet, observing that (in the simulation) the $CO_2$ velocity was best retained with the 1.5 mm nozzle. In accordance with this observation, when experimentally verified, the 1.5 mm nozzle obtained the greatest spray distance of 40 mm (compared to 18 mm and 10 mm for the 1.0 mm and 0.5 mm nozzles, respectively). With this in mind, Pereira et al. went on to reference the research of Park and colleagues [49], who observed that the maximum area fraction coverage of MQL droplets occurs at 30 mm from the nozzle tip. Given this figure, the authors determined that the most viable nozzle diameter of the three trialled was 1.5 mm owing to the capacity to most closely resemble the optimal distance for MQL coverage and general velocity profile.

Having determined that a nozzle diameter of 1.5 mm was most appropriate in a cryogenic cooling context, the authors went on to manufacture two nozzle adaptors of equal diameter, proximity and general architecture with differing outlets, one of which featured a convergent $CO_2$ outlet whilst the other instead employed a converging diverging nozzle. The convergent nozzle was chosen with simple volumetric continuity in mind, whilst the convergent divergent nozzle was employed with the outlook of exploring the impact, or lack thereof, of compressibility on the fluid dynamics at tool exit. Given the two designs, the authors first inputted the nozzle exit geometries into their CFD simulations followed by experimentally verifying the most suitable nozzle. As part of their simulations, they observed that the use of a convergent divergent nozzle was able to create a much greater exit velocity (475 m/s as opposed to 400 m/s) at the expense of a larger spray spread (15.6 mm as opposed to 9 mm). Consequently, they determined that the impact of increasing MQL concentration by 73% in the cut was of greater importance than increasing exit velocity by 18% (particularly given the threshold velocity of 325 m/s was met in each case); as such, they opted to utilise the convergent nozzle for experimental validation.

In order to verify the performance of the nozzle in a practical context, it was then employed during cryogenic-assisted end milling of a billet of Inconel 718, a material, which previously (Sections 5.3 and 6.1) has been shown to exhibit poor cryogenic machinability. The authors elected to use TiN-coated carbide inserts, a cutting speed of 120 m/min and a feed rate of 0.12 mm/tooth, ultimately indexing tool life performance against dry machining, MQL, $CO_2$ in isolation and emulsion cooling (Figure 9). As part of their research, the authors observed that, of the dry and near dry coolant strategies, $CO_2$ + MQL (as delivered through their bespoke nozzle) led to the longest tool life, such that it yielded an approximate 100% increase relative to dry machining. Of the other near dry strategies, $CO_2$ cooling in isolation performed markedly poorer than MQL in isolation. This observation can be taken as inference of the relative heightened importance of lubricity, relative to cooling in the milling of Inconel 718. Despite the relative success of the nozzle in its ability to compete with other dry, or near dry, strategies, it is noteworthy that $CO_2$ + MQL failed to generate comparative tool life to that of the conventional emulsion-cooled strategy.

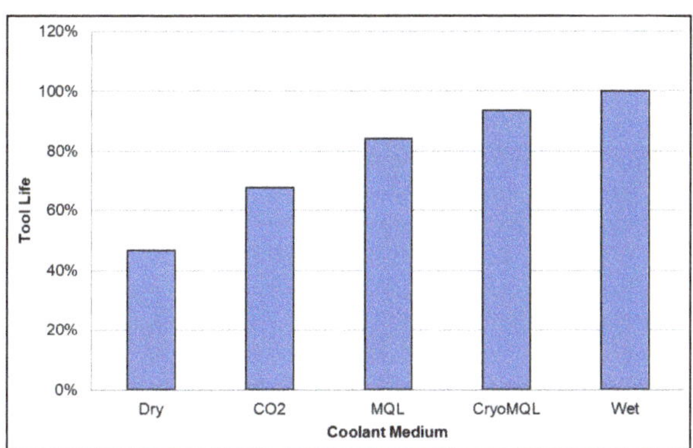

**Figure 9.** Adapted bar chart to show relative tool life during the cryogenic end milling of Inconel 718, adapted from [48].

Although Pereira et al. were able to successfully optimise a nozzle for the delivery of $CO_2$ and MQL, it is worthwhile to remark that emulsion, as has been shown elsewhere in the literature, remains the most suitable option for the machining of Inconel 718. Whilst this observation is clearly of great value in that it laments the conclusions made earlier in this review, it does further constrain the scope of cryogenic coolants to appropriate materials. Moreover, it is equally important to note the limitations in the experimental procedure employed by Pereira and colleagues. For one, this work is similarly limited to the work of Shokrani and Newman, in the sense that the nozzle exit diameter was shown to exhibit a wholly positive correlation with output velocity and, ultimately, performance. For this reason, it would almost certainly be valuable to explore the extent to which this phenomenon persists to assure that the upper limit of nozzle diameter when increased performance is applied. In addition to this limitation, it remains unclear as to whether the parameters with which the nozzle was optimised, namely exit velocity and droplet dispersion, are the most suitable metrics of nozzle performance. In this sense, it would undoubtedly be valuable to assess each of the nozzles in a practical context, although this of course would be cost restrictive.

Another paper that considers nozzle design and optimization for cryogenic machining was published by Gross et al. [50], who employed a similar nozzle optimization focus to the earlier work of Pereira and colleagues. In their research Gross and colleagues undertook preliminary $CO_2$ + MQL nozzle optimization testing followed by cryogenic CNC milling trials on Ti-6Al-4V. In their preliminary testing, the authors first measured the temperature in the $CO_2$ jet stream over a range of distances with four nozzle diameters: 0.5, 0.3 and 0.2 mm. In order to realise this goal, the authors utilised a type K thermocouple, which was vertically fixtured to coincide with the mid-stream of the $CO_2$ jet. As part of their research, the authors trialled both smooth jet nozzles (SJN) and plastic tube nozzles (PT), wherein a $CO_2$ pressure of 56 bar was generally employed, other than in one ancillary trial wherein a higher pressure of 71 bar was chosen. In addition to the temperature profiles developed as part of this research, Gross and colleagues also later went on to test the relative spread of their MQL by placing blotting paper in front of the MQL jet and qualitatively analysing the profile made by the lubricant. The experimental set-up employed in their research is outlined in Figure 10.

Having undertaken the aforementioned tests, the authors made a series of interesting observations on the ramifications of varying nozzle diameter, distance and $CO_2$ pressure (Figure 11). Foremost, Gross and colleagues note that the enlargement of nozzle diameter from 0.2 mm to 0.3 mm (and thus, increase in $CO_2$ mass flow rate) does not reduce the minimum recorded temperature; however, it does serve to increase the size of the low temperature area at the nozzles outlet. They also note that this observation equally applies to the use of a higher-pressure $CO_2$ supply, wherein, again, no significant change is made to the minimum recorded temperature, despite an increased low-temperature area. The temperature data also suggest that the use of each of the SJN strategies leads to a more gradual reduction in temperature from the nozzle's tip onwards than was experienced with the PT nozzle. It is, however, difficult to ascertain with any reliability the extent to which this is a consequence of nozzle construction, as the plastic tubing nozzle featured a 0.5 mm diameter in comparison to the 0.2 and 0.3 mm smooth jet nozzles employed in the tests. Additionally, in the lubricant blot paper tests, the authors observed a significant number of lubricant splashes at the oil trace when MQL was applied with compressed air. By contrast, when the MQL was used in conjunction with $CO_2$, no such splashing existed. The authors note this as an expected consequence of the increased flow speed of the $CO_2$ jet (relative to the compressed air) focusing the oil droplets.

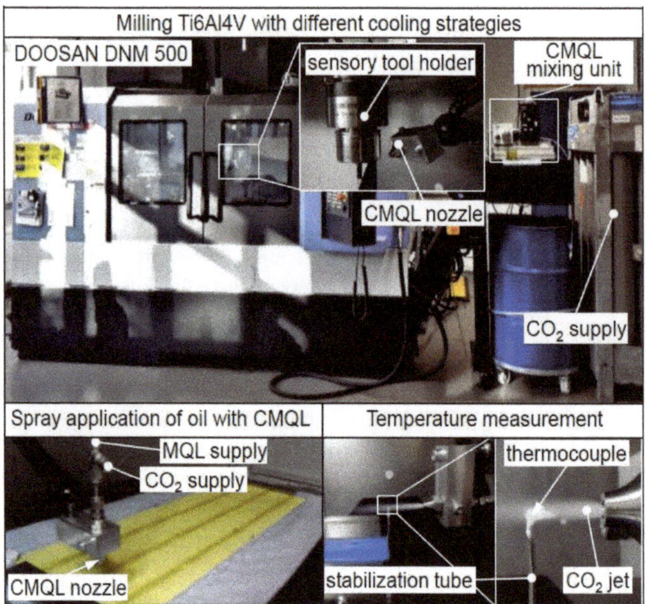


**Figure 10.** Experimental set-up employed during temperature measurement, MQL dispersion and milling trials. Reprinted with permission from Daniel Gross (2019), Copyright 2019 MM Science Journal [50].

Whilst many of the prior observations are valuable for the optimization of cryogenic machining research, one of the most important observations made during the trials of Gross et al. was that temperature reduces exponentially (rather than linearly) as the nozzle is moved further away from the tool. As the relative success of an MWF strategy is highly contingent upon the ability of the coolant to dispel the heat generated in the machining process (and thus cool the tool), this observation is incredibly important to the relative success of cryogenic MWFs. These data thus clearly infer the likely importance of the nozzle's proximity to the tool and thus provide motivation to revisit and optimise trials in which $CO_2$ cooling has in the past proven to be ineffective. In many cases, it may be true that where the poor performance of $CO_2$ coolants was previously assumed to be a consequence of inappropriate material selection, poor selection of feeds and speeds or inadequate tooling, the adverse results rather may have been a consequence of the nozzle's lack of proximity to the tool.

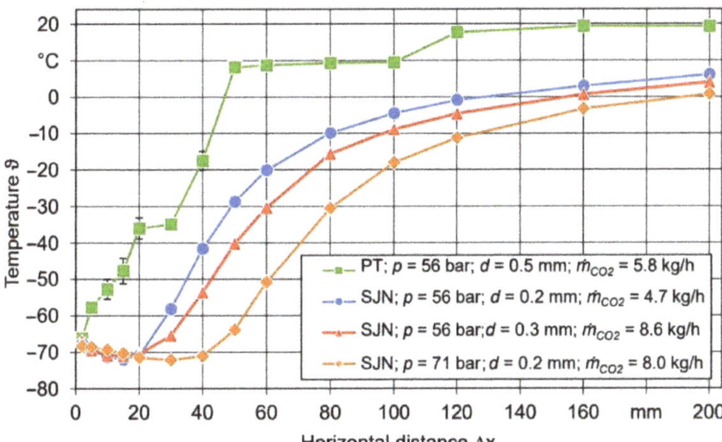


**Figure 11.** Temperature distribution for four nozzle/pressure configurations. Reprinted & adapted with permission from Daniel Gross (2019), Copyright 2019 MM Science Journal [50].

## 7. Future Work and Conclusions

It is clear that $CO_2$ MWFs are one of the best candidates to replace conventional MWFs in many machining scenarios. Despite this promise, this review has clearly outlined the limitations of the technology, wherein the performance of a given MWF strategy, i.e., its ability to drive improved machinability outcomes, is highly contingent upon several factors, ranging from the material species being machined (Sections 5 and 6.1) to the nozzle set-up employed (Section 6.3). Further to this point, it is not immediately clear as to the interplay between these variables, and the dependencies that very likely would present as a greater volume of data is accumulated. This complexity creates uncertainty around the technology, which impedes uptake and creates hesitancy. It is thus the task of researchers in the field to alleviate the unknown, and thereby reduce the risk associated with early adoption. It is the opinion of the author that there are two primary forms of data that must be gathered to facilitate this:

### 7.1. Machinability Data

The pure gathering of machinability data can be thought of as a full factorial approach to acquisition. In order to have true confidence in adopting a $CO_2$ coolant strategy for a given context, prior experimental data supporting its use, in said context, must be available. By focusing on conducting a multitude of machining trials across a broad parameter set, we are able to produce data with a high degree of specificity; if the process outlined in a trial is replicated in industry with equivalent experimental control, the industrial adopter can reasonably expect similar performance outcomes. Unfortunately, however, whilst this approach informs us as to whether a specific process will or will not be successful, it does little to increase understanding regarding the physics of the process.

### 7.2. Explanatory Data

Whilst understanding the physics of a $CO_2$ MWF strategy does not remove the necessity of machining trials, it does increase the probability that the resources used during those trials are wisely allocated. Knowledge of the science informs an engineer as to which variables would be prudent to manipulate in order to drive improved performance outcomes, and as such, data that explains the fundamental physics of a process are crucial. In the context of $CO_2$ MWFs, knowledge of the thermo/fluid dynamics of the coolant or the tribology of the cut allows a machinist or project engineer to make on-the-spot adjustments based on feedback received during the machining operation. As an example, knowledge of the machinability data informs an engineer that $CO_2$ MWF strategies perform poorly in the context of machining Inconel (Section 6); however, knowledge of the phenomena is needed to explain why that is the case.

Undoubtedly, both machinability and explanatory data are crucial to encourage the uptake of $CO_2$ MWF strategies; however, it is clear that the majority of the research outlined in this review has been focused upon developing a broad machinability data set, with much less emphasis on understanding the science of the process. This is a reasonable strategy during the genesis of a new technology, as it explores whether a technology has the scope to be of value; this, in turn, paves the way for significant research commitment in the future. Nonetheless, as the prospective benefits of $CO_2$ MWF strategies are made clear (in this review and elsewhere), it becomes apparent that explanatory experimentation should become the focus of future research efforts, not least to allow for more direct and effective machinability research going forward.

To summarise, this review was produced with the outlook of assessing the operational implications of cryogenic cooling, as well as the practicalities of employing cryogenic MWFs. The review also served to assess the sustainability of $CO_2$ MWF strategies by considering the social, operational, and economic implications of conventional MWFs and contrasting that against $CO_2$ cooling. Further, this work begins to highlight the contexts within which cryogenic coolants are most suitably applied in lieu of the current industrial standard of conventional emulsion-based cooling. In this sense, a compelling case for cryogenic-assisted machining is made and, as such, it is shown that there is clearly a significant motivation for future research, and undoubtedly a great deal of remaining scope for novelty within the field.

**Author Contributions:** Conceptualization, L.P. and N.T.; methodology, T.S.; investigation, L.P.; writing—original draft preparation, L.P.; writing—review and editing, N.T and T.S.; visualization, N.T.; supervision, N.T. and T.S. All authors have read and agreed to the published version of the manuscript.

**Funding:** This research was funded by the EPSRC Industrial Doctorate Centre in Machining Science (EP/L016257/1).

**Institutional Review Board Statement:** Not applicable.

**Informed Consent Statement:** Not applicable.

**Conflicts of Interest:** The authors would like to declare no conflict of interest.

## References

1. Key to Metals AG. High Temperature Nickel-Based Superalloys for Turbine Discs: Part One. Total Materia. 2011. Available online: https://www.totalmateria.com/ (accessed on 21 December 2021).
2. Okoshi, M. *Minutes of General Meeting of the Japan Society of Precision Engineering*; Tokyo, Japan, unpublished work; 1967.
3. Uehara, K.; Kumagai, S. Chip Formation, Surface Roughness and Cutting Force in Cryogenic Machining. *Ann. CIRP* **1969**, *17*, 409–416.
4. Uehara, K.; Kumagai, S. Characteristics of tool wear in cryogenic machining. *CIRP Ann.* **1970**, *18*, 273–277.
5. Yildiz, Y.; Nalbant, M. A review of cryogenic cooling in machining processes. *Int. J. Mach. Tools Manuf.* **2008**, *48*, 947–964. [CrossRef]
6. Shokrani, A.; Dhokia, V.; Muñoz-Escalona, P.; Newman, S. State-of-the-art cryogenic machining and processing. *Int. J. Comput. Integr. Manuf.* **2013**, *26*, 616–648. [CrossRef]
7. Tapoglou, N.; Lopez, M.I.A.; Cook, I.; Taylor, C.M. Investigation of the Influence of $CO_2$ Cryogenic Coolant Application on Tool Wear. *Procedia CIRP* **2017**, *63*, 745–749. [CrossRef]
8. Witkowski, A.; Majkut, M.; Rulik, S. Analysis of pipeline transportation systems for carbon dioxide sequestration. *Arch. Thermodyn.* **2014**, *35*, 117–140. [CrossRef]
9. IPCC. Special Report: Global Warming of 1.5 °C, UN.org. 2020. Available online: https://www.ipcc.ch/sr15 (accessed on 21 December 2021).
10. Grossi, M.; Riccò, B. Metalworking fluid degradation assessment by measurements of the electrical parameters at different temperatures. *Procedia Environ. Sci. Eng. Manag.* **2017**, *4*, 59–68.
11. Passman, F.J. Microbial problems in metalworking fluids. *Tribol. Lubr. Technol.* **1988**, *65*, 14–17.

12. Health and Safety Executive (HSE) and UKLA. Good Practice Guide for Safe Handling and Disposal of Metalworking Fluids. 2018. Available online: https://www.ukla.org.uk/wp-content/uploads/UKLA-HSE-Good-Practice-Guide-for-Safe-Handling-and-Disposal-of-Metalworking-Fluids.pdf (accessed on 10 January 2021).
13. Chipasa, K. Best practice guide for the disposal of water-mix metalworking fluids, UKLA. 2011. Available online: http://www.ukla.org.uk/wp-content/uploads/UKLA-PERA-Best-Practice-Guide-for-the-Disposal-of-Water-mix-Metalworking-Fluids.pdf (accessed on 10 January 2021).
14. Bennett, E.O.; Bennett, D.L. Minimizing human exposure to chemicals in metalworking fluids. *Lubr. Eng.* **1987**, *43*, 167–175.
15. Kumar, P.; Jafri, S.A.H.; Bharti, P.K.; Siddiqui, M.A. Study of Hazards Related To Cutting Fluids and Their Remedies. *Int. J. Eng. Res. Technol.* **2014**, *3*, 1225–1229.
16. Kimberly-Clark. Alarming Number of Workers Fail to Wear Required Protective Equipment (NYSE:KMB). 2012. Available online: http://investor.kimberly-clark.com/ (accessed on 10 January 2021).
17. Hodges, A. COSHH regulations and chemical hazards associated with metal-working fluids. *Tribol. Int.* **1992**, *25*, 135–139. [CrossRef]
18. Jonsson, M.; Thelin, J. Toolholders with fluid supply conduits for supercritical $CO_2$. European Patent EP3219421A1, 9 September 2016.
19. Musil, J.C.; Bukvic, C. Apparatus and method for cooling a cutting tool using supercritical carbon dioxide. U.S. Patent 10052694, 21 August 2016.
20. Fernández, D.; Sandá, A.; Bengoetxea, I. Cryogenic milling: Study of the effect of $CO_2$ cooling on tool wear when machining Inconel 718, grade EA1N steel and Gamma TiAl. *Lubricants* **2019**, *7*, 10. [CrossRef]
21. Busch, K.; Hochmuth, C.; Pause, B.; Stoll, A.; Wertheim, R. Investigation of Cooling and Lubrication Strategies for Machining High-temperature Alloys. *Procedia CIRP* **2016**, *41*, 835–840. [CrossRef]
22. Wang, Z.; Rajurkar, K. Cryogenic machining of hard-to-cut materials. *Wear* **2000**, *239*, 168–175. [CrossRef]
23. Rahim, E.A.; Ibrahim, M.; Mohid, Z. Experimental Investigation of Supercritical Carbon Dioxide (SCCO2) Performance as a Sustainable Cooling Technique. *Procedia CIRP* **2016**, *40*, 637–641. [CrossRef]
24. Rahim, E.A.; Rahim, A.A.; Ibrahim, M.R.; Mohid, Z. Performance of turning operation by using supercritical carbon dioxide (SCCO2) as a cutting fluid. *ARPN J. Eng. Appl. Sci.* **2016**, *11*, 8609–8612.
25. Mulyana, T.; Rahim, E.A.; Yahaya, S.N.M. The influence of cryogenic supercritical carbon dioxide cooling on tool wear during machining high thermal conductivity steel. *J. Clean. Prod.* **2017**, *164*, 950–962. [CrossRef]
26. Cordes, S.; Hübner, F.; Schaarschmidt, T. Next Generation High Performance Cutting by Use of Carbon Dioxide as Cryogenics. *Procedia CIRP* **2014**, *14*, 401–405. [CrossRef]
27. Wika, K.; Gurdal, O.; Litwa, P.; Hitchens, C. Influence of supercritical $CO_2$ cooling on tool wear and cutting forces in the milling of Ti-6Al-4V. *Procedia CIRP* **2019**, *82*, 89–94. [CrossRef]
28. NRoss, N.S.; Sheeba, P.T.; Jebaraj, M.; Stephen, H. Milling performance assessment of Ti-6Al-4V under $CO_2$ cooling utilizing coated AlCrN/TiAlN insert. *Mater. Manuf. Process.* **2021**, *36*, 1–15.
29. Tapoglou, N.; Taylor, C.; Makris, C. Milling of aerospace alloys using supercritical $CO_2$ assisted machining. *Procedia CIRP* **2021**, *101*, 370–373. [CrossRef]
30. Cai, C.; Liang, X.; An, Q.; Tao, Z.; Ming, W.; Chen, M. Cooling/Lubrication Performance of Dry and Supercritical $CO_2$-Based Minimum Quantity Lubrication in Peripheral Milling Ti-6Al-4V. *Int. J. Precis. Eng. Manuf. Technol.* **2021**, *8*, 405–421. [CrossRef]
31. Supra Alloys. Titanium Grade Overview. Supra Alloys. 2020. Available online: http://www.supraalloys.com/titanium-grades.php (accessed on 21 April 2020).
32. Sadik, M.I.; Isakson, S.; Malakizadi, A.; Nyborg, L. Influence of Coolant Flow Rate on Tool Life and Wear Development in Cryogenic and Wet Milling of Ti-6Al-4V. *Procedia CIRP* **2016**, *46*, 91–94. [CrossRef]
33. An, Q.; Cai, C.; Zou, F.; Liang, X.; Chen, M. Tool wear and machined surface characteristics in side milling Ti6Al4V under dry and supercritical $CO_2$ with MQL conditions. *Tribol. Int.* **2020**, *151*, 106511. [CrossRef]
34. Kaynak, Y.; Gharibi, A. Cryogenic Machining of Titanium Ti-5553 Alloy. *J. Manuf. Sci. Eng.* **2019**, *141*, 1–23. [CrossRef]
35. MacHai, C.; Biermann, D. Machining of β-titanium-alloy Ti-10V-2Fe-3Al under cryogenic conditions: Cooling with carbon dioxide snow. *J. Mater. Process. Technol.* **2011**, *211*, 1175–1183. [CrossRef]
36. Kale, A.; Khanna, N. A Review on Cryogenic Machining of Super Alloys Used in Aerospace Industry. *Procedia Manuf.* **2017**, *7*, 191–197. [CrossRef]
37. Patil, N.; Asem, A.; Pawade, R.; Thakur, D.; Brahmankar, P. Comparative Study of High Speed Machining of Inconel 718 in Dry Condition and by Using Compressed Cold Carbon Dioxide Gas as Coolant. *Procedia CIRP* **2014**, *24*, 86–91. [CrossRef]
38. Pusavec, F.; Hamdi, H.; Kopac, J.; Jawahir, I. Surface integrity in cryogenic machining of nickel based alloy—Inconel 718. *J. Mater. Process. Technol.* **2011**, *211*, 773–783. [CrossRef]
39. Cui, P.; Wang, S. Application of microfluidic chip technology in pharmaceutical analysis: A review. *J. Pharm. Anal.* **2019**, *9*, 238–247. [CrossRef]
40. Aldwell, B.; O'Mahony, J.; O'Donnell, G.E. The Effect of Workpiece Cooling on the Machining of Biomedical Grade Polymers. *Procedia CIRP* **2015**, *33*, 305–310. [CrossRef]
41. Kakinuma, Y.; Kidani, S.; Aoyama, T. Ultra-precision cryogenic machining of viscoelastic polymers. *CIRP Ann.* **2012**, *61*, 79–82. [CrossRef]

42. Pu, Z.; Outeiro, J.C.; Batista, A.C.; Dillon, O.W., Jr.; Puleo, D.A.; Jawahir, I.S. Enhanced surface integrity of AZ31B Mg alloy by cryogenic machining towards improved functional performance of machined components. *Int. J. Mach. Tools Manuf.* **2012**, *56*, 17–27. [CrossRef]

43. Yang, S.; Umbrello, D.; Dillon, O.W.; Puleo, D.A.; Jawahir, I. Cryogenic cooling effect on surface and subsurface microstructural modifications in burnishing of Co–Cr–Mo biomaterial. *J. Mater. Process. Technol.* **2015**, *217*, 211–221. [CrossRef]

44. Morkavuk, S.; Köklü, U.; Bağcı, M.; Gemi, L. Cryogenic machining of carbon fiber reinforced plastic (CFRP) composites and the effects of cryogenic treatment on tensile properties: A comparative study. *Compos. Part B Eng.* **2018**, *147*, 1–11. [CrossRef]

45. LittleMachineShop. Cutting Speeds. 2020. Available online: https://littlemachineshop.com/reference/cuttingspeeds.php (accessed on 21 December 2021).

46. Shokrani, A.; Newman, S.T. A New Cutting Tool Design for Cryogenic Machining of Ti–6Al–4V Titanium Alloy. *Materials* **2019**, *12*, 477. [CrossRef]

47. Williams, J. Engineering surfaces. *Eng. Tribol.* **1995**, *25*, 22–28.

48. Pereira, O.; Rodríguez, A.; Barreiro, J.; Fernández-Abia, A.I.; de Lacalle, L.N.L. Nozzle design for combined use of MQL and cryogenic gas in machining. *Int. J. Precis. Eng. Manuf. Green Technol.* **2017**, *4*, 87–95. [CrossRef]

49. Park, K.-H.; Olortegui-Yume, J.; Yoon, M.-C.; Kwon, P. A study on droplets and their distribution for minimum quantity lubrication (MQL). *Int. J. Mach. Tools Manuf.* **2010**, *50*, 824–833. [CrossRef]

50. Gross, D.; Appis, M.; Hanenkamp, N. Investigation on the productivity of milling Ti6al4v with cryogenic minimum quantity lubrication. *MM Sci. J.* **2019**, *2019*, 3393–3398. [CrossRef]

 *metals*

*Review*

# Green Metalworking Fluids for Sustainable Machining Operations and Other Sustainable Systems: A Review

Muhammad Azhar Ali Khan [1,*], Muzafar Hussain [2], Shahrukh Khan Lodhi [3], Bouchaib Zazoum [1], Muhammad Asad [1,*] and Abdulaziz Afzal [1]

[1] Mechanical Engineering Department, College of Engineering, Prince Mohammad Bin Fahd University, Al Khobar 31952, Saudi Arabia

[2] Department of Mechanical Engineering, Khalifa University of Science and Technology, Abu Dhabi 127788, United Arab Emirates

[3] Department of Mathematics and Computer Science, Illinois Institute of Technology, Chicago, IL 60616, USA

* Correspondence: mkhan6@pmu.edu.sa (M.A.A.K.); masad@pmu.edu.sa (M.A.)

**Abstract:** Many efforts have been made over the years to minimize the usage of mineral oil-based MWFs. This includes the trail of its alternatives, such as vegetable oil-based MWFs, nanofluids, etc. These alternatives have shown comparable results to mineral oil-based MWFs in producing a better surface finish and machining efficiency. Apart from the conventional flooding of MWFs, several alternative techniques have been developed by researchers to minimize or eliminate the usage of MWFs, including dry machining, high pressure coolant technique, minimum quantity lubrication, etc. which have also demonstrated promising results. This review attempts to highlight the drawbacks of mineral oil-based MWFs and to assess the applicability of vegetable oil-based MWFs in machining applications. Furthermore, other sustainable machining techniques are discussed in the literature review section, which highlight the main issues associated with the mentioned machining operations and their shortcomings based on the most recent literature. From the comprehensive and critical review that was performed, we inferred that the alternative methods are not mature enough at this stage and that they fall behind in some associated outcomes, some of which may be the tribological properties, surface finish or surface roughness, the cutting forces, the amount of working fluid consumed, etc. More efforts are still needed to fully eliminate the use of MWFs. Moreover, the applications of nanofluids in machining operations have been reviewed in this paper. We concluded from the critical review that nanofluids are an emerging technology which have found their place in machining applications due to their excellent thermophysical properties, but are still in their developmental stage, and more detailed studies are needed to make these a cost-effective solution.

**Keywords:** metalworking fluids; sustainability; machining; nanofluid; vegetable oil; mineral oil

**Citation:** Khan, M.A.A.; Hussain, M.; Lodhi, S.K.; Zazoum, B.; Asad, M.; Afzal, A. Green Metalworking Fluids for Sustainable Machining Operations and Other Sustainable Systems: A Review. *Metals* 2022, *12*, 1466. https://doi.org/10.3390/met12091466

Academic Editors: Francisco J. G. Silva and Jorge Salguero

Received: 30 May 2022
Accepted: 17 August 2022
Published: 31 August 2022

**Publisher's Note:** MDPI stays neutral with regard to jurisdictional claims in published maps and institutional affiliations.



## 1. Introduction

Sustainable machining is being adopted all over the world in manufacturing units as a common practice, as all economic and business activities demand sustainability. It would not be wrong for sustainable manufacturing to be characterized as a branch or extension of the sustainable development philosophy [1]. The sustainable manufacturing philosophy adds value to the final product while keeping the quality environment for the upcoming generations [2]. A wide range of parameters are included in sustainable manufacturing, such as the personal health of the workers, environmental issues, and the safety related to machining operation. As all the basic ingredients of sustainability are an integral part of sustainable manufacturing processes, which include the cost associated with machining operation, safety of the environment, and society, it therefore has a broader perspective than just green and eco-friendly machining operation [3]. The beginning of sustainable manufacturing processes start from the selection of the raw materials, into the early process of manufacturing, and until the finishing of the final product, keeping in view the integrity and objectives of the organization and its performance. The major manufacturing activity is machining, which encompasses a wide range of operational variables that have the room or potential for transformation towards sustainable development. These operational variables include but are not limited to the cooling and lubricating fluids used in machining operation, disposal of water or other working fluids, energy conservation, life of the tool, and recycling of the chips [4]. MWFs are generally used to cool the workpiece during the machining process and serve to lubricate the workpiece from the beginning, and it is well-known that these fluids are generally required to achieve a high quality output as well as a smoother and higher efficiency in the machining process. Additionally, MWFs are used to decrease the friction between the tool and the workpiece during machining operation, thereby reducing the potential for detrimental effects such as adhesion, galling, and welding; they remove the heat generated at the interface and carry away the chips and other debris that are generated during the machining operation [5,6].

The widely used mineral-based MWFs are the primary cause of many diseases in the machine operators such as skin infections, lung problems, and may also lead to the development of cancer. In addition, studies have found that they are not biodegradable, therefore it is required to treat them before disposing them off into the environment. Otherwise, they may cause serious issues to the environment [7]. In order to achieve sustainability in machining operation, several improvements are needed in this regard, such as developing new materials and applications methods; newer technologies are also needed to dispose-off MWFs [8]. Furthermore, green MWF development will also allow for cutting-edge technology to make processes more sustainable and ensure the safety of the workers and environment. The opportunities for performing sustainable machining are illustrated in Figure 1, and these opportunities can be used in order to address the issues pertaining to MWFs that are based on mineral oil. The most important aspect, in terms of the quality and economical perspectives, is the dimensional exactness of the workpiece [9,10]. Therefore, the machining operators should be able to identify the conditions which result in the precise dimensions for most of the used working materials [11–13].

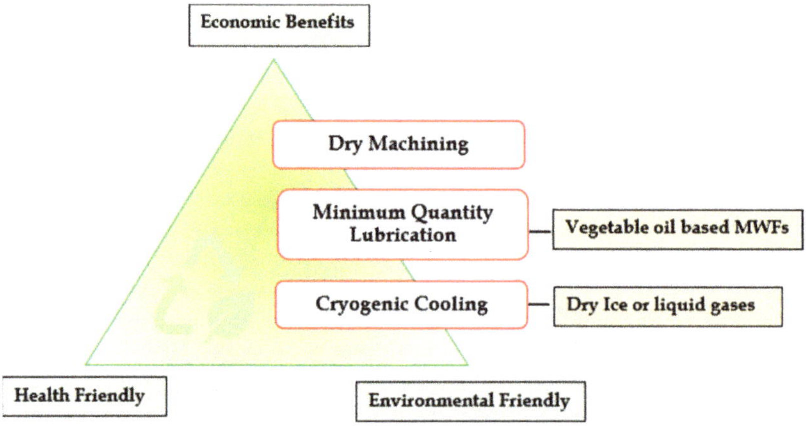



**Figure 1.** The opportunities for performing sustainable machining.

MWFs hold a major percentage of the effluents that are disposed into the environment [14], and in a study by Cheng et al. [15], it was quoted that the volume of MWF waste had been estimated to be more than 20 billion liters. To curb this issue, environmental regulation authorities have been urging companies to adopt or develop new ways of controlling and discharging the industrial MWFs to mitigate their detrimental effects to the environment and natural habitats. Consequently, there is a need for environmentally friendly MWFs to achieve sustainability in machining operations [16]. New MWFs such as the ones based on vegetable oil provide better results than mineral oil-based MWFs. This is because of the fact that a far more effective layer of lubricant is formed between the tool and workpiece, developed by the saturated fatty acids present in vegetable oil [17–19]. The vegetable oil-based MWFs have shown an enhanced performance compared to the mineral oil-based MWF for the drilling operation performed on AISI 316L steel, increasing tool life up to 177% and reducing the thrust force up to 7%. It was also demonstrated by Lawal et al. [12] that the presence of triglycerides in the vegetable oil gives better properties that are needed in the lubricants.

Several studies have been conducted to assess the economic impact of MWFs. Adler et al. [20] provided figures that over two billion gallons of machining fluids were consumed by manufacturers in North America in the year 2002. Similarly, another research by Marksberry and Jawahir [6] showed that the total annual consumption of MWFs was 640 million gallons globally in 2007, whereas around 100 million gallons were utilized in US manufacturing sectors; the actual consumption was far larger than this figure, according to other sources. Lawal et al. [21] revealed that in 2005, the global consumption of MWFs was quite high, i.e., more than 1200 million gallons, and the projected increase over the decade was 1.2 percent. The actual estimate was not possible, due to the pervasive nature of filed processes. Pusavec et al. [22] revealed that 15% to 20% of the overall cost of machining processes is due to the MWFs utilized for cooling and lubrication purposes. Replacing the cutting fluids with sustainable machining processes so that it can save up to 20% of overall machining costs would be a huge achievement for manufacturers. King et al. [23] also discussed that about 7% to 17% of the total manufacturing costs is related to the cutting fluids, and 4% is related to tooling expenses. Fluid expenses in industries include the purchase of fluids, setup of a fluid dispensing system, maintenance, waste treatment, and fluid disposal [10]. Brinksmeier et al. [24] showed that MWFs have expenditures of around 16.9% of the overall manufacturing sectors in European automotive industries. Hence, it is obvious from all of these studies that the cost for the handling of MWFs is almost 17–20% of the total manufacturing cost.

## 2. Scientometric Analysis

Scientometric analysis [25–28] is usually carried out after importing the databases from authentic libraries. Usually, the Scopus and Web of Science databases are selected for the analysis, but it has been reported and observed that Scopus provides a wider and more inclusive coverage of content. The access to profiles of all authors, institutions, serial sources, and the availability of the interrelated databases interface makes the use of Scopus more convenient and comfortable for practical use [29]. Therefore, the Scopus database has been selected for analysis.

Scientometric analysis usually starts by selecting some of the most frequent or widely used keywords on the topic. Therefore, some relevant keywords were used to start the analysis after a preliminary literature review. A total of 1834 documents were filtered out and only the published articles were selected. Articles that were in press were omitted from the analysis. After the search was complete, the database was exported to the commercially available integrated development environment (IDE) R Studio [30], which was used to analyze the database.

### 2.1. Annual Scientific Publication

Figure 2 shows the annual scientific publications, which range from 1975 to 2021. It can be seen from the figure that research on metalworking fluids and sustainable machining started from 1975 and only had a few articles published until the early 2000s. However, a spike was observed in the research from 2003 onwards, where the number of annually published papers increased and in the past seven years, a substantial advancement has been made in the research area of sustainable MWFs and sustainable machining operations. Therefore, the prime focus of this article was to review the papers published in the past

10 years, but for the sake of establishing some basic concepts and forming the bases, some earlier literature has also been cited.

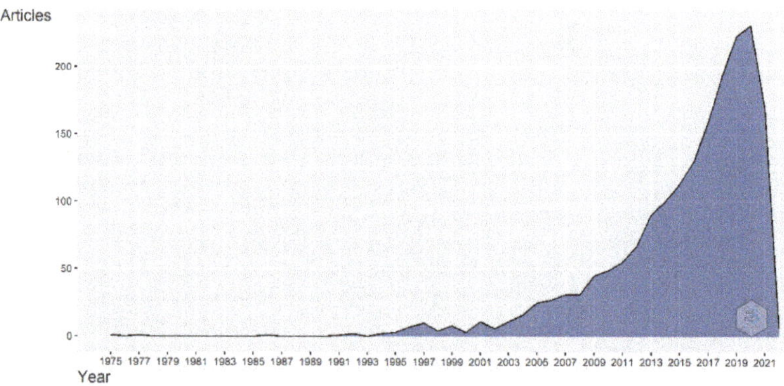

**Figure 2.** Annual scientific publication on metalworking fluids and sustainable machining operations.

*2.2. Sources of Documents*

The comparison of different journals and the number of documents in the journals can be seen in Table 1. Among other journals, the highest number of papers have been published in the *International Journal of Advanced Manufacturing Technology*, with a total of 106 articles. After that, the second highest number of publications has been in the *Journal of Cleaner Production*, with a total of 101 articles; the lowest number of articles has been in the *Wear* journal, i.e., 15 articles. It can be deduced from the analysis that most of the articles targeted the sustainability and machinability aspects of the different metalworking fluids, and therefore a limited number of articles have been published that examine the wear characteristics.

**Table 1.** Most relevant sources and their number of published articles.

| Sources | Number of Articles |
| --- | --- |
| *International Journal of Advanced Manufacturing Technology* | 106 |
| *Journal of Cleaner Production* | 101 |
| *MATERIALS TODAY: PROCEEDINGS* | 73 |
| *PROCEDIA CIRP* | 64 |
| *Journal Of Manufacturing Processes* | 38 |
| *Advanced Materials Research* | 32 |
| *Lecture Notes in Mechanical Engineering* | 32 |
| *Proceedings of the Institution of Mechanical Engineers Part B: Journal of Engineering Manufacture* | 32 |
| *IOP Conference Series: Materials Science and Engineering* | 29 |
| *Journal Of Materials Processing Technology* | 29 |
| *Procedia Manufacturing* | 29 |
| *AIP Conference Proceedings* | 25 |
| *Materials And Manufacturing Processes* | 24 |
| *International Journal of Machining and Machinability of Materials* | 21 |
| *Tribology International* | 21 |
| *Key Engineering Materials* | 20 |
| *Applied Mechanics and Materials* | 17 |
| *Journal of the Brazilian Society of Mechanical Sciences and Engineering* | 15 |
| *Machining Science and Technology* | 15 |
| *Wear* | 15 |

## 2.3. Word Cloud

To pictorially illustrate the representation of the different keywords used in the different articles related to metalworking fluids and sustainable machining operation, a word cloud highlighting the different keywords is shown in Figure 3. It can be seen from the figure that the most frequently used keywords are "cutting tools", "minimum quantity lubrication", "sustainable development", etc. Therefore, it can be inferred by the word cloud that a considerable amount of research work has been conducted on transforming conventional machining operation to a more sustainable machining operation through the adoption of different strategies such as nanofluids, minimum quantity lubrication, dry machining, and vegetable oils, among others.

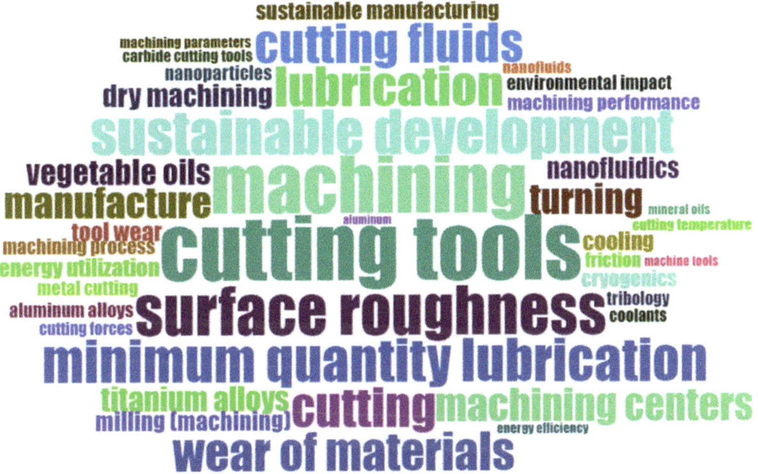


**Figure 3.** Pictorial representation of the most frequently used keywords.

From the scientometric analysis, it is evident that a considerable amount of research has been carried out in the area of metalworking fluids for the development of machining and different sustainable techniques and is still in progress. Therefore, considering the substantial amount of research output, a comprehensive review is needed which summarizes the impact of mineral oil-based MWFs, comparing it with its counterparts (i.e., vegetable oil-based MWFs) and also shedding light on the different sustainable machining techniques available in the market.

This paper reviews the adverse effects of mineral oil-based MWFs and compares it with vegetable oil-based MWFs, highlighting the tribological performances of the vegetable oil-based MWFs. The paper also highlights the potential of other sustainable machining operations such as dry machining, high pressure coolant technique, minimum quantity lubrication, and the potential use of nanofluids in machining operations.

## 3. Relevant Literature

### 3.1. Adverse Effects of Mineral Oil-Based MWFs

Almost all of the available MWFs are derived from petroleum products, and the elements that are present in mineral oil-based MWFs are the major cause of the moisture and oil smoke observed during machining operation, which causes an uncomfortable environment for machine operators [31]. The health and environmental aspects are of the utmost importance in most countries, and regulations pertaining to the use of MWFs have aimed to guarantee the health of workers and to protect the environment [32,33].

There are several methods to apply MWFs onto the interface of the workpiece and the tool. This can be in the form of flooding, through a jet, or through mist in the several directions, as illustrated in Figure 4 [34]. In 1987, the International Agency for Research on Cancer (IARC) declared that the mineral oil-based MWFs, which were widely used in machining operation, were carcinogenic [31]. In an experimental investigation conducted on laboratory animals to investigate the toxicity of water-based MWFs by Bennett [35], it was reported that the specific additives and surfactants present in the MWFs caused cancer to the animals. In a review study by Park et al., the authors reported that nitrosamine and other amines in MWFs were carcinogenic, which are formed by the nitrates and are also used as corrosion inhibitors [33]. In 1984, the US Environmental Protection Agency (USEPA) fully banned the usage of nitrites that contained alkanol amines for cutting fluid, due to the detrimental effects they have on human health [31]. It was also reported in a review article that the mineral oil-based MWFs are composed of constituents which are suspected to be carcinogenic, and which favor the spread of tumors [33]. Any combination of sulfur, nitrosamines, long chain aliphatic compounds, formaldehydes, and Polycyclic Aromatic Hydrocarbons (PAHs) release biocide contaminants, which are also regarded as carcinogenic in nature, thereby posing a serious threat to machine operators [36,37]. The use of acid-refined MWFs results in the development of skin cancer. In order to decrease the PAHs present in crude oil, refining is performed. However, acid-refined MWFs contain a substantial amount of PAHs, which are a cause of skin cancer. Skin irritation is thought to be the most common health related issue resulting from the use of mineral oil-based MWFs during machining operation. These can be caused by the direct contact of the operator with MWFs [33,38]. It has been found that almost all of the mineral oil-based used in metalworking are found to have pH levels ranging from 9.5 to 11.0, where the higher acidic MWFs cause skin-related problems and, in the worst case scenario, can lead to skin diseases. Therefore, researchers around the globe are working to develop MWFs that can ensure the safety of workers and avoid any undesirable outcomes for machine operators [31].

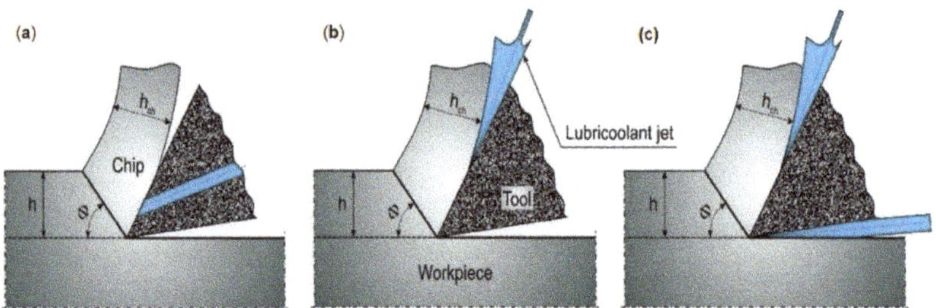


**Figure 4.** The methods of supplying a lubricating medium to the work–tool interface, inserted from: (**a**) from the side of the rake face; (**b**) on the rake face; (**c**) on the flank and rake face [39].

Certain elements are added to enhance the properties of the mineral oil-based MWFs, e.g., sulfur, which increases the heat capacity of MWFs and also increases their ability to lubricate under extreme pressure conditions [40]. Another problem related to the health of machine operators is linked to the inhalation of MWF vapors, which has also increased since the increase in machining speed. The inhalation of mineral oil-based MWFs may lead to digestive problems and respiratory system diseases and may also lead to the development of cancers. Choi et al. [41] reported that the presence of dissolved ions of Co, Cr, and Ni in mineral oil-based MWFs are the potential source of skin disorders, and that many skins reactions occur when neat mineral oil-based MWFs are used.

### 3.2. Vegetable Oil-Based MWFs

A considerable amount of research around the world has been aimed at developing alternatives of the harmful mineral oil-based MWF in order to make machining processes sustainable. Recent studies on sustainable machining have revealed that the vegetable oil-based MWFs have shown a better performance [42–48]. The vegetable oil-based MWFs have demonstrated better cooling and lubrication characteristics when used during machining operation compared with the mineral oil-based MWFs. As a result, they have gained much attention, and they have been a topic of interest for many researchers. Over the years, it has been the practice to choose the MWFs based on the cutting process, the tool material, the work material, and the operation conditions [49,50]. This was the old trend, but as the research is progressing in this area, the selection of MWFs is also changing. Now, the selection of MWFs is more concerned about their impact on the environment and on the health of the machine operators, in addition to other process requirements.

One of the most important advantages of a vegetable oil-based MWF is that it can easily be broken down into eco-friendly species with the aid of enzymes or chemical reactions. The residue can easily be disposed-off in an environmentally friendly manner without posing any serious challenge to the environment, therefore maintaining sustainability. Furthermore, the toxicity level of the vegetable oil-based MWFs is considerably less than that of the mineral oil-based MWFs [51]. Vegetable oil-based MWFs are also less severe than mineral oil-based MWFs when machine operators become exposed to the bio-degradable vegetable oil-based MWF. Another advantage of vegetable oil-based MWFs is that filtration is not required before it is disposed-off, which considerably reduces the costs associated with it. The environmental and economic benefits of vegetable oil-based MWFs, compared with the mineral oil-based MWFs, is shown in Figure 5. The figure illustrates that the environmental impact of mineral oil-based MWFs is considerably lower than that of vegetable oil-based MWFs; because different additives are also added to minimize the environmental impact of mineral oil-based MWFs, they prove to be less economical than vegetable oil-based MWFs. In a study by John et al. [52], it was concluded that by using a vegetable oil-based MWF, better cooling rates were achieved and improved lubrication characteristics were observed due to their higher retention time. In a study by Mannekote and Kailas [53] on the effect of oxidation on the tribological properties of vegetable oil-based MWFs, they reported that when compared to mineral oil-based MWFs, the vegetable oil-based MWFs had a higher tendency to oxidize when exposed to oxygen, and they can easily be converted to compounds like $H_2O$, $CO_2$, and $CH_4$. On the other side, Erhan et al. [54] showed that vegetable oil-based MWFs have a lower ability to maintain their characteristics in high temperature and high humidity environments, which are properties that are needed to perform cooling and lubrication operations. One of the solutions to address the shortcomings of vegetable oil-based MWFs is through the formulation of water-soluble MWFs, where the surfactants are other introduced additives, resulting in the modification of the chemical structure; this method makes the MWF capable of operating satisfactorily in extreme conditions without jeopardizing its lubrication and cooling characteristics, and it has been confirmed in different studies [55,56].

One of the most successful methods for the formulation of vegetable oil-based, water-soluble MWFs is the process of emulsification. In this process, the aquatic and oleic phases are mixed and are rigorously shaken to disperse oil droplets in water and vice versa. The addition of water plays a crucial role in altering the properties of the MWF. Water acts as the cooling agent as it possesses a higher specific heat capacity [57]. However, one of the challenges associated with emulsification is effective mixing or, in other words, the homogenization. The main reason for this is the dispersion resistance of the vegetable oil droplets during the mixing process. As a result, ultrasonic technology was introduced into the market to obtain effective homogenization and thus obtain stable emulsified products [39,58]. In the criterion for determining the stability of the emulsion, one of the parameters used is the hydrophilic-lipophilic (HL) value. The values of the HL can be used to identify whether or not the surfactants or additives have a higher inclination towards the vegetable oils [14,59]. To highlight the basic components of the emulsifier, it should be noted here that it consists of the hydrophilic group in the case of water and lipophilic group for oil. The hydrophilic group has a stronger affinity towards water, whereas the lipophilic chain has a higher proclivity towards oil [60]. The emulsifiers can be classified based on their hydrophilic and/or their lipophilic value [61].

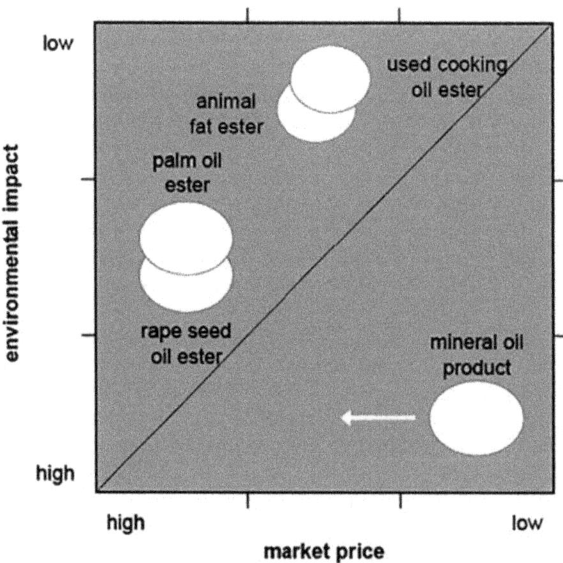


**Figure 5.** The economic and environmental impact of vegetable oil-based MWFs and mineral oil-based MWFs [51].

In order to represent the relative composition of the hydrophilic group and the lipophilic group, a parameter known as the hydrophilic-lipophilic balance (HLB) is used. The HLB scale ranges from 1 to 20, and it shows the affinity of the emulsifier towards water or oil. One way to explain the parameter is that emulsifiers with higher HLB values are more effective for oil in water emulsions, and less useful for water in oil emulsions [62]. The HLB value plays a significant role in the synthesis of the vegetable oil-based MWFs. The surfactants or additives that are to be added and the base oil can be selected based on the HLB values, which are a good indicator of solubility during the preparation of stable emulsions for bio-lubrication purposes [63]. The hydrophilic head points out towards the water phase, while the hydrophobic tail points out towards the oleic phase [64].

### 3.3. Characteristics of Vegetable Oil-Based MWFs in Machining Applications

It has been shown that vegetable oil-based MWFs have shown superior cooling and lubrication properties compared with mineral oil-based MWFs. This is mainly because of the fact that the presence of saturated fatty acids in vegetable oil aid the formation of the lubricant layer at the work–tool interface, and the structure of the triglycerides provides the desired lubrication characteristics [65]. In a study by Sani et al. [66], it was shown that by using modified jatropha oil with ionic liquid, the cutting energy was reduced. Ionic liquids consist of acidic ionic liquids (AIL) and protic ionic liquids (PIL). The authors also reported that better results were obtained when the mixture consisted of 10% AIL with jatropha oil and 1% PIL with jatropha oil. They highlighted that the specific cutting energy was reduced around 4 to 5%, the cutting temperatures were reduced by 7 to 10%, the friction coefficient was reduced by 2 to 3%, and the tool–cup contact's length was reduced by 8 to 11% when the results were compared with the reference mineral oil-based MWF. Their results are also shown in graph form in Figure 6.

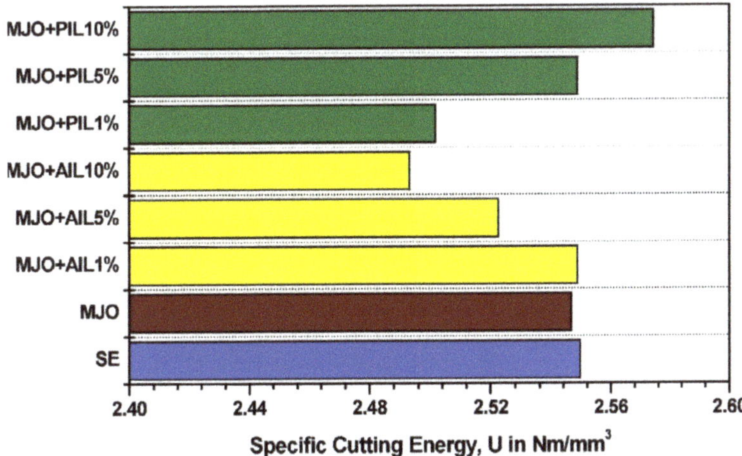



**Figure 6.** The variation of the specific cutting energy with respect to the mixture of modified jatropha oil (MJO) and AIL/PIL [66].

Vamsi Krishna et al. [67] showed in their research that better surface quality was obtained when using nano-boric acid in coconut oil compared with the surface quality obtained from the industrial lubricant SAE 40, and their results are shown in Figure 7. It can be seen from the figure that the coconut-based oil resulted in lower values of surface roughness while also changing the cutting speed.

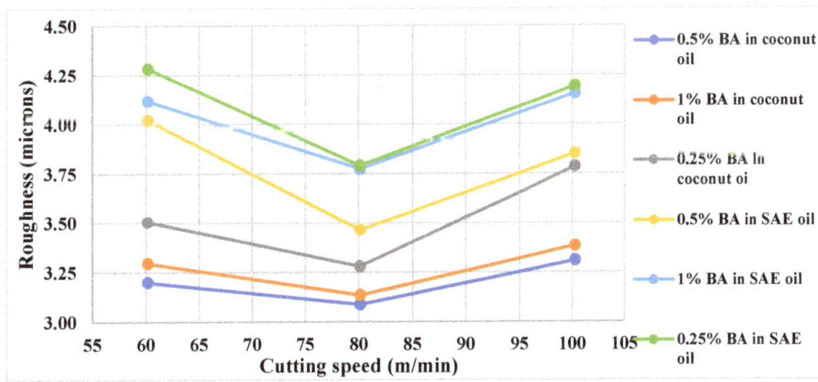



**Figure 7.** The variation of the surface roughness with the cutting speed [67].

### 3.4. Sustainable Machining Techniques

As in the machining process, the heat produced is a major problem, and it can incur economic and technical costs either directly or indirectly [68]. Taylor [69] in the early 1900s pointed out that heat generated in the cutting zone plays a significant role in the cutting process. MWFs were used to address this, which imposes a serious challenge to the environment and the machine operators as discussed above in detail. In order to minimize the use and side effects of MWFs, several potential methods are available, including dry machining, machining with minimum quantity lubrication, machining with high-pressure jet assistance, and machining with alternative fluids such as gas, vapor, and solid lubricants.

### 3.4.1. Dry Machining

As the name suggests, dry machining does not make use of conventional cutting fluids during the machining process. This process is only accepted by companies if it makes sure that the quality of the product is better or at least the same as when cutting fluids are used [70]. Several techniques are adopted to improve the dry machining process, such as the tool material and tool coating. In terms of the tool material, it is important to optimize the flute width, number of flutes, and margin size to have an extended tool life. Since cutting fluids are absent in this process, different methodologies are adopted to achieve the desirable finish of the workpiece, of which include the use of diamond-like carbon (DLC) coatings on the surface of tools, among others. In an investigation by Fukui et al. [71], the tribological behavior and performance of the DLC-coated tools working on the aluminum alloy workpiece were assessed. They reported that the DLC coatings on the surface of the tool resulted in improved tool life when compared with uncoated tools during the dry machining process. The comparison of the surface roughness in both cases is shown in Figure 8. It can be seen from the figure that when a DLC coating is applied, the surface roughness is lower.

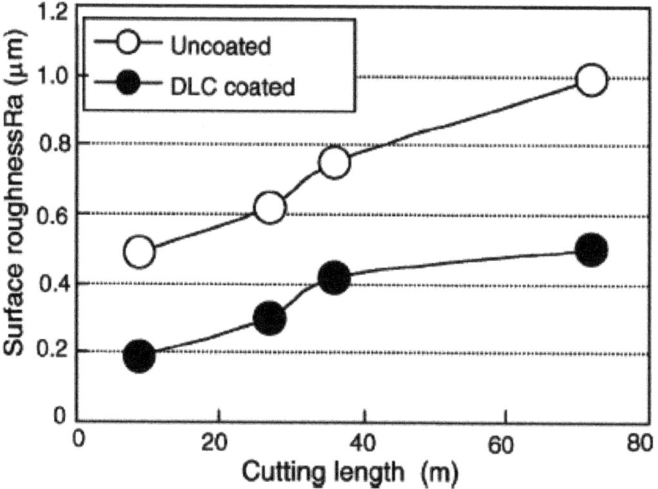


**Figure 8.** The variation of the surface roughness with along the cutting length [71].

Klocke and Eisenblätter [72] carried out several investigations to implement the dry machining process in the production of cast iron, steel, aluminum, and some other materials. They reported that for the case of uncoated tools, some unwanted built-up edges were formed, and that the surface quality was also disturbed. The improvements in the dry machining process were further discussed by Sreejith and Ngoi [73]. According to their study, the dry machining process cannot match the wet machining process in many aspects, and it is only acceptable if the surface finish and other desired properties are equivalent to that obtained by the wet machining process; the authors stated that if it was to be employed, several improvements were necessary. A new system was proposed by Vereschaka et al. [74] in which the cutting tool was coated with a multi-layered, nano-scale coating, along with an ionized gas dispensing system and exciting system. They reported an improved performance in terms of the surface finish when cutting titanium alloys, steel, and nickel-based alloys. In a study by Devillez et al. [75] concerning the successful implementation of dry machining processes for Inconel®718 using a coated carbide tool, they reported that a reasonable surface finish and micro-hardness was observed, and that the values were comparable to the ones obtained in the flooded conditions. Additionally, no severe changes were observed in the microstructure. The authors had merged different cutting techniques to reduce the cutting forces and surface roughness.

From several studies on the dry machining process, it can be inferred that although many researchers have reported successful implementations of the dry machining process for different materials such as cast iron, steel, and aluminum, etc., major technological improvements are still necessary in order to minimize the cutting forces and cutting temperatures. Furthermore, improved methods are still needed to flush out the chips that are formed during the machining process. Researchers have also reported the improved performance of coated tools such as coated carbide tools and DLC-coated tools when it comes to the surface roughness, but a high tool wear rate was still reported by many researchers.

### 3.4.2. High Pressure Coolant Technique

Out of the several techniques to increase the machining efficiency, one of the techniques is the high-pressure coolant technique [76]. High-pressure coolant refers to the pumping of coolant at pressures exceeding 300 psi. In general, the pressures are in the range of 1000 psi. In some ultra-high-pressure coolants, the pressure reaches up to 3000 psi and therefore, solely depend on the requirements of surface being machined. There are several advantages to using of the high-pressure coolant technique, such as optimal chip control, which is accomplished by virtue of the coolant at high pressure breaking the chips into smaller pieces, preventing the chips from wrapping around the workpiece and chuck. High-pressure coolant evacuates the chips from the work area before the cutting tool gets into contact with them, therefore resulting in a better surface finish. Because of the above two benefits, the high-pressure coolant technique allows machine operators to work at increased feed rates, resulting in faster cycle times. Dahlman and Escursell [77] reported in their study that when the high-pressure coolant technique was applied, the chip control and reduction in the amount of built-up edges considerably improved during the turning process of decarburized steel. They also reported that the surface roughness was reduced as much as 80%, and that the tool wear was significantly reduced, of which the tools were prone to high temperature cracking. Ezugwu et al. [78] investigated the high-pressure coolant technique for the machining of hard metal alloys, such as Inconel 718, AISI 1045, and Tie6Ale4V steel; they also used different tool materials, such as cubic boron nitride (cBN) and TiAlN-coated carbide tools. They reported that by increasing the supply pressure of the coolant, the cooling and lubrication conditions were enhanced, along with a reduction of the cutting forces. This also resulted in the improved separation of chips and improved the surface roughness values. Kramar et al. [79] experimentally investigated different machining techniques, including the dry machining, conventional flooded machining, and the high-pressure cooling techniques for performing turning operations on piston rods which were already surface-hardened. They reported that out of all of techniques, the high-pressure cooling technique showed promising results, as the chip deformation was enhanced and the fluid consumption was reduced. However, the only shortcoming of the high-pressure coolant technique as reported by them was in its inability to reduce the depth of cut notches. The graphical interpretation of their results is shown in Figure 9. Pusavec et al. [22] conducted an experimental investigation on the cost analysis of the high-pressure cooling, conventional flood machining, and cryogenic machining techniques. They reported that the high-pressure cooling technique was 30% less costly compared to the other two techniques. Ayed et al. [80] experimentally investigated the tool deterioration and wear patterns on uncoated WC inserts, employing the conventional flooded machining and high-pressure water-jet-assisted machining. They reported promising results when compared with flooded machining with respect to the plastic deformation and flank wear of the cutting tool. They also reported a drawback of this technique, which was in its inability to reduce abrasion and adhesion wear, which led to notch wear. In a study by da Silva et al. [81], they studied the effect of the high-pressure coolant technique while machining a Ti-6Al-4V alloy with a polycrystalline diamond under high-speed conditions.

They reported that by increasing the fluid pressure, the tool life increased, and the adhesion was considerably reduced, specifically at 20.3 MPa.

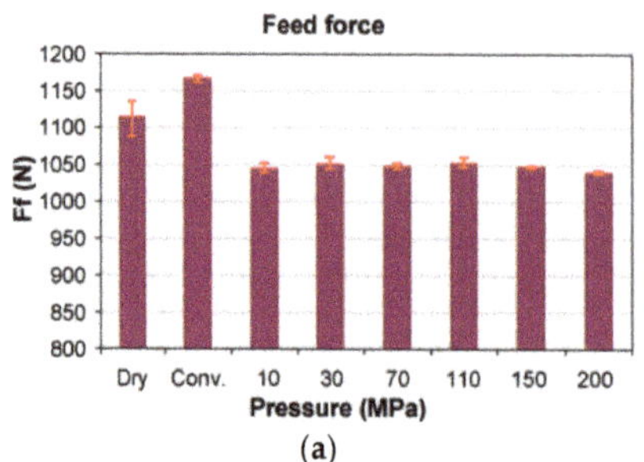
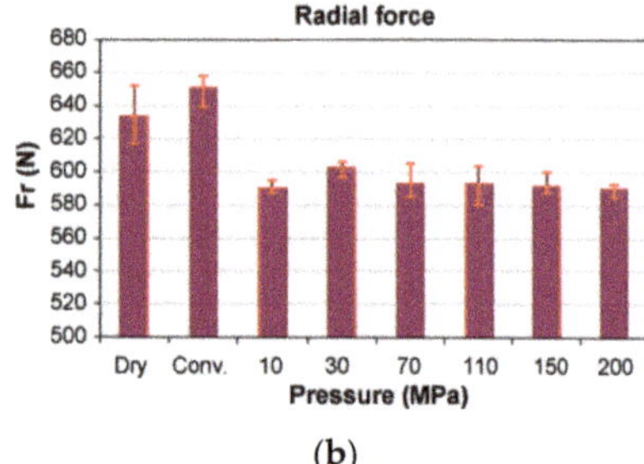

(a)

(b)

**Figure 9.** The effect of coolant pressure on (**a**) the feed force, and (**b**) the radial force [79].

### 3.4.3. Minimum Quantity Lubrication (MQL)

As has been discussed, efforts are being made by many researchers to achieve manufacturing goals that are eco-friendly in nature due to the polices regulated by governments for preventing pollution globally, with the long-term aspects of the environment in mind [82]. There are many examples of such countries, including the USA, EU, China, and Malaysia, in particular which is clearly shown by Figure 10 that the number of publications are increasing gradually on yearly basis. [83]. As machining is one of the main processes in manufacturing sectors, it is therefore considered to have a significant process and important role to play in the context of green metalworking and sustainability, as it has a direct impact over the cost, life, and performance quality of so many components [84–86]. Therefore, minimum quantity lubrication (MQL) is one of the highlighted techniques that is playing a key role in sustainable machining in the last two decades, and the research needs to work more on this process to make manufacturing environmentally friendly as per the demand of industrial sectors.

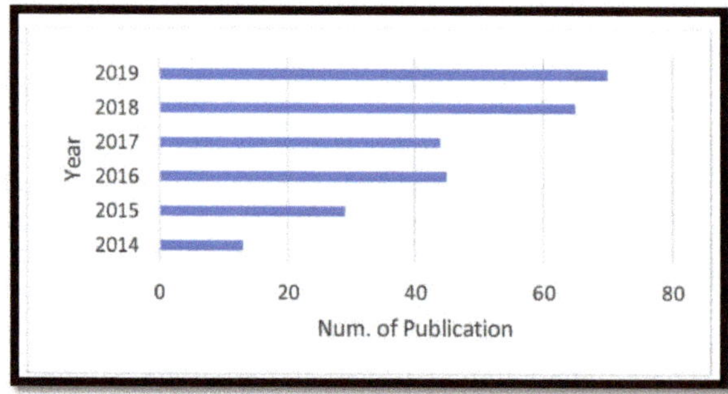

**Figure 10.** Research articles published about the advancements of MQL from 2014 to 2019 [83].

There are many researchers that are in agreement with the argument that MQL has the potential to replace the conventional methods of flooding which are used for machining processes i.e., grinding, milling, drilling, and turning [87–90]. Najiha et al. [91] reported that the MQL technique is one of the practical ways for a green manufacturing process, as it is one of the most cost-efficient techniques and also guarantees both sustainability and worker health. This claim has also been supported by many other scientists and researchers who believe that the minimum quantity of cutting fluid should be consumed in this way [88,92,93]. All of these studies depict that the MQL technique has importance in the emerging efficient and eco-friendly manufacturing techniques of the modern era. It can be seen in Figure 11 that around 7% to 17% of the cost of the manufacturing process constitutes cutting fluid, and if replaced by minimum quantity lubrication, it will save a substantial amount of budget. Hence, a substantial amount could be saved by switching the conventional methods with the MQL technique in industries to reduce budget costs.

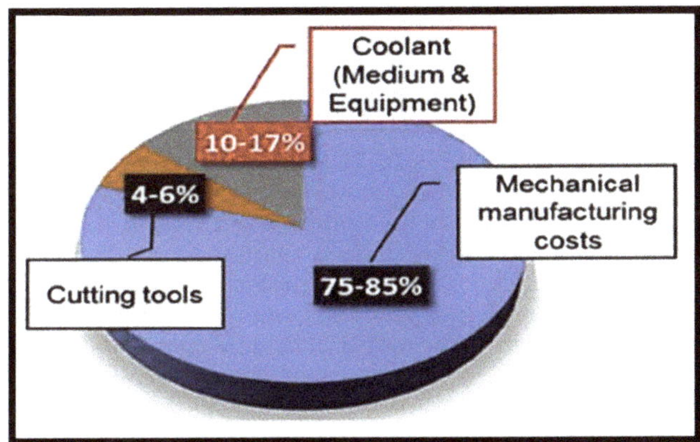


**Figure 11.** The quantitative distribution of manufacturing costs in industries [94].

Khan and Dhar [95] studied the benefits of using vegetable-based oil in manufacturing instead of cutting fluids, as they are very good pressure absorbents, have the ability to accelerate the material removal rate (MRR), and provide very minimum loss due to vaporization, misting, among other reasons. Moreover, many other researches have supported these advantages of MQL, especially in studies focusing on milling, drilling, and turning [19,96–98]. Dixit et al. [99] observed that synthetic oils are also very effective for machining and have similar properties to vegetable oil-based MWFs, having high boiling temperatures, low viscosities and better flash points. Moreover, some studies have revealed that synthetic oil machining is far better than both vegetable- and mineral-based oils [100].

Commercially, the MQL technique comprises five major parts, which are the cutting fluid tank, air compressor, flow control system, tubes, and spray nozzle [90]. It generally uses an atomizing method and a minimum amount of spraying, composed of an oil mixture and pressurized air sent at a flow rate below 1000 mL/h, and it directly sprays the mixture into cutting zone as has been described in many studies [101–103]. This consumes 10,000 times less cutting fluid volume as compared to the flooding technique. Furthermore, the MQL system is categorized into internal and external applications, as shown in Figure 12.

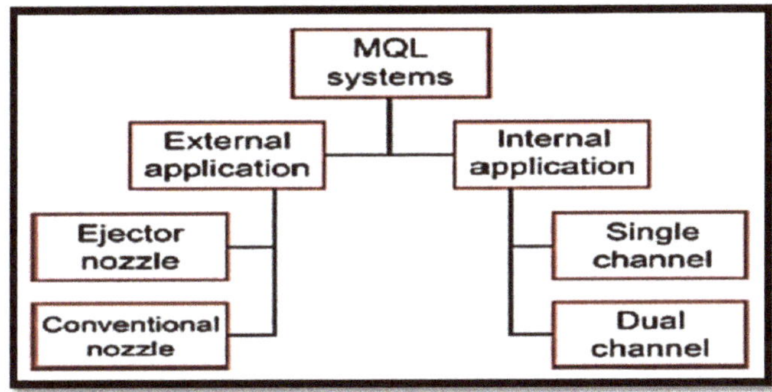


**Figure 12.** Minimum quantity lubrication delivery system categorizations [95].

All of these review studies indicate that the use of both vegetable oil-based MWFs and synthetic esters are safe to use for machining in place of conventional techniques and cutting fluids, as they are non-toxic and sustainable; MQL is a more feasible choice for machining applications due to being risk-free for the health of workers and environment.

### 3.5. Nanofluids

When the fluids are suspended with nano-sized particles, they are referred to as nanofluids. Nanofluids are the colloidal dispersion of the nanometer-sized particles, which are referred to as the nanoparticles in the base fluids [104]. The base fluids may include water, ethylene glycol, engine oil, or any other cutting fluid. The current recent advances in nanotechnology allows us to use nanofluids as conventional MWFs in conjunction with the minimum quantity lubrication technique in machining processes [105]. The inherent properties that nanofluids offer, such as enhanced heat transfer and improved tribological properties, allow them to be used in applications where better cooling and lubrication are required during the machining process, thus making the machining process more viable. Therefore, using nanofluids as an alternative to the conventional MWFs is one of the novel technological approaches in machining. Based on their heat transfer and tribological characteristics, nanoparticles that comprise $MoS_2$, $CuO$, $ZnO$, diamond, $Ag$, and titanium have been investigated for their use in machining operation. A considerable amount of research is being conducted to investigate the feasibility of nanofluids prepared from the colloidal dispersion of nanoparticles in the base fluid for machining operation. The decision to use nanofluids as a coolant is solely due to the enhanced thermal conductivity characteristics of the nanoparticles that are suspended in the base fluids [106]. It has been reported that the size of the nanoparticles also play an important role in determining the thermal conductivity of the nanofluids [107–110]. Furthermore, it has been reported that the nanofluids with smaller-sized nanoparticles have more enhanced thermal conductivity due to their extended specific surface area. Other factors which affect the thermal conductivity of nanofluids include the temperature of the nanofluid and the concentration of nanoparticles in the base fluid [111–115]. The thermal conductivity of nanoparticles and the base fluid also considerably affect the thermal conductivity of the nanofluid. The higher the thermal conductivity of the nanoparticles and thermal conductivity ratio i.e., the higher the ratio of the thermal conductivity of the nanoparticles and thermal conductivity of the base fluid, the higher the thermal conductivity of the resulting nanofluid will be [116,117]. Adding to the thermal conductivity, the pH of the base fluids and additives also play an important role. It has been observed that an increase in the pH of the base fluids and additives increase the thermal conductivity of the nanofluids. This is because of the fact that an increase in the pH value of the base fluids and additives results in the prevention of agglomeration and the improvement of the nanoparticle suspension [118–122].

In addition to thermal conductivity, other factors which affect the performance of nanofluids include the stability and viscosity of the nanofluids. The stability of the nanofluid is very important for improved heat transfer and thus the stability depends on various factors, such as the characteristics of the nanoparticles themselves, the methods of preparation, ultrasonication, stirring, etc. [123] Moreover, the viscosity of the nanofluids plays an important role in the performance of the nanofluids. Viscosity is defined as the internal resistance of the fluid to flow, i.e., the fluid's internal friction to flow, expressed as the force per unit area, which resists the flow; this property is widely affected by external physical parameters, such as temperature. Therefore, viscosity is an important parameter to be considered in thermal and fluid flow applications. Several investigations have been carried out to investigate the viscosity of nanofluids, and these investigations have reported an increase in the viscosity of the nanofluids with an increase in the volume fraction. Additionally, the size of the nanoparticles was seen to have a minimal effect on the viscosity of the nanofluids [124–127].

Nanofluids are widely used in the MQL technique to minimize the amount of lubricant used, and numerous attempts have been made to perform machining operation using nanofluids under the MQL method. Various studies have reported better surface finish, lower cutting forces, lower power consumption, and a higher tool life wen nanofluids were used, compared with dry machining and machining using flooded cooling methods. Prasad and Srikant [128] performed an experimental investigation on the turning of AISI 1040 using nanographite particles mixed with cutting fluid using the MQL method. They reported that as the concentration of the nanoparticles was increased, there was a spike in the values of the pH, viscosity, and thermal conductivity, in addition to lower tool wear, surface roughness, nodal temperatures, and cutting forces. They also observed a better machining performance at 0.3% nanoparticle concentration and a flowrate of 15 mL/min. Rahmati et al. [129] performed the slot milling of Al6061-T6, allowing the use of nanoparticles under the MQL approach. They reported that at 1% nanoparticle concentration in the mineral oil, the lowest cutting forces were observed; the lowest cutting temperature was observed at 0.5% nanoparticle concentration. Similarly, Sarhan et al. [130] reported a considerable reduction in the coefficient of friction at the tool–chip interface and consequently a decrease in the cutting forces, specific energy, and power when using $SiO_2$ in tandem with mineral oil under the MQL method. Yücel et al. [131] performed a turning operation on the AA 2024 T3 aluminum alloy, using the MoS2-based nanofluid and the MQL technique, in order to investigate the tribological and machining characteristics. They reported that significant improvements were achieved in the surface roughness, surface topography, and maximum temperature. They also added that by using the nanofluid-based MQL, the built-up edges were eliminated, and they obtained less damaged edges compared with dry machining.

Recently, Şirin and Kivak [132] performed a milling operation on the Inconel X-750 superalloy to see the effects of hybrid nanofluids using the MQL technique. They investigated the combination of different nanofluids, cutting speeds, and feed rates, and reported that using the hexagonal boron nitride (hBN)/graphite nanofluids resulted in a better performance compared to their counterparts under all criteria. They also added that hBN/graphite nanofluids achieved 36.17% and 6.08% improvements in tool life, respectively, compared to the graphite/$MoS_2$ and hBN/$MoS_2$ nanofluids. Junankar et al. [133] conducted a performance evaluation of a Cu nanofluid in a turning operation of bearing steel using the MQL approach. They analyzed the effect of the cutting sped, feed rate, and depth of the cut to perform a multi-objective optimization; this analysis was conducted using the grey relational analysis technique, and it was performed to obtain the optimum conditions of operation and their impact on the surface roughness and the cutting zone temperature. They reported that Cu nanofluid in conjunction with MQL resulted in the most significant cooling environment compared with vegetable oil MWFs. They also reported that the surface roughness and the cutting zone temperature were considerably reduced when the machining operation was performed using a Cu nanofluid under the MQL method. Haq et al. [134] evaluated the effects of a nanofluid-based MQL technique while performing a milling operation on the Inconel 718 superalloy, and compared the results of the simple MQL and nanofluid-based MQL approaches. They investigated the effect of feed rate, speed, flow rate, depth of the cut on the material removal rate, and the surface roughness, and conducted the optimization using the response surface methodology. They reported that the nanofluid-based MQL approach was better as compared with the simple MQL method, and resulted in decreased surface roughness, temperature, and power. Barewar et al. [135] investigated the sustainable machining of the Inconel 718 superalloy using an Ag/ZnO-based hybrid nanofluid and the MQL method, and performed the optimization using the Taguchi method with the grey relational analysis. They reported that the nanofluid-based MQL method resulted in an improved surface finish, minimum tool wear, and lower cutting temperature when compared with the simple MQL method and dry machining. Tiwari et al. [136] performed a computational analysis to see the characteristics of the surfaces of different concentrations of different nanofluids in conjunction with the MQL technique. They analyzed different nanofluids such as $Al_2O_3$, CuO, and $TiO_2$ at different concentrations (1% to 6%, at an interval of 1%) through the MATLAB software. From their analysis, they reported that the nanofluid-based MQLs resulted in intermittent chips which were easy to remove in contrast to the normal MQLs, which resulted in continuous chips. They also added that by using the nanofluid-based MQL method, cutting power was reduced, and a better surface finish was obtained. Mohana Rao et al. [137] performed an experimental investigation to observe the effects of cutting parameters on the tuning of EN-36 steel using both dry MQL and nanofluid-based MQL methods. They performed the investigations at 6% and 8% volume concentration of $Al_2O_3$ nanofluid and used the Taguchi analysis to optimize the process. They reported that at the 8% volume concentration, the surface roughness, temperature, cutting forces, and tool wear was lower compared with the 6% volume concentration and compared with dry machining.

Khanafer et al. [138] investigated the micro-drilling of the Inconel®718 superalloy using a MQL-$Al_2O_3$ nanofluid, and reported that the thrust forces were lower in the case of MQL-based nanofluid cooling compared with simple MQL cooling and flood cooling. They also reported that burr formation, tool wear, and cooling rates were improved in the case of the MQL-$Al_2O_3$ nanofluid. Sharma et al. [139] compared three different types of nanofluids, namely $Al_2O_3$, $TiO_2$, and $SiO_2$, with varying volume fractions to be utilized in metal cutting fluids. They concluded that the $Al_2O_3$ nanofluid exhibited better thermal properties compared with $SiO_2$ and $TiO_2$. Sharma et al. [140] experimentally investigated the turning operation of AISI 1040 steel using $Al_2O_3$ nanoparticle-based cutting fluids and the MQL approach. They reported that the performance of $Al_2O_3$ nanofluids were better in terms of the surface roughness, tool wear, cutting force, and chip morphology when compared with dry machining and wet machining with conventional cutting fluid. Minh et al. [141] investigated the performance of 0.5% (by volume concentration) $Al_2O_3$ nanofluids in MQL in the hard milling of $60Si_2Mn$ steel using cemented carbide tools. They reported that the tool life was considerably improved, and they observed a reduction in the roughness and cutting forces in the range of 35–60% under the MQL conditions. They added that it could be attributed to the improved tribological behavior as well as the cooling and lubricating effect of the nanoparticles.

The above analyses indicate that cutting fluid applications, as well as cooling and lubrication media, can be customized by using properly selected nanofluids in varying amounts. For an enhanced cooling effect, i.e., for an enhanced heat removal rate, nanofluids can be tailored to meet the requirements. When the objective is to obtain more lubrication, nanofluids can be used as a cutting medium in the form of droplets with the MQL technique. From the above analyses, we inferred that the $Al_2O_3$ nanoparticles have shown promising results compared with their counterparts. However, nanofluids are still in the developmental phase, but the applications of nanofluids in machining have promising prospects compared to nano-coolants.

## 4. Conclusions

In this review, we attempted to highlight the properties and associated drawbacks of the mineral oil-based MWFs and to perform a comprehensive literature review on the potential alternatives to mineral oil-based MWFs, such as vegetable oil-based MWFs; and we attempted to investigate other sustainable machining operations, including the high-pressure coolant, dry machining, MQL, and the nanofluid methods. The pros and cons of all of the associated alternatives were critically reviewed in terms of their applicability, adaptability, cost-effectiveness, and environmental impact so that an unbiased analysis could be performed for the use of MWFs in machining applications. As seen from the comprehensive literature review, machining is one of the main parts of every manufacturing plant and it cannot be ignored, and MWFs play a key role in the cost of machining. Cutting fluids have widely been used in conventional machining process for cooling applications, but they are a main factor of increased costs along with adverse effects on the environment and health of workers/operators. Therefore, it is necessary to minimize the utilization of mineral oil-based MWFs, and instead adopt vegetable oil-based MWFs and other sustainable methods such as the MQL, dry machining, high pressure coolant, and nanofluid methods, among other alternatives. These alternative techniques not only have been shown to reduce the cost of manufacturing by 10% to 17%, but also have a more positive impact on the environment by eliminating issues concerning cutting fluid disposal and by minimizing the contact of the workers with the MWFs, contact that may result in severe diseases. These techniques demonstrated very promising results as compared with the conventional methods where mineral oil-based MWFs were used in the machining process. By keeping this in view with the latest demands of industrial sectors and modern machining process requirements, these techniques are one of the best options to opt for, allowing for the implementation of sustainable machining processes in replacement of traditional methods.

## 5. Future Recommendations

- It is evident that the utilization of vegetable oil-based MWFs have shown better performance in terms of decreasing the overall cutting temperature, cutting forces, and surface roughness, among other desired properties. They have also proved themselves to be more eco-friendly as well, but there are some shortcomings (which can be further studied), and there is room for improvement in these shortcomings. Little attention was paid to the oxidation and thermal stabilities of the vegetable oil-based MWFs.
- For the vegetable oil-based MWFs, it was seen that most of the research was carried out for ferrous materials and alloys, and little attention was paid to the non-ferrous materials, such as copper, brass, and aluminum.
- These days, super alloys are also being widely used due to their excellent properties. Therefore, consideration should be given in exploring the application of vegetable oil-based MWFs in the case of super alloys and other mentioned materials.
- Nanofluids have become an emerging technology due to their excellent thermophysical properties and they have proven themselves to be an excellent candidate in machining applications, offering desired properties such as decreased interface temperature, lower cutting forces, lower power consumption, and improved surface finish. However, the properties of the nanofluids can be further enhanced by tweaking different parameters such as the size of the nanoparticles, shape of the nanoparticles, volumetric concentration, and spray nozzle angle, among other parameters.
- The application of nanofluids has not been cost-effective up to this point, and some studies have reported a negative impact on the environment. Therefore, efforts can be made to develop novel nanofluids which are more eco-friendly and provide cost-effective solutions.
- A limited number of research has been done on hybrid nanofluids, i.e., the combination of different nanoparticles and their properties; therefore, efforts can be made to test different hybrid nanoparticles and their performance under different conditions, in terms of the thermal conductivity, stability, viscosity, material removal rate, cutting forces, cutting temperatures, and power consumption, among other attributes.

**Author Contributions:** Conceptualization, M.H. and M.A.A.K.; methodology, M.H. and S.K.L.; writing—original draft preparation, M.H. and M.A.A.K.; writing—review and editing, B.Z., M.A. and A.A. All authors have read and agreed to the published version of the manuscript.

**Funding:** The APC was funded by Prince Mohammad Bin Fahd University, Saudi Arabia.

**Institutional Review Board Statement:** Not applicable.

**Informed Consent Statement:** Not applicable.

**Data Availability Statement:** Not applicable.

**Acknowledgments:** The authors would like to acknowledge the supportive environment provided by Prince Mohammad Bin Fahd University to facilitate the preparation process of the manuscript and its publication.

**Conflicts of Interest:** The authors declare no conflict of interest.

# References

1. Leahu-Aluas, S. *Sustainable Manufacturing—An Overview for Manufacturing Engineers*; Sustainable Manufacturing Consulting, 2010.
2. Skerlos, S.J.; Hayes, K.F.; Clarens, A.F.; Zhao, F. Current advances in sustainable Metalworking Fluids research. *Int. J. Sustain. Manuf.* **2008**, *1*, 180. [CrossRef]
3. Lozano, R. Envisioning sustainability three-dimensionally. *J. Clean. Prod.* **2008**, *16*, 1838–1846. [CrossRef]
4. Günay, M.; Yücel, E. An evaluation on machining processes for sustainable manufacturing. *Gazi Univ. J. Sci.* **2013**, *26*, 241–252.
5. Nachtman, E.S.; Kalpakjian, S. *Lubricants and Lubrication in Metalworking Operations*; Marcel Dekker, Inc.: New York, NY, USA, 1985; p. 215.
6. Marksberry, P.W.; Jawahir, I.S. A comprehensive tool-wear/tool-life performance model in the evaluation of NDM (near dry machining) for sustainable manufacturing. *Int. J. Mach. Tools Manuf.* **2008**, *48*, 878–886. [CrossRef]
7. Wickramasinghe, K.C.; Perera, G.I.P.; Herath, H.M.C.M. Formulation and performance evaluation of a novel coconut oil–based metalworking fluid. *Mater. Manuf. Process.* **2017**, *32*, 1026–1033. [CrossRef]
8. Talib, N.; Rahim, E.A. Performance of modified jatropha oil in combination with hexagonal boron nitride particles as a bio-based lubricant for green machining. *Tribol. Int.* **2018**, *118*, 89–104. [CrossRef]
9. Wickramasinghe, K.C.; Herath, H.M.C.M.; Perera, G.I.P. Empirical investigation of surface quality and temperature during turning AISI 304 steel with vegetable oil based metalworking fluids. In Proceedings of the 2016 Manufacturing & Industrial Engineering Symposium (MIES), Colombo, Sri Lanka, 22 October 2016; pp. 1–4. [CrossRef]
10. Paul, S.; Pal, P.K. Study of Surface Quality During High Speed Machining Using Eco-Friendly Cutting Fluid. *Int. J. Mach. Mach. Mater.* **2011**, *11*, 24–28. Available online: http://mech-ing.com/journal/Archive/2011/11/121_Swarup%20Paul.pdf (accessed on 19 May 2022).
11. Çiçek, A.; Kıvak, T.; Samtaş, G. Application of Taguchi Method for Surface Roughness and Roundness Error in Drilling of AISI 316 Stainless Steel. *Stroj. Vestn. J. Mech. Eng.* **2012**, *58*, 165–174. [CrossRef]
12. Lawal, S.A.; Choudhury, I.A.; Nukman, Y. Application of vegetable oil-based metalworking fluids in machining ferrous metals—A review. *Int. J. Mach. Tools Manuf.* **2012**, *52*, 1–12. [CrossRef]
13. Panda, A.; Duplák, J.; Vasilko, K. Analysis of Cutting Tools Durability Compared with Standard ISO 3685. *Int. J. Comput. Theory Eng.* **2012**, *4*, 621–624. [CrossRef]
14. Kuram, E.; Ozcelik, B.; Demirbas, E.; Şik, E.; Tansel, I.N. Evaluation of New Vegetable-Based Cutting Fluids on Thrust Force and Surface Roughness in Drilling of AISI 304 Using Taguchi Method. *Mater. Manuf. Process.* **2011**, *26*, 1136–1146. [CrossRef]
15. Cheng, C.; Phipps, D.; Alkhaddar, R.M. Treatment of spent metalworking fluids. *Water Res.* **2005**, *39*, 4051–4063. [CrossRef] [PubMed]
16. Benedicto, E.; Carou, D.; Rubio, E.M. Technical, Economic and Environmental Review of the Lubrication/Cooling Systems Used in Machining Processes. *Procedia Eng.* **2017**, *184*, 99–116. [CrossRef]
17. Lathi, P.; Mattiasson, B. Green approach for the preparation of biodegradable lubricant base stock from epoxidized vegetable oil. *Appl. Catal. B Environ.* **2007**, *69*, 207–212. [CrossRef]
18. Guo, S.; Li, C.; Zhang, Y.; Wang, Y.; Li, B.; Yang, M.; Zhang, X.; Liu, G. Experimental evaluation of the lubrication performance of mixtures of castor oil with other vegetable oils in MQL grinding of nickel-based alloy. *J. Clean. Prod.* **2017**, *140*, 1060–1076. [CrossRef]
19. Belluco, W.; De Chiffre, L. Performance evaluation of vegetable-based oils in drilling austenitic stainless steel. *J. Mater. Process. Technol.* **2004**, *148*, 171–176. [CrossRef]
20. Adler, D.P.; Hii, W.W.-S.; Michalek, D.J.; Sutherland, J.W. Examining the Role of Cutting Fluids in Machining and Efforts to Address Associated Environmental/Health Concerns. *Mach. Sci. Technol.* **2006**, *10*, 23–58. [CrossRef]
21. Lawal, S.A.; Choudhury, I.A.; Nukman, Y. Developments in the formulation and application of vegetable oil-based metalworking fluids in turning process. *Int. J. Adv. Manuf. Technol.* **2013**, *67*, 1765–1776. [CrossRef]
22. Pusavec, F.; Kramar, D.; Krajnik, P.; Kopac, J. Transitioning to sustainable production—Part II: Evaluation of sustainable machining technologies. *J. Clean. Prod.* **2010**, *18*, 1211–1221. [CrossRef]
23. King, N.; Keranen, L.; Gunter, K.; Sutherland, J. *Wet Versus Dry Turning: A Comparison of Machining Costs, Product Quality, and Aerosol Formation*; SAE International: Warrendale, PA, USA, 2001. [CrossRef]
24. Brinksmeier, E.; Heinzel, C.; Wittmann, M. Friction, Cooling and Lubrication in Grinding. *CIRP Ann.* **1999**, *48*, 581–598. [CrossRef]
25. Olawumi, T.O.; Chan, D.W.M. A scientometric review of global research on sustainability and sustainable development. *J. Clean. Prod.* **2018**, *183*, 231–250. [CrossRef]
26. Khalaj, M.; Kamali, M.; Costa, M.E.V.; Capela, I. Green synthesis of nanomaterials—A scientometric assessment. *J. Clean. Prod.* **2020**, *267*, 122036. [CrossRef]
27. Tariq, S.; Hu, Z.; Zayed, T. Micro-electromechanical systems-based technologies for leak detection and localization in water supply networks: A bibliometric and systematic review. *J. Clean. Prod.* **2021**, *289*, 125751. [CrossRef]
28. Huang, L.; Zhou, M.; Lv, J.; Chen, K. Trends in global research in forest carbon sequestration: A bibliometric analysis. *J. Clean. Prod.* **2020**, *252*, 119908. [CrossRef]
29. Pranckutė, R. Web of Science (WoS) and Scopus: The Titans of Bibliographic Information in Today's Academic World. *Publications* **2021**, *9*, 12. [CrossRef]
30. RStudio. Available online: https://www.rstudio.com/ (accessed on 14 April 2022).

31. Li, K.; Aghazadeh, F.; Hatipkarasulu, S.; Ray, T.G. Health Risks from Exposure to Metal-Working Fluids in Machining and Grinding Operations. *Int. J. Occup. Saf. Ergon.* **2003**, *9*, 75–95. [CrossRef]
32. Kawatra, S.K.; Hess, M.J. Environmental Beneficiation of Machining Wastes—Part III: Effects of Metal Working Fluids on the Spontaneous Heating of Machining Swarf. *J. Air Waste Manag. Assoc.* **1999**, *49*, 588–593. [CrossRef] [PubMed]
33. Park, D.; Stewart, P.A.; Coble, J.B. A Comprehensive Review of the Literature on Exposure to Metalworking Fluids. *J. Occup. Environ. Hyg.* **2009**, *6*, 530–541. [CrossRef] [PubMed]
34. Çolak, O. Investigation on Machining Performance of Inconel 718 under High Pressure Cooling Conditions. *Stroj. Vestn. J. Mech. Eng.* **2012**, *58*, 683–690. [CrossRef]
35. Bennett, E.O. Water based cutting fluids and human health. *Tribol. Int.* **1983**, *16*, 133–136. [CrossRef]
36. Park, D.; Stewart, P.; Coble, J.B. Determinants of Exposure to Metalworking Fluid Aerosols: A Literature Review and Analysis of Reported Measurements. *Ann. Occup. Hyg.* **2009**, *53*, 271–288. [CrossRef] [PubMed]
37. Wu, C.-C.; Liu, H.-M. Determinants of Metals Exposure to Metalworking Fluid Among Metalworkers in Taiwan. *Arch. Environ. Occup. Health* **2014**, *69*, 131–138. [CrossRef] [PubMed]
38. Suuronen, K. Metalworking Fluids—Allergens, Exposure, and Skin and Respiratory Effects Katri Suuronen People and Work. Ph.D. Thesis, University of Eastern Finland, Kuopio, Finland, 2014.
39. Krolczyk, G.M.; Maruda, R.W.; Krolczyk, J.B.; Wojciechowski, S.; Mia, M.; Nieslony, P.; Budzik, G. Ecological trends in machining as a key factor in sustainable production—A review. *J. Clean. Prod.* **2019**, *218*, 601–615. [CrossRef]
40. Raynor, P.C.; Kim, S.W.; Bhattacharya, M. Mist generation from metalworking fluids formulated using vegetable oils. *Ann. Occup. Hyg.* **2005**, *49*, 283–293. [CrossRef]
41. Choi, U.S.; Ahn, B.G.; Kwon, O.K.; Chun, Y.J. Tribological behavior of some antiwear additives in vegetable oils. *Tribol. Int.* **1997**, *30*, 677–683. [CrossRef]
42. D'Amato, R.; Wang, C.; Calvo, R.; Valášek, P.; Ruggiero, A. Characterization of vegetable oil as cutting fluid. *Procedia Manuf.* **2019**, *41*, 145–152. [CrossRef]
43. Pal, A.; Chatha, S.S.; Sidhu, H.S. Experimental investigation on the performance of MQL drilling of AISI 321 stainless steel using nano-graphene enhanced vegetable-oil-based cutting fluid. *Tribol. Int.* **2020**, *151*, 106508. [CrossRef]
44. Osama, M.; Singh, A.; Walvekar, R.; Khalid, M.; Gupta, T.C.S.M.; Yin, W.W. Recent developments and performance review of metal working fluids. *Tribol. Int.* **2017**, *114*, 389–401. [CrossRef]
45. Kumar Gajrani, K.; Ravi Sankar, M. Past and Current Status of Eco-Friendly Vegetable Oil Based Metal Cutting Fluids. *Mater. Today Proc.* **2017**, *4*, 3786–3795. [CrossRef]
46. Fernando, W.L.R.; Sarmilan, N.; Wickramasinghe, K.C.; Herath, H.M.C.M.; Perera, G.I.P. Experimental investigation of Minimum Quantity Lubrication (MQL) of coconut oil based Metal Working Fluid. *Mater. Today Proc.* **2020**, *23*, 23–26. [CrossRef]
47. Saikiran, M.; Kumar, P. An investigation on the effects of vegetable oil based cutting fluids in the machining of copper alloys. *Mater. Today Proc.* **2019**, *19*, 455–461. [CrossRef]
48. Rapeti, P.; Pasam, V.K.; Rao Gurram, K.M.; Revuru, R.S. Performance evaluation of vegetable oil based nano cutting fluids in machining using grey relational analysis-A step towards sustainable manufacturing. *J. Clean. Prod.* **2018**, *172*, 2862–2875. [CrossRef]
49. Mamidi, V.; Xavior, A. A review on selection of cutting fuids. *J. Res. Sci. Technol.* **2012**, *1*, 1174–2277.
50. Odi-Owei, S. Tribological properties of some vegetable oils and fats. *Lubr. Eng* **1989**, *45*, 685–690.
51. Talib, N.; Rahim, E.A. The Performance of Modified Jatropha-Oil Based Trimethylolpropane (TMP) Ester on Tribology Characteristic for Sustainable Metalworking Fluids (MWFs). *Appl. Mech. Mater.* **2014**, *660*, 357–361. [CrossRef]
52. John, J.; Bhattacharya, M.; Raynor, P.C. Emulsions containing vegetable oils for cutting fluid application. *Colloids Surf. A Physicochem. Eng. Asp.* **2004**, *237*, 141–150. [CrossRef]
53. Mannekote, J.K.; Kailas, S.V. The Effect of Oxidation on the Tribological Performance of Few Vegetable Oils. *J. Mater. Res. Technol.* **2012**, *1*, 91–95. [CrossRef]
54. Erhan, S.Z.; Sharma, B.K.; Perez, J.M. Oxidation and low temperature stability of vegetable oil-based lubricants. *Ind. Crops Prod.* **2006**, *24*, 292–299. [CrossRef]
55. Perera, G.I.P.; Herath, H.M.; Perera, I.S.J.; Medagoda, M.M.P. Investigation on white coconut oil to use as a metal working fluid during turning. *Proc. Inst. Mech. Eng. Part B J. Eng. Manuf.* **2015**, *229*, 38–44. [CrossRef]
56. Nune, M.M.R.; Chaganti, P.K. Development, characterization, and evaluation of novel eco-friendly metal working fluid. *Measurement* **2019**, *137*, 401–416. [CrossRef]
57. Coker, A.K. Physical properties of Liquids and Gases. In *Ludwig's Applied Process Design for Chemical and Petrochemical Plants*; Elsevier: Amsterdam, The Netherlands, 2007; pp. 103–132.
58. Arora, G.; Kumar, U.; Bhowmik, P. Vegetable oil based cutting fluids—Green and sustainable machining—II. *J. Mater. Sci. Mech. Eng.* **2015**, *2*, 1–5.
59. Doll, K.M.; Sharma, B.K. Emulsification of Chemically Modified Vegetable Oils for Lubricant Use. *J. Surfactants Deterg.* **2011**, *14*, 131–138. [CrossRef]
60. Byers, J.P. *Metalworking Fluids*; CRC Press: Boca Raton, FL, USA, 2017; ISBN 1498722237.
61. Fernandes, C.P.; Mascarenhas, M.P.; Zibetti, F.M.; Lima, B.G.; Oliveira, R.P.R.F.; Rocha, L.; Falcão, D.Q. HLB value, an important parameter for the development of essential oil phytopharmaceuticals. *Rev. Bras. Farmacogn.* **2013**, *23*, 108–114. [CrossRef]

62. Genot, C.; Berton, C.; Ropers, M.-H. The Role of the Interfacial Layer and Emulsifying Proteins in the Oxidation in Oil-in-Water Emulsions. In *Lipid Oxidation*; Elsevier: Amsterdam, The Netherlands, 2013; pp. 177–210.
63. Noor El-Din, M.R.; Mishrif, M.R.; Kailas, S.V.; Suvin, P.S.; Mannekote, J.K. Studying the lubricity of new eco-friendly cutting oil formulation in metal working fluid. *Ind. Lubr. Tribol.* **2018**, *70*, 1569–1579. [CrossRef]
64. Gacek, M.M.; Berg, J.C. Effect of surfactant hydrophile-lipophile balance (HLB) value on mineral oxide charging in apolar media. *J. Colloid Interface Sci.* **2015**, *449*, 192–197. [CrossRef] [PubMed]
65. Abdalla, H.S.; Baines, W.; McIntyre, G.; Slade, C. Development of novel sustainable neat-oil metal working fluids for stainless steel and titanium alloy machining. Part 1. Formulation development. *Int. J. Adv. Manuf. Technol.* **2007**, *34*, 21–33. [CrossRef]
66. Abdul Sani, A.S.; Rahim, E.A.; Sharif, S.; Sasahara, H. Machining performance of vegetable oil with phosphonium- and ammonium-based ionic liquids via MQL technique. *J. Clean. Prod.* **2019**, *209*, 947–964. [CrossRef]
67. Vamsi Krishna, P.; Srikant, R.R.; Nageswara Rao, D. Experimental investigation on the performance of nanoboric acid suspensions in SAE-40 and coconut oil during turning of AISI 1040 steel. *Int. J. Mach. Tools Manuf.* **2010**, *50*, 911–916. [CrossRef]
68. Trent, E.M.; Wright, P.K. *Metal Cutting*; Butterworth-Heinemann: Oxford, UK, 2000; ISBN 075067069X.
69. Taylor, F.W. *On the Art of Cutting Metals: An Address Made at the Opening of the Annual Meeting in New York, December, 1900 [ie., 1906]*; American Society of Mechanical Engineers: New York, NY, USA, 1907.
70. Najiha, M.S.; Rahman, M.M.; Yusoff, A.R. Environmental impacts and hazards associated with metal working fluids and recent advances in the sustainable systems: A review. *Renew. Sustain. Energy Rev.* **2016**, *60*, 1008–1031. [CrossRef]
71. Fukui, H.; Okida, J.; Omori, N.; Moriguchi, H.; Tsuda, K. Cutting performance of DLC coated tools in dry machining aluminum alloys. *Surf. Coat. Technol.* **2004**, *187*, 70–76. [CrossRef]
72. Klocke, F.; Eisenblätter, G. Dry Cutting. *CIRP Ann.* **1997**, *46*, 519–526. [CrossRef]
73. Sreejith, P.; Ngoi, B.K. Dry machining: Machining of the future. *J. Mater. Process. Technol.* **2000**, *101*, 287–291. [CrossRef]
74. Vereschaka, A.A.; Vereschaka, A.S.; Grigoriev, S.N.; Kirillov, A.K.; Khaustova, O.U. Development and Research of Environmentally Friendly Dry Technological Machining System with Compensation of Physical Function of Cutting Fluids. *Procedia CIRP* **2013**, *7*, 311–316. [CrossRef]
75. Devillez, A.; Le Coz, G.; Dominiak, S.; Dudzinski, D. Dry machining of Inconel 718, workpiece surface integrity. *J. Mater. Process. Technol.* **2011**, *211*, 1590–1598. [CrossRef]
76. Çolak, O. Optimization of Machining Performance in High-Pressure Assisted Turning of Ti6Al4V Alloy. *Stroj. Vestn. J. Mech. Eng.* **2014**, *60*, 675–681. [CrossRef]
77. Dahlman, P.; Escursell, M. High-pressure jet-assisted cooling: A new possibility for near net shape turning of decarburized steel. *Int. J. Mach. Tools Manuf.* **2004**, *44*, 109–115. [CrossRef]
78. Ezugwu, E.O.; Bonney, J.; Fadare, D.A.; Sales, W.F. Machining of nickel-base, Inconel 718, alloy with ceramic tools under finishing conditions with various coolant supply pressures. *J. Mater. Process. Technol.* **2005**, *162–163*, 609–614. [CrossRef]
79. Kramar, D.; Krajnik, P.; Kopac, J. Capability of high pressure cooling in the turning of surface hardened piston rods. *J. Mater. Process. Technol.* **2010**, *210*, 212–218. [CrossRef]
80. Ayed, Y.; Germain, G.; Ammar, A.; Furet, B. Degradation modes and tool wear mechanisms in finish and rough machining of Ti17 Titanium alloy under high-pressure water jet assistance. *Wear* **2013**, *305*, 228–237. [CrossRef]
81. da Silva, R.B.; Machado, Á.R.; Ezugwu, E.O.; Bonney, J.; Sales, W.F. Tool life and wear mechanisms in high speed machining of Ti–6Al–4V alloy with PCD tools under various coolant pressures. *J. Mater. Process. Technol.* **2013**, *213*, 1459–1464. [CrossRef]
82. Sayuti, M.; Sarhan, A.A.D.; Salem, F. Novel uses of $SiO_2$ nano-lubrication system in hard turning process of hardened steel AISI4140 for less tool wear, surface roughness and oil consumption. *J. Clean. Prod.* **2014**, *67*, 265–276. [CrossRef]
83. Hamran, N.N.N.; Ghani, J.A.; Ramli, R.; Haron, C.H.C. A review on recent development of minimum quantity lubrication for sustainable machining. *J. Clean. Prod.* **2020**, *268*, 122165. [CrossRef]
84. Goindi, G.S.; Sarkar, P.; Jayal, A.D.; Chavan, S.N.; Mandal, D. Investigation of ionic liquids as additives to canola oil in minimum quantity lubrication milling of plain medium carbon steel. *Int. J. Adv. Manuf. Technol.* **2018**, *94*, 881–896. [CrossRef]
85. Goindi, G.S.; Chavan, S.N.; Mandal, D.; Sarkar, P.; Jayal, A.D. Investigation of Ionic Liquids as Novel Metalworking Fluids during Minimum Quantity Lubrication Machining of a Plain Carbon Steel. *Procedia CIRP* **2015**, *26*, 341–345. [CrossRef]
86. Najiha, M.S.; Rahman, M.M. Experimental investigation of flank wear in end milling of aluminum alloy with water-based $TiO_2$ nanofluid lubricant in minimum quantity lubrication technique. *Int. J. Adv. Manuf. Technol.* **2016**, *86*, 2527–2537. [CrossRef]
87. Marques, A.; Paipa Suarez, M.; Falco Sales, W.; Rocha Machado, Á. Turning of Inconel 718 with whisker-reinforced ceramic tools applying vegetable-based cutting fluid mixed with solid lubricants by MQL. *J. Mater. Process. Technol.* **2019**, *266*, 530–543. [CrossRef]
88. Osman, K.A.; Ünver, H.Ö.; Şeker, U. Application of minimum quantity lubrication techniques in machining process of titanium alloy for sustainability: A review. *Int. J. Adv. Manuf. Technol.* **2019**, *100*, 2311–2332. [CrossRef]
89. Paturi, U.M.R.; Maddu, Y.R.; Maruri, R.R.; Narala, S.K.R. Measurement and Analysis of Surface Roughness in WS2 Solid Lubricant Assisted Minimum Quantity Lubrication (MQL) Turning of Inconel 718. *Procedia CIRP* **2016**, *40*, 138–143. [CrossRef]
90. Sharif, M.N.; Pervaiz, S.; Deiab, I. Potential of alternative lubrication strategies for metal cutting processes: A review. *Int. J. Adv. Manuf. Technol.* **2017**, *89*, 2447–2479. [CrossRef]
91. Najiha, M.S.; Rahman, M.M.; Kadirgama, K. Parametric optimization of end milling process under minimum quantity lubrication with nanofluid as cutting medium using pareto optimality approach. *Int. J. Automot. Mech. Eng.* **2016**, *13*, 3345–3360. [CrossRef]

92. Boswell, B.; Islam, M.N.; Davies, I.J.; Ginting, Y.R.; Ong, A.K. A review identifying the effectiveness of minimum quantity lubrication (MQL) during conventional machining. *Int. J. Adv. Manuf. Technol.* **2017**, *92*, 321–340. [CrossRef]
93. Eltaggaz, A.; Hegab, H.; Deiab, I.; Kishawy, H.A. Hybrid nano-fluid-minimum quantity lubrication strategy for machining austempered ductile iron (ADI). *Int. J. Interact. Des. Manuf.* **2018**, *12*, 1273–1281. [CrossRef]
94. Tai, B.; Stephenson, D.; Furness, R.; Shih, A. Minimum Quantity Lubrication for Sustainable Machining. In *Encyclopedia of Sustainable Technologies*; Elsevier: Amsterdam, The Netherlands, 2017; pp. 477–485.
95. Khan, M.M.A.; Dhar, N.R. Performance evaluation of minimum quantity lubrication by vegetable oil in terms of cutting force, cutting zone temperature, tool wear, job dimension and surface finish in turning AISI-1060 steel. *J. Zhejiang Univ. A* **2006**, *7*, 1790–1799. [CrossRef]
96. Ginting, Y.R.; Boswell, B.; Biswas, W.; Islam, N. Advancing Environmentally Conscious Machining. *Procedia CIRP* **2015**, *26*, 391–396. [CrossRef]
97. Islam, M.N. Effect of additional factors on dimensional accuracy and surface finish of turned parts. *Mach. Sci. Technol.* **2013**, *17*, 145–162. [CrossRef]
98. Sales, W.; Becker, M.; Barcellos, C.S.; Landre, J.; Bonney, J.; Ezugwu, E.O. Tribological behaviour when face milling AISI 4140 steel with minimum quantity fluid application. *Ind. Lubr. Tribol.* **2009**, *61*, 84–90. [CrossRef]
99. Dixit, U.S.; Sarma, D.K.; Davim, J.P. Machining with Minimal Cutting Fluid. In *Environmentally Friendly Machining*; Springer: Berlin, Germany, 2012; pp. 9–17.
100. Ramana, M.V.; Mohan Rao, G.K.; Rao, D.H. Experimental Investigations and Selection of Optimal Cutting Conditions in Turning Of Ti-6al-4v Alloy With Different Cutting Fluids By Minimum Quantity Lubrication (MQL) Methodology. *i-Manag. J. Mech. Eng.* **2012**, *2*, 45–52. [CrossRef]
101. Banerjee, N.; Sharma, A. Improving machining performance of Ti-6Al-4V through multi-point minimum quantity lubrication method. *Proc. Inst. Mech. Eng. Part B J. Eng. Manuf.* **2019**, *233*, 321–336. [CrossRef]
102. Fitrina, S.; Kristiawan, B.; Surojo, E.; Wijayanta, A.T.; Miyazaki, T.; Koyama, S. Influence of minimum quantity lubrication with $Al_2O_3$ nanoparticles on cutting parameters in drilling process. *AIP Conf. Proc.* **2018**, *1931*, 030056. [CrossRef]
103. Paul, S.; Ghosh, A. An Experimental Evaluation of Solid Lubricant Based Nanofluids in Small Quantity Cooling and Lubrication during Grinding. *Mater. Sci. Forum* **2017**, *890*, 98–102. [CrossRef]
104. Beck, M.P.; Yuan, Y.; Warrier, P.; Teja, A.S. The effect of particle size on the thermal conductivity of alumina nanofluids. *J. Nanopart. Res.* **2009**, *11*, 1129–1136. [CrossRef]
105. Srikant, R.R.; Rao, D.N.; Subrahmanyam, M.S.; Krishna, V.P. Applicability of cutting fluids with nanoparticle inclusion as coolants in machining. *Proc. Inst. Mech. Eng. Part J J. Eng. Tribol.* **2009**, *223*, 221–225. [CrossRef]
106. Rao, S.N.; Satyanarayana, B.; Venkatasubbaiah, K. Experimental estimation of tool wear and cutting temperatures in MQL using cutting fluids with CNT inclusion. *Int. J. Eng. Sci. Technol.* **2011**, *3*, 2928–2931.
107. Xie, H.; Wang, J.; Xi, T.; Liu, Y.; Ai, F.; Wu, Q. Thermal conductivity enhancement of suspensions containing nanosized alumina particles. *J. Appl. Phys.* **2002**, *91*, 4568–4572. [CrossRef]
108. Chon, C.H.; Kihm, K.D.; Lee, S.P.; Choi, S.U.S. Empirical correlation finding the role of temperature and particle size for nanofluid ($Al_2O_3$) thermal conductivity enhancement. *Appl. Phys. Lett.* **2005**, *87*, 153107. [CrossRef]
109. Mintsa, H.A.; Roy, G.; Nguyen, C.T.; Doucet, D. New temperature dependent thermal conductivity data for water-based nanofluids. *Int. J. Therm. Sci.* **2009**, *48*, 363–371. [CrossRef]
110. Chopkar, M.; Das, P.K.; Manna, I. Synthesis and characterization of nanofluid for advanced heat transfer applications. *Scr. Mater.* **2006**, *55*, 549–552. [CrossRef]
111. Das, S.K.; Putra, N.; Thiesen, P.; Roetzel, W. Temperature Dependence of Thermal Conductivity Enhancement for Nanofluids. *J. Heat Transf.* **2003**, *125*, 567–574. [CrossRef]
112. Choi, S.U.S.; Zhang, Z.G.; Yu, W.; Lockwood, F.E.; Grulke, E.A. Anomalous thermal conductivity enhancement in nanotube suspensions. *Appl. Phys. Lett.* **2001**, *79*, 2252–2254. [CrossRef]
113. Ding, Y.; Alias, H.; Wen, D.; Williams, R.A. Heat transfer of aqueous suspensions of carbon nanotubes (CNT nanofluids). *Int. J. Heat Mass Transf.* **2006**, *49*, 240–250. [CrossRef]
114. Yu, W.; Xie, H.; Li, Y.; Chen, L. Experimental investigation on thermal conductivity and viscosity of aluminum nitride nanofluid. *Particuology* **2011**, *9*, 187–191. [CrossRef]
115. Eastman, J.A.; Choi, S.U.S.; Li, S.; Yu, W.; Thompson, L.J. Anomalously increased effective thermal conductivities of ethylene glycol-based nanofluids containing copper nanoparticles. *Appl. Phys. Lett.* **2001**, *78*, 718–720. [CrossRef]
116. Chopkar, M.; Sudarshan, S.; Das, P.K.; Manna, I. Effect of Particle Size on Thermal Conductivity of Nanofluid. *Met. Mater. Trans. A* **2008**, *39*, 1535–1542. [CrossRef]
117. Wang, X.; Xu, X.; Choi, S.U.S. Thermal Conductivity of Nanoparticle—Fluid Mixture. *J. Thermophys. Heat Transf.* **1999**, *13*, 474–480. [CrossRef]
118. Lee, D.; Kim, J.-W.; Kim, B.G. A New Parameter to Control Heat Transport in Nanofluids: Surface Charge State of the Particle in Suspension. *J. Phys. Chem. B* **2006**, *110*, 4323–4328. [CrossRef] [PubMed]
119. Assael, M.J.; Metaxa, I.N.; Arvanitidis, J.; Christofilos, D.; Lioutas, C. Thermal Conductivity Enhancement in Aqueous Suspensions of Carbon Multi-Walled and Double-Walled Nanotubes in the Presence of Two Different Dispersants. *Int. J. Thermophys.* **2005**, *26*, 647–664. [CrossRef]

120. Wang, X.; Zhu, D.; Yang, S. Investigation of pH and SDBS on enhancement of thermal conductivity in nanofluids. *Chem. Phys. Lett.* **2009**, *470*, 107–111. [CrossRef]

121. Murshed, S.M.S.; Leong, K.C.; Yang, C. Characterization of Electrokinetic Properties of Nanofluids. *J. Nanosci. Nanotechnol.* **2008**, *8*, 5966–5971. [CrossRef]

122. Yu, W.; France, D.M.; Routbort, J.L.; Choi, S.U.S. Review and Comparison of Nanofluid Thermal Conductivity and Heat Transfer Enhancements. *Heat Transf. Eng.* **2008**, *29*, 432–460. [CrossRef]

123. Babita; Sharma, S.K.; Gupta, S.M. Preparation and evaluation of stable nanofluids for heat transfer application: A review. *Exp. Therm. Fluid Sci.* **2016**, *79*, 202–212. [CrossRef]

124. Prasher, R.; Song, D.; Wang, J.; Phelan, P. Measurements of nanofluid viscosity and its implications for thermal applications. *Appl. Phys. Lett.* **2006**, *89*, 133108. [CrossRef]

125. Anoop, K.B.; Sundararajan, T.; Das, S.K. Effect of particle size on the convective heat transfer in nanofluid in the developing region. *Int. J. Heat Mass Transf.* **2009**, *52*, 2189–2195. [CrossRef]

126. He, Y.; Jin, Y.; Chen, H.; Ding, Y.; Cang, D.; Lu, H. Heat transfer and flow behaviour of aqueous suspensions of $TiO_2$ nanoparticles (nanofluids) flowing upward through a vertical pipe. *Int. J. Heat Mass Transf.* **2007**, *50*, 2272–2281. [CrossRef]

127. Anoop, K.B.; Kabelac, S.; Sundararajan, T.; Das, S.K. Rheological and flow characteristics of nanofluids: Influence of electroviscous effects and particle agglomeration. *J. Appl. Phys.* **2009**, *106*, 034909. [CrossRef]

128. Prasad, M.; Srikant, R. Performance Evaluation of Nano Graphite Inclusions in Cutting Fluids with Mql Technique in Turning of Aisi 1040 Steel. *Int. J. Res. Eng. Technol.* **2013**, *02*, 381–393. [CrossRef]

129. Rahmati, B.; Sarhan, A.A.D.; Sayuti, M. Investigating the optimum molybdenum disulfide (MoS2) nanolubrication parameters in CNC milling of AL6061-T6 alloy. *Int. J. Adv. Manuf. Technol.* **2014**, *70*, 1143–1155. [CrossRef]

130. Sarhan, A.A.D.; Sayuti, M.; Hamdi, M. Reduction of power and lubricant oil consumption in milling process using a new $SiO_2$ nanolubrication system. *Int. J. Adv. Manuf. Technol.* **2012**, *63*, 505–512. [CrossRef]

131. Yücel, A.; Yıldırım, Ç.V.; Sarıkaya, M.; Şirin, Ş.; Kıvak, T.; Gupta, M.K.; Tomaz, Í.V. Influence of MoS2 based nanofluid-MQL on tribological and machining characteristics in turning of AA 2024 T3 aluminum alloy. *J. Mater. Res. Technol.* **2021**, *15*, 1688–1704. [CrossRef]

132. Şirin, Ş.; Kıvak, T. Effects of hybrid nanofluids on machining performance in MQL-milling of Inconel X-750 superalloy. *J. Manuf. Process.* **2021**, *70*, 163–176. [CrossRef]

133. Junankar, A.A.; Yashpal; Purohit, J.K.; Gohane, G.M.; Pachbhai, J.S.; Gupta, P.M.; Sayed, A.R. Performance evaluation of Cu nanofluid in bearing steel MQL based turning operation. *Mater. Today Proc.* **2021**, *44*, 4309–4314. [CrossRef]

134. Ul Haq, M.A.; Hussain, S.; Ali, M.A.; Farooq, M.U.; Mufti, N.A.; Pruncu, C.I.; Wasim, A. Evaluating the effects of nano-fluids based MQL milling of IN718 associated to sustainable productions. *J. Clean. Prod.* **2021**, *310*, 127463. [CrossRef]

135. Barewar, S.D.; Kotwani, A.; Chougule, S.S.; Unune, D.R. Investigating a novel Ag/ZnO based hybrid nanofluid for sustainable machining of inconel 718 under nanofluid based minimum quantity lubrication. *J. Manuf. Process.* **2021**, *66*, 313–324. [CrossRef]

136. Tiwari, A.; Agarwal, D.; Singh, A. Computational analysis of machining characteristics of surface using varying concentration of nanofluids ($Al_2O_3$, CuO and $TiO_2$) with MQL. *Mater. Today Proc.* **2021**, *42*, 1262–1269. [CrossRef]

137. Mohana Rao, G.; Dilkush, S.; Sudhakar, I.; Anil babu, P. Effect of Cutting Parameters with Dry and MQL Nano Fluids in Turning of EN-36 Steel. *Mater. Today Proc.* **2021**, *41*, 1182–1187. [CrossRef]

138. Khanafer, K.; Eltaggaz, A.; Deiab, I.; Agarwal, H.; Abdul-latif, A. Toward sustainable micro-drilling of Inconel 718 superalloy using MQL-Nanofluid. *Int. J. Adv. Manuf. Technol.* **2020**, *107*, 3459–3469. [CrossRef]

139. Sharma, A.K.; Tiwari, A.K.; Dixit, A.R. Characterization of $TiO_2$, $Al_2O_3$ and $SiO_2$ Nanoparticle based Cutting Fluids. *Mater. Today Proc.* **2016**, *3*, 1890–1898. [CrossRef]

140. Sharma, A.K.; Singh, R.K.; Dixit, A.R.; Tiwari, A.K. Characterization and experimental investigation of $Al_2O_3$ nanoparticle based cutting fluid in turning of AISI 1040 steel under minimum quantity lubrication (MQL). *Mater. Today Proc.* **2016**, *3*, 1899–1906. [CrossRef]

141. Minh, D.T.; The, L.T.; Bao, N.T. Performance of $Al_2O_3$ nanofluids in minimum quantity lubrication in hard milling of 60Si2Mn steel using cemented carbide tools. *Adv. Mech. Eng.* **2017**, *9*, 1–9. [CrossRef]

*Article*

# Build-Up an Economical Tool for Machining Operations Cost Estimation

Francisco J. G. Silva [1,2,*], Vitor F. C. Sousa [1,2], Arnaldo G. Pinto [1,2], Luís P. Ferreira [1,2] and Teresa Pereira [1,2]

[1]  Mechanical Engineering Department, ISEP-School of Engineering, Polytechnic of Porto, 4200-072 Porto, Portugal; vcris@isep.ipp.pt (V.F.C.S.); agp@isep.ipp.pt (A.G.P.); luispintoferreira@eu.ipp.pt (L.P.F.); mtp@isep.ipp.pt (T.P.)

[2]  INEGI-Instituto de Ciência e Inovação em Engenharia Mecânica e Engenharia Industrial, 4200-465 Porto, Portugal

*  Correspondence: fgs@isep.ipp.pt

**Abstract:** Currently, there is a lack of affordable and simple tools for the estimation of these costs, especially for machining operations. This is particularly true for manufacturing SMEs, in which the cost estimation of machined parts is usually performed based only on required material for part production, or involves a time-consuming, non-standardized technical analysis. Therefore, a cost estimation tool was developed, based on the calculated machining times and amount of required material, based on the final drawing of the requested workpiece. The tool was developed primarily for milling machines, considering milling, drilling, and boring/threading operations. Regarding the considered materials, these were primarily aluminum alloys. However, some polymer materials were also considered. The tool first estimates the required time for total part production and then calculates the total cost. The total production time is estimated based on the required machining operations, as well as drawing, programming, and machine setup time. A part complexity level was also introduced, based on the number of details and operations required for each workpiece, which will inflate the estimated times. The estimation tool was tested in a company setting, comparing the estimated operation time values with the real ones, for a wide variety of parts of differing complexity. An average error of 14% for machining operation times was registered, which is quite satisfactory, as this time is the most impactful in terms of machining cost. However, there are still some problems regarding the accuracy in estimating finishing operation times.

**Keywords:** cost estimation; budgeting; machining; operation times; operation costs

**Citation:** Silva, F.J.G.; Sousa, V.F.C.; Pinto, A.G.; Ferreira, L.P.; Pereira, T. Build-Up an Economical Tool for Machining Operations Cost Estimation. *Metals* **2022**, *12*, 1205. https://doi.org/10.3390/met12071205

Academic Editor: Shoujin Sun

Received: 7 May 2022
Accepted: 13 July 2022
Published: 15 July 2022

**Publisher's Note:** MDPI stays neutral with regard to jurisdictional claims in published maps and institutional affiliations.

## 1. Introduction

Having good budgeting tools and methods is crucial for the future success of a company [1,2], and is also useful for smaller-to-medium enterprises (SME). Moreover, correct budgeting can solve common problems, such as poor material/resource management, especially in manufacturing companies, as analyzed by Siyanbola et al. [3] in their study of the impact of budgeting operations on the performance of a manufacturing company. Usually, in these companies, particularly SMEs, the provided budget is based on the required workpiece material coupled with the empirical knowledge acquired by each company, not following a standardized procedure. Indeed, this was also registered by the previously mentioned authors, who state that it is common for the production team (machine operators) to know and have an influence on the budgetary process. Furthermore, as stated by Nikitina et al. [1], there is a need for communication within the company, especially between the production department and the financial one, to perform budgets and cost estimations. As these budgets are made with the knowledge of the production team (operators/workers), this makes the budgeting process a random one, prone to mistakes and cost miscalculations. Moreover, as the budgeting process largely depends on the empirical knowledge acquired over time, if there is a staff change in the company, this can cause adaptation problems for the new budgeter, resulting in budget errors. These problems can cause order cancelations, especially due to the delay in budget delivery, injuring the companies' competitiveness, as miscalculations in cost can drive clients away to the competition (if the budget value is too high) or result in revenue losses (if the budget value is too low). There is a lack of estimation tools for machining times and cost, especially for cases where the parts are usually produced in small series, with varying geometries and machining operations involved, as seen in SME manufacturing companies. Silva et al. [4] reported that there is a growing interest in outsourcing machining operations to these SMEs; however, this results in requests that have high part variability (in terms of dimensions, detail, and geometry), as well as being requested in small series/quantities. Therefore, the budgeting process for these companies is quite hard, requiring careful analysis of each part, even resulting in a need to perform multiple budgets. This high part variability and order amount hinders the accuracy of budgets and makes the cost estimation process quite a time-consuming one. This, coupled with the intricacies of the machining process, such as the influence of tools, material, and process parameters on the overall performance of the machining process, induces even more budgeting errors.

Machining processes are still the most used to produce high-precision parts for the manufacturing industries, and due to the popularity of these processes, there is a large amount of research performed about them, focused on studying the influence of process parameters and developing ways to optimize them [5,6]. There is, also, a lot of research conducted about the use of coated tools, that improve the overall tool's life by reducing the amount of sustained wear, usually by employing coatings with high wear resistance, as reported by Martinho et al. [7], these coating extend tool-life. Studies around this subject are usually focused on hard to machine materials, evaluating the tested tools' wear behavior, as seen in this study by Gouveia et al. [8] where a comparative study of various machining tools is made, when machining a duplex stainless-steel alloy. Studies such as these offer a valuable insight on the influence of cutting geometry, tool coating, and machining parameters on tool wear [9]. Parent et al. [10] mention that the machining parameters also have quite a relevant impact on the performance of a certain machining operation, being tightly related with process optimization, especially regarding machining cost optimization. These studies are important when trying to optimize the machining process, also having the possibility of registering the cutting forces developed during these processes, allowing for further optimization, as these are strongly related to the overall process' stability, efficiency, and even energy consumption [11]. The study of the machining processes and their optimization may prove quite useful for cost estimation, as it provides ways to best manage material/resources [12] and machining operations times. Choosing tools, coatings, and even more efficient machining strategies, induces an increase in productivity reducing the overall part production cost by, primarily, reducing the machining time. This was reported by Huang et al. [13], where the authors devise a new machining strategy, where the cutting length and machining time are promoted for pocket milling operations. A model was successfully developed, able to generate a spiral toolpath that can be applied for a multitude of pocket milling operations, in which the material removal rate, cycle time, and tool-path length were optimized. The employment of lubricants can also be beneficial. These are known to improve machining times, as they allow for higher feeds and cutting speeds, as reported by Agarwal et al. [14], where the authors employ a solid lubricant in the machining of AISI 304 stainless-steel alloy and then compare the results to dry and wet machining. The authors report that even when compared to wet machining (commonly used for part production when valuing machine surface quality), the use of solid lubricant improves the produced surface quality, reduces the cutting forces, and improves the material removal rate. The use of these solid lubricants not only shows advantages in terms of machining performance (for some alloys), but also shows promissory results in terms of sustainability [15]. Cryogenic machining is also a popular and promising lubrication/refrigeration method, as reported by Agrawal et al. [16]. The authors analyze the tool-wear, tool-life, machined surface roughness and overall process cost of machining operations of Ti6Al4V titanium alloy, using cryogenic machining, and then comparing the results with wet-machining operations. The authors have reported that for lower speeds, the cryogenic turning of this alloy does not present considerable advantages over wet turning.

From a cost estimation standpoint, the selection of an adequate machining strategy for part production is crucial, requiring knowledge about the machining parameters and their influence [1,4]. A careful planning of the machining operations required for a determined part is very beneficial, for example, optimizing the material consumption for a set of operations, by grouping up similar shapes that require the same machining operation [17]. Correct operation sequence planning is also beneficial, especially when producing a series of parts. This is true for a wide variety of processes, such as additive manufacturing [18] or even machine assembly processes [19]. Plaza et al. [20], propose a decision system for optimizing machining operations by selecting an appropriate strategy based on the part request. The authors relate the correct strategy selection to a reduction in tool wear, machining forces, and overall machining cost. Machining parameters have a great influence on the process [21], affecting factors such as tool wear [22], surface quality [23], and material removal rate [24], which affect the total operation time and, thus, the cost of the machining operations. Zhang et al. [25], study the reduction of energy consumption for micro-milling processes, by proposing an energy model. The authors successfully developed a mechanistic model for the prediction of energy consumption. The optimization method was put into practice, and, with the proposed methodology, the authors were able to reduce energy consumption by almost 8%. Still, regarding the machining process optimization, this time regarding the production of better surface quality, Mersni et al. [26], have studied the optimization of machined surface quality, for ball-end milling operations of Ti6Al4V titanium alloy. The authors have employed the Taguchi method and analysis of variance, to determine the best set of machining parameters to obtain the best possible surface roughness quality. In another interesting study, by Narita [27], a method to minimize machining costs is proposed. The method consists of analyzing the most influential parameters on the overall machining cost and then determining the best set of parameters to minimize this. Cost optimization is a common research topic, either by the implementation of optimal parameters or by monitoring tool behavior, such as a monitoring system [28], which can also be used to determine the economic impact of the process itself [29]. To optimize the cost of the machining processes, there have been some applications developed for this purpose, with cost estimations based on machining times, as proposed by Ben-Arieh and Li [30], where a web-based application, based on the Java 2 Enterprise Edition was developed. The prototype was successfully developed and able to predict the machining time of rotational parts, based on the machining parameters that were used. The work proposes the linking of multiple design stations inside a manufacturing shop, to provide these cost estimations in a faster manner; however, there was no practical validation presented for this work. Machining times are usually acquired from empirical knowledge, obtained from years of working at a certain company, making it hard to use the application for different machining processes. Energy consumption also impacts machining costs. In fact, the optimization of energy consumption is quite a popular research topic. The most influential parameters on machining cost are the toolpath, cutting tool selection, and tool sequence [31–33]. Machining parameters also influence the machine's energy consumption during milling. In fact, tests were conducted on milling titanium alloy by Tlhabadira et al. [34], concluding that increases in cutting speed and depth of cut produce an increase in energy consumption. Still, regarding energy consumption, in companies with multiple machines that produce a high number of parts, there is a need to properly schedule the production orders, with a correct machine selection being important [35,36].

Cycle times significantly influence the overall efficiency and cost of a process, being tied closely to productivity [37]. Estimation of these times is important when wanting to reduce/predict the cost of a determined operation. This is especially true for machining operations, where the total part production cost is largely dependent on this factor. In terms of cost estimation for machining processes, it is very important to have accurate methods to predict these times, either by acquired empirical knowledge or by the development and implementation of methods that seek to estimate and optimize these times. Regarding optimization based on empirical knowledge, Pal and Saini [38], propose the optimization of cycle time when machining a forged crankshaft. A study of the process was performed, identifying possible improvements in terms of actions performed during machining and setup operations. The authors were able to improve the cycle time by 4.42%, resulting in a reduction in overall machining cost of about 7%. However, these empirical studies are quite time-consuming and expensive, as they require the use of consumables to perform the required tests. In terms of machining simpler parts, the machining time prediction is much more straightforward, when compared to parts with more organic shapes. However, there are some methods that can be used for these complex shapes, as proposed by Timar et al. [39], by optimizing the tool path for the machining of curved surfaces, determining the best strategy, and set of machining parameters to perform the task in the least amount of time. Regarding cycle time optimization, the Taguchi method can be successfully employed to reduce machining times, while maintaining productivity requirements, as studied by Sakidaze et al. [40], where the authors use this method to reduce the cycle time in plateau honing of a diesel engine cylinder. Still, regarding machining parameter optimization for reducing the operation time, Cafieri et al. [41] propose an approach for the optimization of plunge milling time is presented, based on mixed-integer nonlinear programming. The authors optimized the machining parameters and validated the obtained results from tests performed on CNC machines, finding that they could reduce the operation times by 55%. This highlights the importance of machine selection in the machining process' performance [42,43]. Still, regarding the development of algorithms for cycle time optimization/prediction for milling operations, these are usually developed based on the parameters used during the process. However, there are some methods that can be coupled with this simple calculation, especially for complex parts. One of these methods is toolpath evaluation [44]. This data can be used to predict the machining times, with some authors creating methods that use this stored information and apply it to new processes, where information regarding outputs such as machining times and surface roughness can be obtained [45], even offering process cost estimations as proposed by Ning et al. [46]. In that study, the authors propose a process for machining cost estimation based on convolutional neural network part feature recognition. The model was successfully developed by the authors, offering a fast and accurate way of determining machining costs. However, this has a quite complex implementation, requiring constant learning of new parts for an accurate estimation.

Regarding the overall production time for a certain machined part, there is also the need to consider the preparation times of the machined parts on the budget, especially machine setup and part design times [47,48]. Some machined parts undergo multiple machining operations, needing to be extracted from the machine to be re-adjusted or placed in a different machine. These preparation times can cause problems from a cost estimation standpoint, as these setup times are not always well defined. The use of optimized jigs enables the fixturing of multiple parts, which undergo different machining operations, or by producing systems that are simpler to operate, resulting in faster setup times [49]. Kumar et al. [50], present a study on the development of a fixture that is meant to reduce operation time for a machined part that undergoes a variety of machining processes, including turning, milling, and drilling. The authors developed a fixture that was able to hold the part in place for different operations without requiring extracting, which resulted in a gain of 4 min per produced component. In a similar study, by Kumar et al. [51], a modular jig for machining parts was designed. This jig enabled the performance of machining operations on more parts simultaneously, registering a reduction of up to 32% in part production time [51]. These studies highlight the importance of machine setup, as well as its influence on the overall production time.

There are a lot of factors influencing the machining processes, from parameters, lubrication methods, and even the material's machinability. All these factors have an impact on the overall machining cost and machining times. Determination of the machining cost and cost estimation is critical for the success of a manufacturing company [48]. Some methods have been developed for the direct cost estimation to produce some parts [52,53], with some recent studies using deep learning methods to predict the manufacturing cost of a part by using 3D CAD models. This enables the optimization of the part's production in the design stage [54].

There is little recent research on the prediction of machining costs based on calculated machining times. These times have the greatest influence on the overall machining cost, due to the cost per hour of the machining operator, equipment amortization, and machine energy consumption. Other factors, such as machine consumables and material quantity also influence the production cost. The analyzed models and methods for cost estimation and optimization are quite complex, showing low adaptability for other applications, especially in the machining of parts. As such, in this paper, the development of an affordable and simple cost estimation tool for machined parts, based on the machining times and required material is presented. The tool was developed to be quite flexible, with easier adaptation for different machining processes. An MS Excel® interface was designed, enabling the fast configuration as estimation of part production times, from preparation to finishing operations. These times are then used to calculate total production costs, which can be used to create and supply accurate budgets to clients, in a short amount of time. The developed method and tool would benefit the budgeting process of part manufacturing companies, mainly SMEs, that see many budgeting problems, mainly associated with high part variability. The budgeting process for these SMEs is usually performed based on the amount of required material for part production, or by involving a careful, non-standardized analysis from the operators that have acquired empirical knowledge over a period (working on the area). As such, the SME budgeting process is quite time-consuming and prone to mistakes, lacking standardization. Furthermore, SMEs lack a vast number of resources, not being able to implement complicated or costly solutions for these problems. Due to these aspects, the developed cost estimation tool has the potential to be used by these SMEs, although it can be employed by any enterprise/user that seeks to perform cost estimation of machined parts.

The present study is divided into five main sections (including the present section), in the following subsection, the background and contextualization for the developed cost estimation tool will be presented. In Section 2, Methodology, the considerations made for the development of the tool will be presented, namely the milling machining centers, types of material, and machining parameters. Furthermore, the working principle and operation time calculation method will be presented. Finally, the validation method that was adopted is presented at the end of Section 2. In Section 3, the results regarding the development of the cost estimation tool are going to be presented, namely the input and output sheets of the developed tool, as well as the implementation results for two case studies. Section 4 offers a discussion of the obtained results and, finally, in Section 5, the concluding remarks about the developed work are given.

*Background–Development of an Affordable and Simple Cost Estimation Tool*

The development of the cost estimation tool was made based on an SME manufacturing company that produces machined parts, primarily by the milling process. The analyzed company followed a conventional cost estimation process, shown in Figure 1. However, due to the high variability and small series of requested parts, the created budgets had some errors, usually resulting in over-estimated production costs and, in some cases, underestimated production costs, which resulted in company revenue loss. This was the case, especially for more complex parts. Thus, there is an opportunity to develop a tool that can be useful for this sector but also provide the necessary knowledge to be adapted and adopted by other kinds of industries.

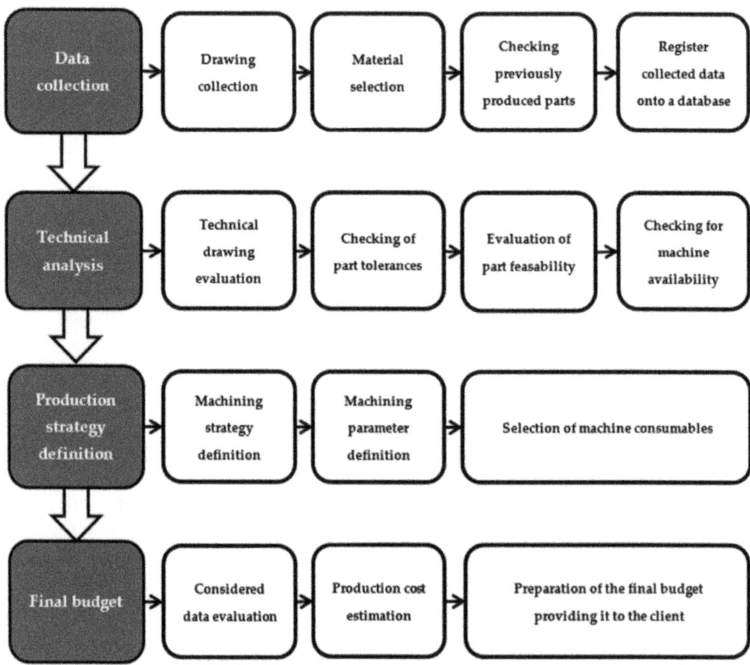

**Figure 1.** Schematic representation of the conventional budgeting process.

To speed up the budgeting process and reduce the errors associated with it, a cost estimation tool based on the analyzed company's resources and conditions, such as machines, workforce, and client requests, was developed. In the subsequent sections, the methodology used for the development and validation of this tool, and the results obtained from this validation, are going to be presented. Furthermore, a discussion of the obtained data is going to be made, analyzing the advantages/disadvantages of the developed tool.

## 2. Materials and Methodology

To develop the cost estimation tool, it was decided that an approach based on the calculation of machining times and founded on the final part's dimension would be the best choice. The calculations were developed for each of the milling machine types considered for validation of the model. The different milling machines that were considered can be observed in Table 1, where the different specifications of each machine are presented.

**Table 1.** Considered CNC milling machining centers.

| Machine | Number of Axes | Workspace Volume | Part Fixation Method | Type |
|---|---|---|---|---|
| M1 | 3 (+2) | $1620 \times 810 \times 760 \ mm^3$ | Mechanical | Vertical |
| M2 | 4 | $1000 \times 450 \times 550 \ mm^3$ | Mechanical | H |
| M3 | 5 | $800 \times 650 \times 550 \ mm^3$ | Mechanical | Vertical |
| M4 | 4 | $4000 \times 2100 \times 275 \ mm^3$ | Vacuum table | Vertical |

As observed in Table 1, the main difference between machines is the workspace volume, the amount of axis, and the workpiece fixation method. These machines were considered as they are selected based on the requested final workpiece (size, tolerances, number of needed axes, etc. . . . ). The machines with vacuum tables are mainly used for the machining of parts with small heights or thickness. In Figure 2, some parts being produced on the mentioned machined can be observed.

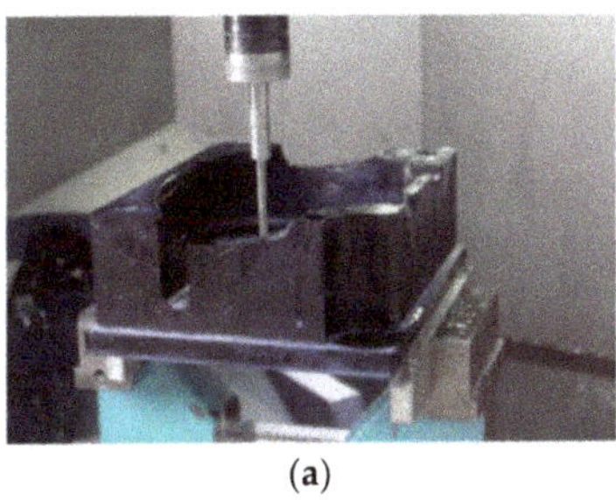


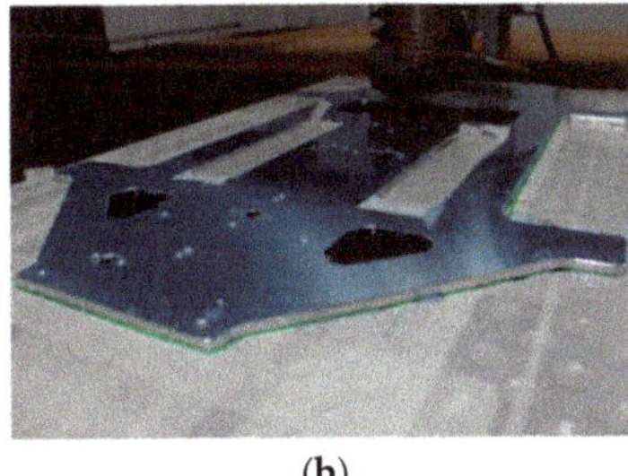


**Figure 2.** The part being machined in an M1 milling machine (**a**), and an M4 milling machine (**b**).

The developed tool's working principle is shown in Figure 3.

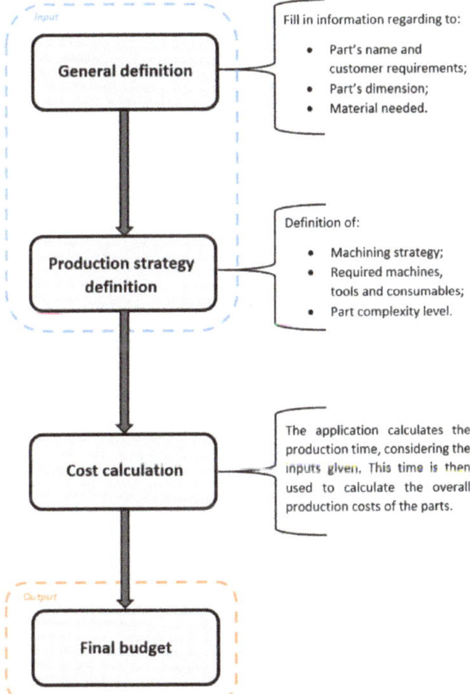


**Figure 3.** Schematic representation of the working principle of the developed tool, for the cost estimation of machined parts.

The inputs for the tool, as observed in Figure 2, are the part's material, the initial required material amount (based on the part's dimensions), machining strategy (parameters and operations), and required machines to obtain the workpiece, and part complexity level definition. Based on all this information, the tool performs the calculation of machining time, which can be used to estimate the overall production cost for the machined parts.

Total production time and, consequently, the total production cost are obtained by determining the operation time for the five production steps that each part undergoes, as follows:

(1)  CAD (2D/3D): The 2D technical drawings are needed for part production; additionally, the 3D drawings can be used to perform the CAM software. If these are not provided, they need to be made.

(2)    CAM: The execution and introduction of the CAM software that is required for part production.

(3)    Machine setup: This step encompasses all the required machine preparation steps for machining, including machine cleaning, tool preparation, tool and holders exchange, and jig placement.

(4)    Machining operations: The different machining operations that the parts are subject to.

(5)    Finishing operations: The operations required to finish the part according to specifications, including machining and manual finishing operations, such as surface roughness improvement (finishing passes) or manual deburring.

These five steps are applied for every part that is produced in the milling machines; however, due to the existing variability from part to part in terms of geometry complexity and required details, the determination of the operation times is insufficient to provide an accurate prediction. Thus, a part complexity level was created to be applied to each of the parts that are being analyzed in terms of cost, which influences the estimated times for each of the production steps.

In the following subsections, the operation time calculation method for each of the five production steps is going to be presented. Furthermore, the working principle of the model will be described in more detail, including the determination of the part complexity level and its influence on the estimated times are going to be explained. Furthermore, the methodology adopted for the validation tests is going to be presented.

### 2.1. Considered Workpiece Materials

The considered workpiece materials were selected based on the requests that are usually performed to the company where the tool was validated. These are mainly aluminum alloys, although some requests are for parts made in a polymeric material. Regarding the considered aluminum alloys, these can be classified as "hard" aluminum alloys and "soft" aluminum alloys, indicating their hardness relative to one another. As for the "hard" aluminum alloys, AW7050, AW7075, AW2017, and AW2030 alloys were considered. Regarding the "soft" ones, the AW6082, AW6063, AW5083, and AW5724 were considered. The most relevant mechanical properties of these alloys can be observed in Table 2, these properties were taken from the material data sheet, provided by the material supplier.

**Table 2.** Mechanical properties values for the considered aluminum alloys.

| Alloy | Ultimate Yield Strength [MPa] | Ultimate Tensile Strength [MPa] | Hardness (HB) |
|---|---|---|---|
| AW7050 | 465 | 520 | 140 |
| AW7075 | 365 | 450 | 130 |
| AW6063 | 215 | 241 | 73 |
| AW6082 | 250 | 290 | 90 |
| AW5083 | 105 | 250 | 70 |
| AW5724 | 185 | 245 | 63 |
| AW2017 | 240 | 385 | 110 |
| AW2030 | 250 | 370 | 115 |

In addition to the mentioned alloys, some polymers were also considered, as some parts made from these materials are requested by the company. These polymeric materials can also be divided into "hard", (e.g., PET and PVC) and "soft", (e.g., HD-PE and PTFE) plastics. Due to their properties, polymeric materials are usually easier to cut than metals; as such, the machining parameters selected for the machining of these parts are usually higher (namely feed rate and axial depth of cut). Taking these higher values as reference (100% of feed rate value is used for these materials), in Figure 4, the percentual values of feed rate and axial depth of cut can be seen for both "hard" and "soft" aluminum alloys and polymeric materials.

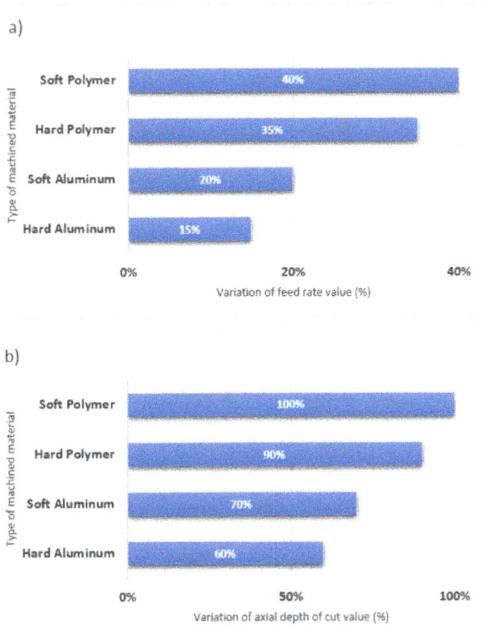

**Figure 4.** Variation of feed rate (**a**); and axial depth of cut (**b**) value, percentual, for the machining of soft and hard aluminum alloys and polymeric materials.

As can be observed in Figure 4, the machining parameter values are higher for softer materials, this will make the machining of these materials faster (given that the workpiece is the same). The variation in axial depth of cut is also dependent on the chosen value of radial depth of cut. This value is divided into three levels:

- First level: the radial depth of cut is equal to 20% of the tool's diameter, mainly selected for contour operations. Enables the selection of higher values of axial depth of cut;
- Second level: radial depth values go from 20% to 45% of the tool's diameter, used for some contour operations, as well as the machining of cavities or slots;
- Third level: radial depth of cut from 45% to 100% of the tool's diameter, used mainly for roughing operations, in cavities or slots. Allows only for low values of axial depth of cut.

### 2.2. Operation Time Estimation

In this subsection, the various methods for the calculation of the operation times for each of the five steps of production are going to be presented, with each of these being divided into one subsubsection. The total operation time estimation is obtained by adding the estimated values obtained for each of the production steps.

### 2.2.1. Operation Time Estimation: CAD 2D/3D

For the estimation of total part production time, the 2D and 3D drawings must be considered. These are a necessity for part production and are not always provided by the client, meaning that in some cases these must be produced. The configuration of the cost estimation tool was made considering the types of drawings for milled parts received. Base times were attributed for 2D and 3D drawings, and are presented in Table 3.

**Table 3.** Base times for the steps regarding the CAD preparation of the parts.

| CAD Step | Time (Min.) |
| --- | --- |
| 2D technical drawings | 5 |
| 3D drawings | 15 |

These base times will be influenced by the part's complexity, which will be explained in more detail in Section 2.2.2.

### 2.2.2. Operation Time Estimation: CAM

Regarding the CAM production step, the estimated based time was defined in a similar way to that determined in the last subsection. This time was determined to be 10 min; however, some parts need to be adjusted inside the machine to be produced, sometimes even requiring multiple clamping operations. These clamping operations would be added to the base time, as they need to be considered in the CAM software. It was determined that each of these clamping operations would add 10 min to the base time.

### 2.2.3. Operation Time Estimation: Machine Setup

The time estimation method for the machine setup production step is the same as the one presented for the CAM production step. The base time for the machine preparation was set to 10 min. Part clamping operation steps were also considered, each of these adding 10 min to the determined base time for this production step.

### 2.2.4. Operation Time Estimation: Machining Operations

For the development of the cost estimation tool, the required machining operations for part production were identified, as follows: side-milling; face-milling; end-milling; drilling, and boring or threading. The equations were obtained from already documented work (such as the Sandvik manual for machining operations), conjugating acquired empirical knowledge to adjust some of these equations, to yield more accurate results in terms of machining time.

Regarding these operations, it is important to note that both part complexity level and production quantity affect the machining times. The influence of the latter will be explained at the end of this subsection.

#### Calculation of Side-Milling Time

Firstly, the part's exterior perimeter ($P_{ext}$) is calculated, considering the part's length and width values. Secondly, the estimated number of roughing passes ($No._{R.P.}$) is defined. This is calculated as shown in Equation (1), considering part thickness ($t$, in mm) and the depth of cut ($a_p$, in mm).

$$No._{R.P.} = \left(\frac{t}{a_e}\right) \tag{1}$$

The value obtained from the calculation of (1) should be rounded up, being equal to an integer. Moreover, the number of finishing passes ($No._{F.P.}$) are also calculated, as shown by the Equation (2), considering part thickness and the tool diameter ($\varnothing_{tool}$).

$$No._{F.P.} = \left(\frac{t}{0.5 \times \varnothing_{tool}}\right) \tag{2}$$

With both the values of the number of finishing ($No._{F.P.}$) and roughing ($No._{R.P.}$) passes, and the values for the exterior perimeter ($P_{ext}$) and the feed rate ($V_f$, in mm/min), the machining time for side-milling ($M.T._{S.M.}$, in minutes) could be calculated, as shown by Equation (3).

$$M.T._{S.M.} = \frac{P_{ext} \times (No._{R.P.} + No._{F.P.})}{V_f} \tag{3}$$

For this kind of operation, the chosen tool diameter usually depends on the thickness of the machined part, as presented in Table 4.

**Table 4.** Different tool diameters chosen in function of part's thickness.

| Part's Thickness [mm] | Tool Diameter [mm] |
| --- | --- |
| 35 | 12 |
| 36–65 | 16 |
| 65–100 | 20 |

Regarding feed rate value, it is chosen based on the machined material and on the current cutting length. For lower cutting lengths the value of feed rate will be lower and, consequently, for higher cutting lengths, the value of feed rate could reach up to 3000 mm/min, for high-performance tools.

Calculation of Face-Milling Time

For this kind of operation, face-mills with diameters between 44 and 64 mm are typically used, using a value of the width of cut ($a_e$) corresponding to 70% of the tool's diameter. The number of facing passes ($No._{FM.P.}$) is calculated by dividing the material's width by the chosen $a_e$ value, and the result should be rounded up, as for Equations (1) and (2).

With the knowledge of the number of facing passes and knowing the length of the part ($P_{length}$) and the tool's diameter ($\varnothing_{tool}$), the facing length ($L_{Facing}$, in mm) can be calculated, as shown in Equation (4).

$$L_{Facing} = (P_{length} + \varnothing_{tool}) \times No._{FM.P.} \tag{4}$$

The value chosen for $a_p$ is usually 1 mm, however, in some cases, face-milling must be performed on the opposite side of the part, requiring its clamping. For the second facing operation the value for $a_p$ is 5 mm. The machining time for the first ($M.T._{1st\ Facing}$) and second machining operation ($M.T._{2nd\ Facing}$) is determined by Equations (5) and (6).

$$M.T._{1st\ Facing} = \frac{L_{Facing}}{V_f} \tag{5}$$

$$M.T._{2nd\ Facing} = \frac{L_{Facing} \times \left(\frac{5}{ae}\right)}{V_f} \tag{6}$$

Calculation of End-Milling Time

End-milling time estimation is performed based on the value of $a_e$, dependent on the tool diameter (40% of this value). The value for end-milling distance per depth increment ($l_{E.M.}$, in mm) needs to first be calculated, and this is dependent on the length and width of the machined cavity. This value is then multiplied by the number of increments (in depth) that will be performed to machine the cavity, obtained by dividing the depth of the cavity ($D_{cavity}$) by the depth of cut value ($a_p$). The obtained value is the total end-milling distance of the operations ($L_{E.M.}$).

The total machining time for the end-milling operations ($M.T._{E.M.}$, in minutes) can be calculated, by using Equation (7), essentially dividing the $L_{E.M.}$ by the feed rate value ($V_f$).

$$M.T._{E.M.} = \frac{L_{E.M.}}{V_f} \tag{7}$$

Regarding finishing operations, these are performed on the interior cavity walls and can be calculated in the same way as the side-milling operations.

Calculation of Drilling Time

Drilling time calculation depends on tool diameter ($\varnothing_{tool}$), chosen according to the desired hole diameter, the depth of the hole ($D_{hole}$, in mm), the value for feed per rotation ($f$, in mm/rotation), and the rotational speed ($N$, in RPM) employed during drilling operations. During the drilling process, the machine performs plunges, quickly retracting and then resuming drilling for a few more millimeters, after retracting again, performing this cycle until the operation is concluded. This promotes a correct chip evacuation from the cutting zone. The number of plunges ($No._{plunges}$) that need to be performed during the process is calculated by Equation (8).

$$No._{plunges} = \frac{D_{hole}}{\varnothing_{tool}} \tag{8}$$

The value for the number of plunges should always be rounded up; next, to this number, one more plunge should be added. After determining the value for $No._{plunges}$, the total drilling length ($L_{drilling}$, in mm) is next. The calculation of this value is determined based on the $No._{plunges}$ needed for the operation and the tool diameter. Equation (9) shows the calculation process for total drilling length.

$$L_{drilling} = \sum_{n=1}^{No._{plunges}} n \times \varnothing_{tool} \times 2 \tag{9}$$

The machining time for drilling operations ($M.T._{Drill}$, in minutes) is calculated according to Equation (10).

$$M.T._{Drill} = \frac{\left(\frac{L_{drilling}}{f}\right)}{N} \tag{10}$$

Calculation of Boring or Threading Times

The method for calculating the machining time of boring and threading ($M.T._{Thread}$, in minutes) operations is shown in Equation (11), where the depth of the hole ($D_{hole}$), the feed per rotation ($f$), and rotational speed ($N$) values are considered.

$$M.T._{Thread} = \frac{\left(2 \times \frac{D_{hole}}{f}\right)}{N} \tag{11}$$

Influence of Production Quantities on the Times

The quantity of requested parts is also considered in the estimation of machining times, as it was found that this factor had an influence on the total production times of machined parts. A larger quantity implies that the worker is more familiar with the production procedures of a certain part. This familiarity causes a slight increase in the production rate. As such, an inflation factor was created for certain quantity levels, which should be multiplied by the estimated machining times, as shown in Table 5.

**Table 5.** Inflation factors applied to machining operation estimated times for different part quantities.

| Part Quantity | Inflation Factor |
| --- | --- |
| 1 | 1.25 |
| 2 to 4 | 1.2 |
| 5 to 8 | 1.2 |
| 9 to 15 | 1.15 |
| 16 to 20 | 1.1 |
| 21 to 25 | 1.08 |
| 26 to 30 | 1.08 |
| $\geq$30 | 1.08 |

### 2.2.5. Operation Time Estimation: Finishing Operations

Expected machining times for finishing operations ($M.T._{Finishing}$, in minutes) were determined, based on the part's dimensions (in mm), namely length ($P_{length}$), width ($P_{width}$), part thickness ($t$), and feed value ($V_f$). Equation (12) shows the way to determine the expected finishing time.

$$M.T._{Finishing} = \frac{(4 \times P_{length}) + (4 \times P_{width}) + (4 \times t)}{V_f} \tag{12}$$

The finishing times are also affected by part complexity level, which will increase with the part's level, as seen for the estimated machining operation times (addressed in detail in Section 2.2.2).

Estimated times obtained from this equation may present deviation, as finishing operations are quite difficult to estimate in this case. As these are dependent on the machining strategy and part geometry, additionally, some parts may require manual finishing, with some of them even needing supplementary clamping to undergo these operations.

### 2.3. Cost Estimation Tool Working Principle

The cost calculation method can be divided into two main steps, one regarding the definition of the initial inputs, such as part's dimensions, client requirements, and needed material volume. The other step encompasses the definition of the part's complexity level and the calculation of the total operation times, based on the conjugation of all the defined parameters and the influence of the complexity level. These steps will be presented in the following Sections 2.2.1 and 2.2.2.

### 2.3.1. Step 1 of Cost Calculation

A flowchart was created to depict the cost estimation model. In Figure 5, the flowchart for the first cost calculation step can be observed.

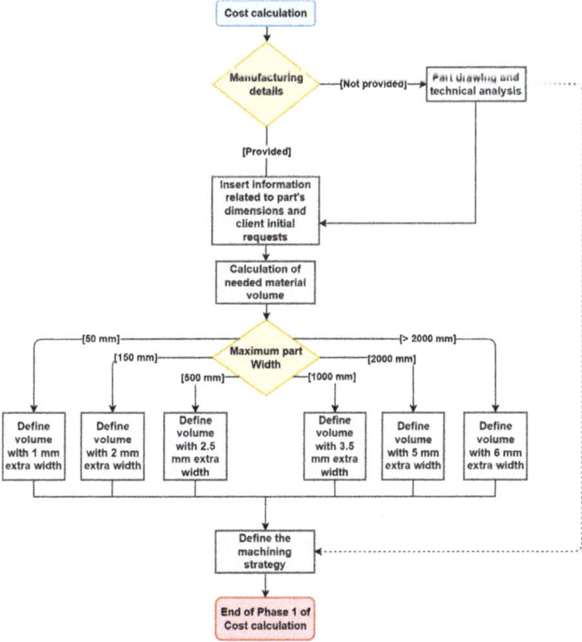


**Figure 5.** Flow chart of the developed model, for the first cost calculation step.

The initial client's requirements are considered, as well as the part's dimensions. The latter is used to determine the amount (in kg) of raw material that will be used in the production process, with a higher volume of material being used for wider parts. With all this information defined, the machining strategy can also be determined, ending the first step of cost calculation.

### 2.3.2. Phase 2 of Cost Calculation

After machining strategy definition, the part complexity level is attributed based on three main parameters of selection:

- Level of detail: Refers to the amount of detail that each of the part's need, based on the number of operations applied for part production. The number of operations that correspond to each level is presented in Table 6.
- Geometry: Regards the complexity of each part, divided into three categories, as seen in Figure 6.
- Machine axis needed: Differentiates the parts that need three or five axes for the machining operations.

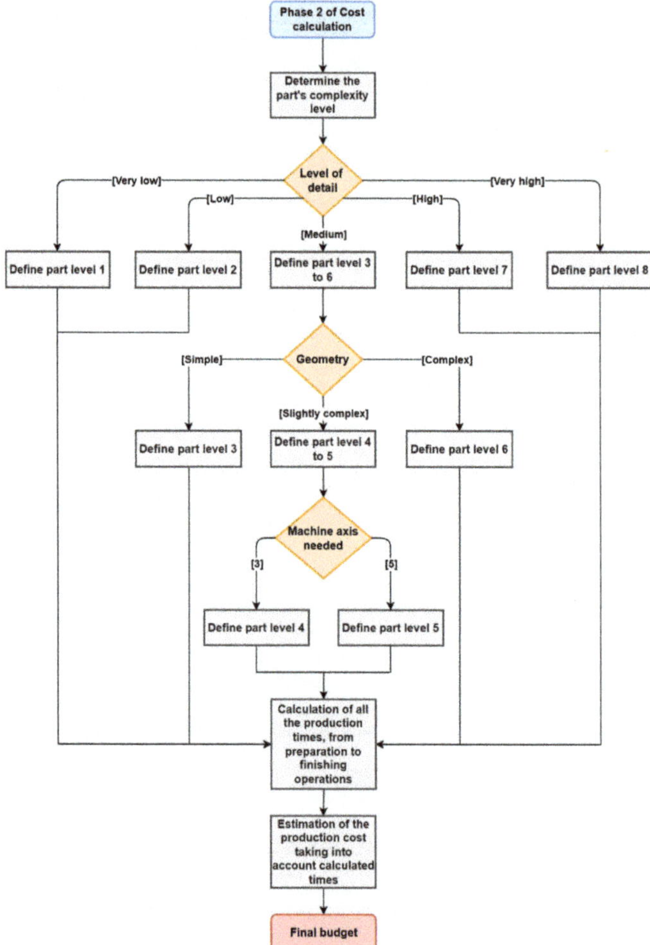


**Figure 6.** Flow chart of the developed model, for the second cost calculation step.

**Table 6.** Number of operations for each level of detail, considered for the definition of part complexity level.

| Level of Detail | Number of Threaded Holes | Number of Simple Holes |
| --- | --- | --- |
| Very low | 0–3 | 0–6 |
| Low | 4–7 | 7–14 |
| Medium | 8–20 | 15–40 |
| High | 21–30 | 41–60 |
| Very high | $\geq$31 | $\geq$61 |

As previously mentioned, the part complexity level influences each of the production steps, from CAD to finishing operations. For the CAD, CAM, and machine setup production steps, the influence of the part complexity level in the estimated time is applied in the same manner. As with an increase in part complexity level, comes an increase in detail and number of machining operations, more operations imply more drawing, programming, and setup steps. The determined times can be observed in Table 7. These should then be added to the base times to obtain the production times for each of mentioned steps and are presented as minutes per detail added to the drawing (min/detail).

**Table 7.** Time increment that should be applied to the production steps for each part complexity level.

| Part Complexity Level | Added Time for Each Production Step (Min/Detail) | | |
| --- | --- | --- | --- |
| | CAD | CAM | Machine Setup |
| 1 | 1.5 | 1 | 1 |
| 2 | 2.5 | 2 | 2 |
| 3 | 3.5 | 3 | 3 |
| 4 | 4.5 | 4 | 4 |
| 5 | 5.5 | 5 | 5 |
| 6 | 6.5 | 6 | 6 |
| 7 | 7.5 | 7 | 7 |
| 8 | 8.5 | 8 | 8 |

Regarding the influence of the part complexity level on the machining and finishing operations estimated time, this is applied in the same manner. An inflation factor was devised for each part complexity level, and this value should be multiplied by the estimated times for these operation steps. Table 8 presents the inflation factor for each of the parts' complexity levels.

**Table 8.** Inflation factors applied to machining and finishing operation estimated times for the different part complexity levels.

| Part Complexity Level | Inflation Factor |
| --- | --- |
| 1 | 1.25 |
| 2 | 1.2 |
| 3 | 1.2 |
| 4 | 1.15 |
| 5 | 1.1 |
| 6 | 1.08 |
| 7 | 1.08 |
| 8 | 1.08 |

With all this defined, the tool can estimate the total production time of a certain part, and these times can then be used to calculate the total operation cost of the process. This value will also be added to the amount of raw material determined in step 1 of cost calculation, to determine the total cost of part production.

*2.4. Validation of the Developed Tool*

The cost estimation tool was subjected to a series of validation tests, being used for two case studies. For Case Study 1, the validation consisted of the analysis of a total of 24 parts, 3 for each different complexity level. The tool was used to estimate the total production time of these parts (for all the mentioned production steps), which were then produced, and their production times were clocked by the machine operators (after each production step the worker would register the time taken up to perform the said task, this would later be compared to the estimated machining times). These times were then compared to the estimated ones, and the percentual deviation from each of the produced parts' production time was registered. The percentual deviations are presented with either a positive or negative value, with it representing a time over-estimation and under-estimation, respectively. This comparison of time estimation is key, as the tool performs the calculation of total operation cost based on these production times.

Regarding Case Study 2, the validation was performed in the same manner; however, the machine types for this case study were slightly different, being CNC milling centers with vacuum tables. These machines performed the same operations as those of Case Study 1; however, the part complexity of the workpieces produced on these machines is somewhat constant, with the parts having low amounts of detail with low complexity (in terms of geometry). Additionally, the finishing operations for these machines are usually performed by the machine itself (differing from the manual finishing operations conducted for parts produced in the machines considered for Case Study 1). For Case Study 2, a total of 10 parts were produced, registering the machining time that was estimated, then, producing the workpiece and timing this manually (as for Case Study 1). These deviations were then averaged and are subsequently presented in the Results section of this study.

## 3. Results

The cost estimation tool was successfully developed and tested, and in the present section the application is going to be presented in Section 3.1, showing the interface while explaining each of the different main interface elements. Furthermore, the accuracy of this tool was tested by estimating and producing various machined parts and then comparing the deviation of the predicted times from the real times. The data obtained from these tests will be presented in Section 3.2.

*3.1. Developed Cost Estimation Tool*

The developed cost estimation tool considers the inputs given in the manner described in the previous section to give a production time estimation for parts produced in CNC milling centers, having an interface in MS Excel®, as seen in Figure 7. The interface is divided into six main sections:

- General data: This is where the information regarding the client, part and project name, and part quantity is defined;
- Material: In this section, the material information is filled, mentioning designation, material properties, raw material tolerances, and dimensions, as well as material price;
- Dimensions: Regarding the part's dimensions;
- Strategy: Here, the part complexity level is set, as well as the variables that directly influence the estimated times, such as low tolerance requirements, as well as cavity and 2D/3D drawing consideration;
- Technical observations: This section should be filled if there is a need to request drawings or drawing corrections for part production.

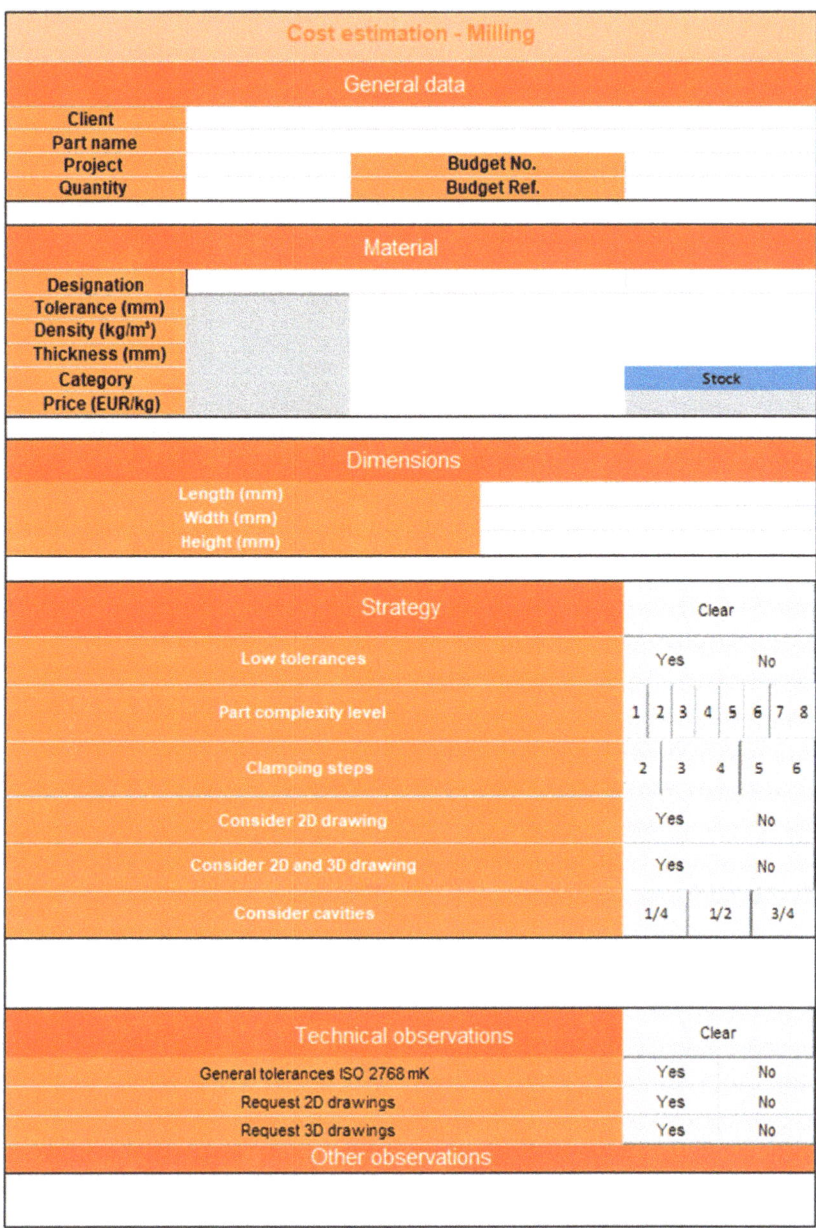

**Figure 7.** MS Excel® interface of the developed cost estimation tool.

To input data in the developed interface requires knowledge of machining processes currently being applied, for example, in the "Strategy" section of the interface, a choice for cavity consideration was added. This should be defined according to the total area of the cavities, in relation to the workpiece area. The chosen value will influence the machining time for end-milling operations; however, this input can be left blank.

After filling the input interface, the cost estimation tool will estimate the operation times for each of the mentioned production steps, exhibiting the results in an output sheet, observed in Figure 8.

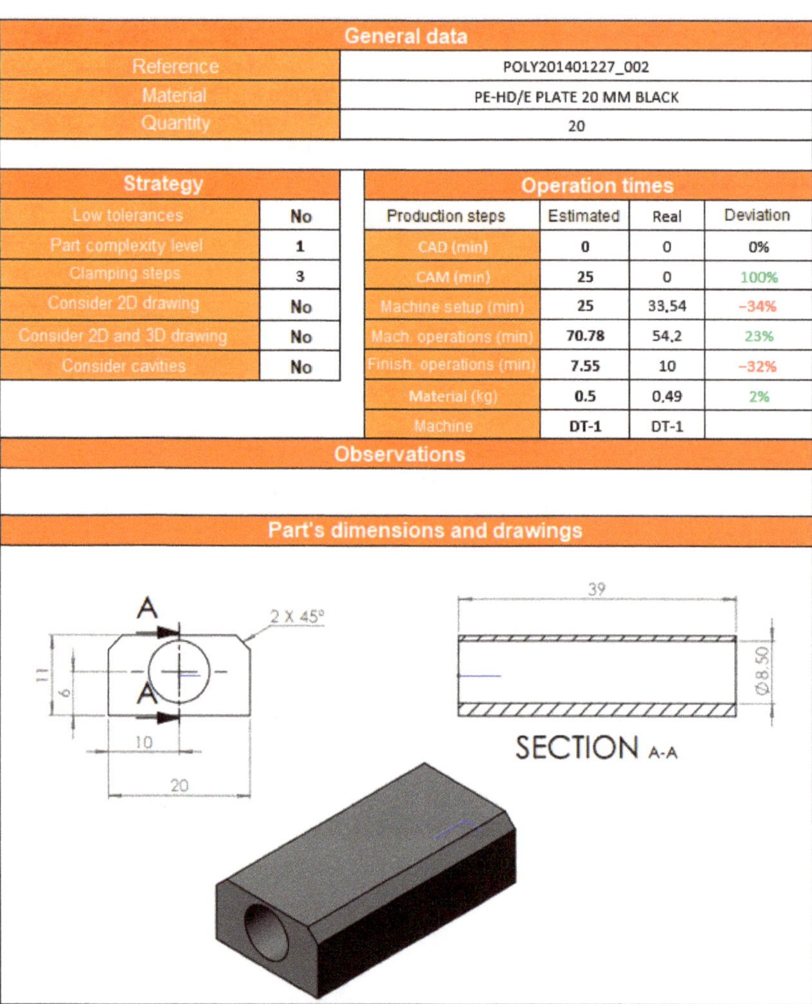



**Figure 8.** Output sheet of a machined polyethylene part produced for Case Study 1.

The output sheet is divided into five main sections, displaying the filled data regarding the material and project information, the adopted strategy, observations, and the part's technical drawing. Estimated times are displayed in the "Operation times" section. Note that in this section there is an input table for the real machining times, which was added for the validation of the cost estimation tool. These real times were registered by the operator after machining.

As can be observed in Figure 6, there are some deviations that originate from the lack of need to perform the CAD drawings of the part. Furthermore, there are some deviations registered for the machine setup times and finishing operations. This can be attributed to the fact that these operations are performed manually, being harder to estimate correctly (highly dependent on the operator). It is also worthy of noting that, although the finishing times have quite a large deviation from the real times, this is since these operations usually have a short duration. This can also be observed in another part, as depicted in Figure 9. Observing the output sheet shown in Figure 9, it can be noted that the highest percentual deviation, in terms of real machining time, is registered for the calculation of finishing operations (−80%). Although this value is considerably high, the difference between estimated and real times is less than one minute. Again, this lack of accuracy in the estimation of these operations is since finishing operations are performed manually (for the parts produced in these machine types). However, analyzing the deviation from all the other production steps, the maximum deviation is +8%, which is incredibly satisfactory. This was registered for the parts of a similar complexity level, with lower deviations being registered.

| General data | | | |
|---|---|---|---|
| Reference | POLY201501578_33 | | |
| Material | AW6082-T651 PLATE 25 MM | | |
| Quantity | 1 | | |

| Strategy | | Operation times | | | |
|---|---|---|---|---|---|
| | | Production steps | Estimated | Real | Deviation |
| Low tolerances | No | CAD (min) | 5 | 5 | 0% |
| Part complexity level | 1 | CAM (min) | 25 | 22.93 | 8% |
| Clamping steps | 3 | Machine setup (min) | 25 | 24.66 | 1% |
| Consider 2D drawing | Yes | Mach. operations (min) | 15.32 | 14.38 | 6% |
| Consider 2D and 3D drawing | No | Finish. operations (min) | 1.11 | 2 | −80% |
| Consider cavities | 1/4 | Material (kg) | 0.46 | 0.47 | −2% |
| | | Machine | DT-1 | DT-1 | |

| Observations |
|---|
| |

| Part's dimensions and drawings |
|---|

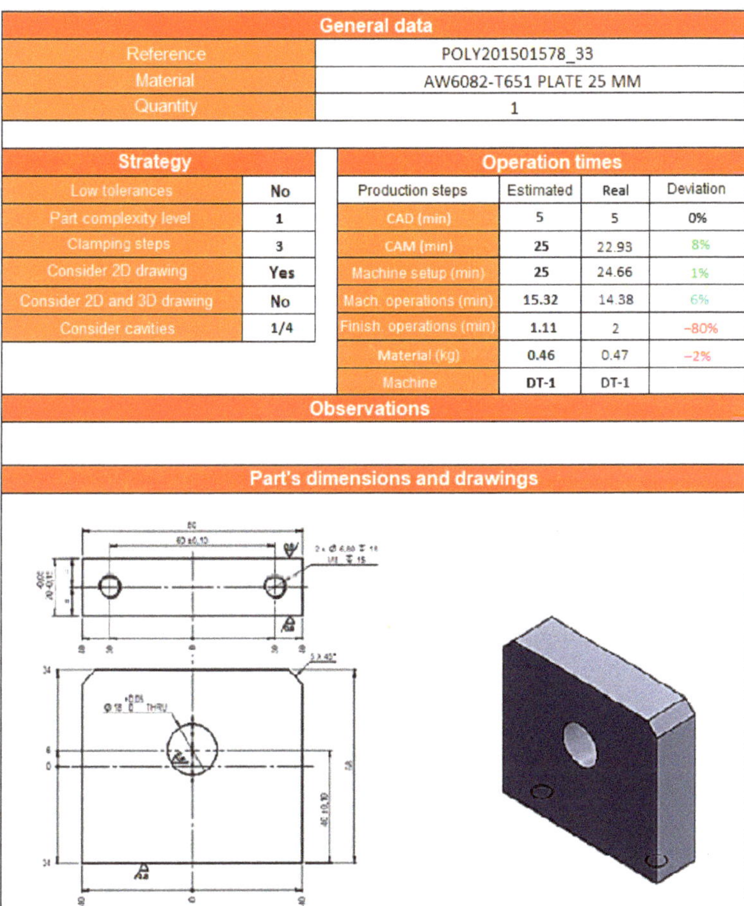

**Figure 9.** Output sheet of a machined aluminum part, with low complexity level.

### 3.2. Cost Estimation Tool Validation

The developed cost estimation tool was validated according to the procedure in Section 2.3. The average percentual deviations were registered for each of the production steps and are presented in Figure 10. Additionally, an influence of part complexity level in these deviations was noticed. Furthermore, this influence behaved slightly differently depending on the analyzed production step. A graph that depicts the variation in absolute percentual deviation of each production step, over the different part complexity levels can be observed in Figure 11.

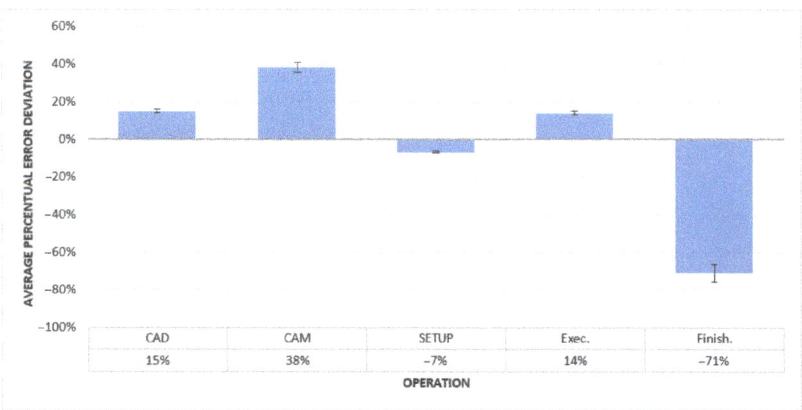

**Figure 10.** Average percentual deviation for each part production step.

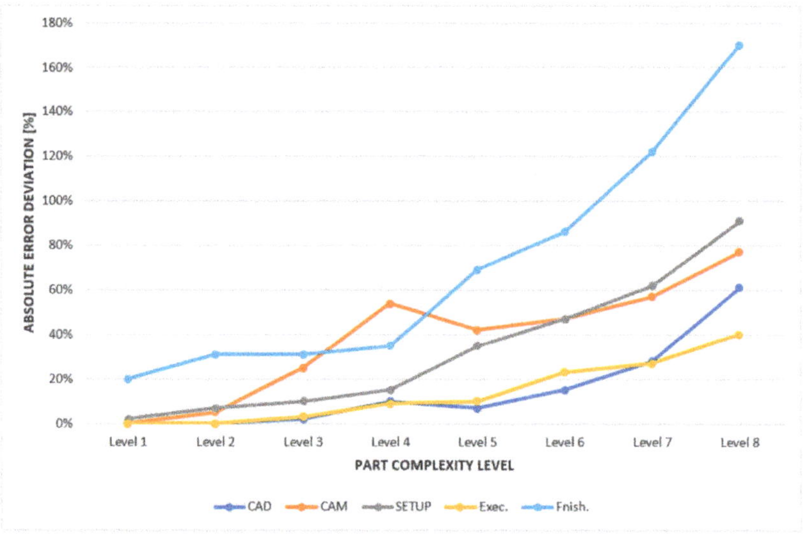

**Figure 11.** Absolute percentual deviation variation of each part production step, over different part complexity levels.

Cost Estimation Tool Validation: Case Study 2

An additional case study was conducted, as the tool can be adapted to different machines. Tests were conducted for CNC milling machines with vacuum tables. These machines conduct mainly: side-milling; end-milling; drilling and face-milling operations, which meant that the equations are presented in Section 2.1 could be used to estimate machining operations. It was noted that the parts usually produced in these machines did not exhibit much variability in terms of shape or complexity. This enabled a more accurate time prediction based on the equations, especially for the step regarding machining and finishing operations, without the need to define part complexity levels. One of the produced parts can be observed in Figure 12, as well as its technical drawing.

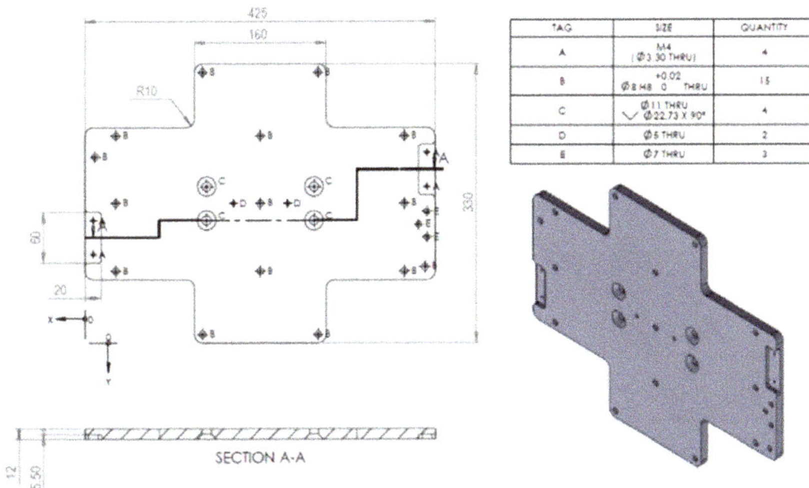

**Figure 12.** Produced part for Case Study 2 in a CNC milling center with vacuum table.

Regarding machining time operation for these parts, it was quite accurate for all the tested ones; however, regarding the performance of CAM software for the machining, it was the step that had the biggest deviation (this is depicted in Figure 13). This is due to the number of holes that these parts have, inducing delays from the developers of the CAM for these parts. Although some of these parts imply complex programming, it was noted that the average percentual deviation registered for this step (−19%) is not as accentuated as that registered for Case Study 1 (38%), this is because the complexity of some the part's machined in machines of this case study is considerably higher, especially in terms of geometry.

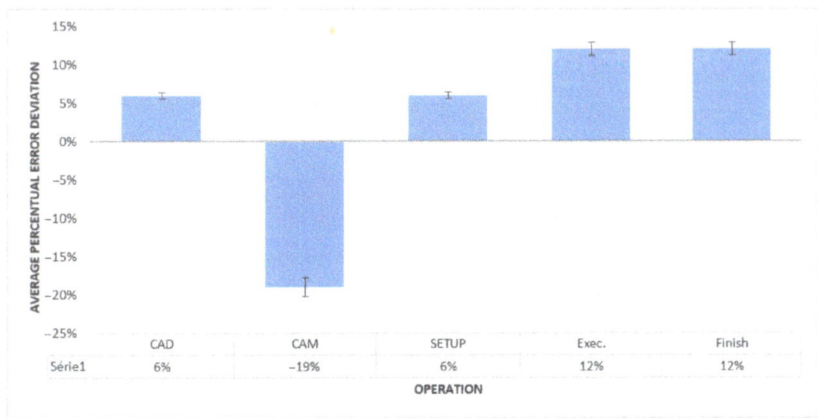

**Figure 13.** Average percentual deviation values for each of the production steps, for parts produced in milling centers with vacuum tables (Case Study 2).

A total of 10 parts were estimated and produced using these machines, registering the average percentual deviation values from the real production times, as presented in Figure 13.

## 4. Discussion

In Figure 10, it can be observed that the percentual deviations are mostly positive. This is quite satisfactory, as a positive percentual deviation is preferred since there will be no direct revenue loss from the production of the parts (associated with negative percentual deviations). For this case, the higher percentual deviation is for the "Finishing operations" step, exhibiting a −71% percentual deviation. This value is quite high, due to the complexity of these types of operations, and the fact that these are usually carefully performed, and are dependent on human work. The second highest percentual deviation is for the "CAM" step, at about 38%. The most influential production step on the overall production cost is the "Machining operations", due to the influence of machining time [34]. This step registered an acceptable percentual deviation of about 14%.

The part's complexity level was also found to influence the percentual deviation error, increasing this error for higher levels. This can be observed in Figure 11, where the highest values for all the considered production steps are registered for higher part complexity levels. Additionally, the error of "Finishing operations" tends to be higher when compared to the other production steps. Again, this is due to the difficulty in predicting these times, as there are many variables that cannot be controlled directly, such as finishing and inspection operations that are performed outside the machine [44]. This is corroborated by the data obtained from Case Study 2, presented in Figure 13. A side-by-side comparison of the average percentual deviation values for each of the production steps, for both case studies, can be observed in Table 9.

**Table 9.** Average percentual deviations for each of the production steps, for both case studies.

| Production Step | Average Percentual Deviation | |
|---|---|---|
| | Case Study 1 | Case Study 2 |
| CAD 2D/3D | +15% | +6% |
| CAM | +38% | −19% |
| Machine setup | −7% | +6% |
| Machining operations | +14% | +12% |
| Finishing operations | −71% | +12% |

In Case Study 2, the produced parts were of similarly low complexity, with finishing operations being conducted inside the machine. This is reflected in the obtained results, as the values for percentual deviation are quite consistent and low when compared to the values for Case Study 1. For the second case study, the highest percentual deviation was −19%, for the "CAM" step, with all the other values being positive deviations.

## 5. Conclusions

The present study presents the development of a cost estimation tool that calculates total operation cost based on the material requirements and machining times while applying a part complexity level as a way of standardization for the budgeting process to expedite it while considering multiple factors that affect the total operation time and thus, the total cost of part production. The cost estimation tool was successfully developed and validated for two machine types; these being milling machines capable of producing parts of different heights.

- Machining time and material are the main factors that influence machining cost;
- The developed tool offers a quite accurate way of predicting machining times and, thus, the operation cost of machined parts obtaining an average percentual deviation of 14% and 12% for Case Study 1 and 2, respectively;
- The tool exhibits high accuracy in predicting CAD 2D/3D drawing times and machine setup times, registering 15% and −7% of percentual deviation, respectively, for Case Study 1, and 6% for both steps for Case Study 2;

- The created part complexity level introduces a level of standardization in the budgeting process, ensuring accurate and fast budgets, especially for parts of lower complexity level;
- The deviation from estimated to real machining times increases with part/complexity level, especially for finishing operations; this is shown by the high deviation registered for the prediction of finishing times, with Case Study 1 exhibiting 71% of average deviation from predicted to real-time values;
- These high deviation values can be attributed to the performance of manual tasks, which are difficult to estimate correctly;
- This is also shown by the CAM production step associated error, which is quite high for Case Study 1 (38%); this is due to the number of operations that need to be programmed by the operator. For simpler parts, such as those of Case Study 2, the deviation drops to 19%;
- Case Study 2 yielded fewer spread results than Case Study 1; this is due to the simpler geometry of the parts considered for this case study and the fact that the number of manual operations is quite reduced (mainly for machine setup operations).

The model can be improved by conducting additional experiments and validation tests, as this tool is easily reprogrammable and adapted to different processes. Furthermore, since the estimation of manual operations can be quite difficult (resulting in deviations from the estimated to the real operation time), these operations should be minimized. Alternatively, the creation of normalized procedures for these operations can improve the accuracy of the time estimations. Despite this, the cost estimation tool can predict the manufacturing times accurately, resulting in the obtention of accurate and fast budgets. This is particularly useful for manufacturing SMEs, as this tool provides a faster and easier alternative to providing budgets for machined parts. The advantages and drawbacks of the developed tool can be observed in Table 10.

**Table 10.** Main advantages and drawbacks of the developed cost estimation tool.

| Advantages | Drawbacks |
| --- | --- |
| Affordable cost estimation tool | Estimation accuracy drops for more complex parts |
| Easy implementation and configuration | Requires some knowledge of the machining processes and operations |
| High adaptability for different processes/machines | Finishing operations are difficult to estimate correctly |
| Fast estimation of operation costs | Accuracy is hindered by the performance of manual operations |
| Simple interface | - |
| Accounts for all steps of part production in its estimation | - |

Regarding further improvements to the developed tool, some other prediction methods can also be employed in conjunction with this tool, to improve prediction accuracy. The average error detected in the model can be smoothly corrected by a determining factor, and the profit yield of the manufacturer can easily accommodate this error in the first stage, being successively corrected through experiments.

**Author Contributions:** F.J.G.S.: conceptualization, work orientation, investigation, supervision, writing—reviewing; V.F.C.S.: investigation, formal analysis, writing—reviewing and editing; A.G.P.: formal analysis, visualization, writing—editing; L.P.F.: formal analysis, writing—editing; T.P.: formal analysis, visualization, writing—reviewing. All authors have read and agreed to the published version of the manuscript.

**Funding:** This research received no external funding.

**Conflicts of Interest:** The authors declare no conflict of interest.

**Nomenclature**

| | |
|---|---|
| $a_e$ | Radial depth of cut (mm); |
| $a_p$ | Axial depth of cut (mm) |
| CAD | Computer-aided design; |
| CNC | Computer numerical control; |
| $D_{cavity}$ | Cavity depth (mm); |
| $D_{hole}$ | Hole depth (mm); |
| $f$ | Feed per rotation (mm/rot); |
| $L_{drilling}$ | Total drilling length (mm); |
| $l_{E.M.}$ | End-milling distance per depth increment (mm); |
| $L_{E.M.}$ | Total end-milling distance (mm) |
| $L_{length}$ | Facing length (mm); |
| $M.T._{1st\ facing}$ | Machining time for the first facing operation (min); |
| $M.T._{2nd\ facing}$ | Machining time for the second facing operation (min); |
| $M.T._{drilling}$ | Machining time for drilling operations (min); |
| $M.T._{E.M.}$ | Machining time for end-milling operations (min); |
| $M.T._{thread}$ | Machining time for threading operations (min); |
| $M.T._{finishing}$ | Machining time for finishing operations (min); |
| $M.T._{S.M.}$ | Machining time for side-milling operations (min); |
| MS Excel | Microsoft Excel® |
| N | Rotational speed (rpm); |
| $No_{Plunges}$ | Number of plunges required; |
| $No._{F.P.}$ | Number of finishing passes; |
| $No._{FM.P.}$ | Number of face-milling passes; |
| $No._{R.P.}$ | Number of roughing passes; |
| $P_{ext}$ | Part's exterior perimeter (mm); |
| $P_{length}$ | Part length (mm); |
| $P_{width}$ | Part width (mm); |
| SME | Smaller-to-medium enterprises; |
| t | Thickness (mm); |
| TiAlN | Titanium aluminum nitride; |
| TiAlSiN | Titanium aluminum silicon nitride; |
| $Vf$ | Feed rate (mm/min); |
| $\varnothing_{tool}$ | Tool diameter (mm); |
| 2D | Two-dimensional; |
| 3D | Three-dimensional. |

## References


1. Nikitina, O.A.; Litovskaya, Y.V.; Ponomareva, O.S. Development of the cost management mechanism for metal products manufacturing on budgeting method. *Acad. Strateg. Manag. J.* **2018**, *17*, 1.
2. Schlegel, D.; Frank, F.; Britzelmaier, B. Investment decisions and capital budgeting practices in German manufacturing companies. *Int. J. Bus. Glob.* **2016**, *16*, 66–78. [CrossRef]
3. Siyanbola, T.T. The Impact of Budgeting and Budgetary Control on The Performance of Manufacturing Company in Nigeria. *J. Bus. Manag. Soc. Sci. Res.* **2013**, *2*, 8–16.
4. Ferreira, V.; Silva, F.J.G.; Martinho, R.P.; Pimentel, C.; Godina, R.; Pinto, B. A comprehensive supplier classification model for SME outsourcing. *Procedia Manuf.* **2019**, *38*, 1461–1472. [CrossRef]
5. Sousa, V.F.C.; Silva, F.J.G. Recent Advances on Coated Milling Tool Technology: A Comprehensive Review. *Coatings* **2020**, *10*, 235. [CrossRef]
6. Sousa, V.F.C.; Silva, F.J.G. Recent Advances in Turning Processes Using Coated Tools: A comprehensive Review. *Metals* **2020**, *10*, 170. [CrossRef]
7. Martinho, R.P.; Silva, F.J.G.; Baptista, A.P.M. Cutting forces and wear analysis of $Si_3N_4$ diamond coated tools in high speed machining. *Vacuum* **2008**, *82*, 1415–1420. [CrossRef]
8. Gouveia, R.M.; Silva, F.J.G.; Reis, P.; Baptista, A.P.M. Machining Duplex Stainless Steel: Comparative Study Regarding End Mill Coated Tools. *Coatings* **2016**, *6*, 51. [CrossRef]
9. Sousa, V.F.C.; Silva, F.J.G.; Alexandre, R.; Fecheira, J.S.; Silva, F.P.N. Study of the wear behaviour of TiAlSiN and TiAlN PVD coated tools on milling operations of pre-hardened tool steel. *Wear* **2021**, *476*, 203685. [CrossRef]
10. Parent, L.; Songmene, V.; Kenné, J.-P. A generalised model for optimising an end milling operation. *Prod. Plan. Control. Manag. Oper.* **2011**, *18*, 319–337. [CrossRef]

11. Sousa, V.F.C.; Silva, F.J.G.; Fecheira, J.S.; Lopes, H.M.; Martinho, R.P.; Casais, R.B.; Ferreira, L.P. Cutting Forces Assessment in CNC Machining Processes: A Critical Review. *Sensors* **2020**, *20*, 4536. [CrossRef] [PubMed]

12. Seong, D.; Suh, M.S. An integrated modelling approach for raw material management in a steel mill. *Prod. Plan. Control. Manag. Oper.* **2012**, *23*, 922–934. [CrossRef]

13. Huang, N.; Jin, Y.; Lu, Y.; Yi, B.; Li, X.; Wu, S. Spiral toolpath generation method for pocket machining. *Comput. Ind. Eng.* **2020**, *139*, 106142. [CrossRef]

14. Agarwal, V.; Agarwal, S. Performance profiling of solid lubricant for eco-friendly sustainable manufacturing. *J. Manuf. Process.* **2021**, *64*, 294–305. [CrossRef]

15. Makhesana, M.A.; Patel, K.M.; Mawandiya, B.K. Environmentally conscious machining of Inconel 718 with solid lubricant assisted minimum quantity lubrication. *Met. Powder Rep.* **2020**, *76* (Suppl. 1), S24–S29. [CrossRef]

16. Agrawal, C.; Wadhwa, J.; Pitroda, A.; Pruncu, C.I.; Sarikaya, M.; Khanna, N. Comprehensive analysis of tool wear, tool life, surface roughness, costing and carbon emissions in turning Ti–6Al–4V titanium alloy: Cryogenic versus wet machining. *Tribol. Int.* **2021**, *153*, 106597. [CrossRef]

17. Silva, J.; Silva, F.J.G.; Campilho, R.D.S.G.; Sá, J.C.; Ferreira, L.P. A model for productivity improvement on machining of components for stamping dies. *Int. J. Ind. Eng. Manag.* **2021**, *12*, 85–101. [CrossRef]

18. Fuchs, C.; Semm, T.; Zaeh, M.F. Decision-based process planning for wire and arc additively manufactured and machined parts. *J. Manuf. Syst.* **2021**, *59*, 180–189. [CrossRef]

19. Maheut, J.; Besga, J.M.; Uribetxebarria, J.; Garcia-Sabater, J.P. A decision support system for modelling and implementing the supply network configuration and operations schedulling problem in the machine tool industry. *Prod. Plan. Control. Manag. Oper.* **2014**, *25*, 679–697. [CrossRef]

20. Plaza, M.; Zebala, W.; Matras, A. Decision system supporting optimization of machining strategy. *Comput. Ind. Eng.* **2019**, *127*, 21–38. [CrossRef]

21. Zubair, A.F.; Mansor, M.S.A. Embedding firefly algorithm in optimization of CAPP turning machining parameters for cutting tool selections. *Comput. Ind. Eng.* **2019**, *135*, 317–325. [CrossRef]

22. Chung, C.; Wang, P.C.; Chinomona, B. Optimization of turning parameters based on tool wear and machining cost for various parts. *Int. J. Adv. Manuf. Technol.* **2022**, *120*, 5163–5174. [CrossRef]

23. Pimenov, D.Y.; Abbas, A.T.; Gupta, M.K.; Erdakov, I.N.; Soliman, M.S.; El Rayes, M.M. Investigations of surface quality and energy consumption associated with costs and material removal rate during face milling of AISI 1045 steel. *Int. J. Adv. Manuf. Technol.* **2020**, *107*, 3511–3525. [CrossRef]

24. Abbas, A.T.; Pimenov, D.Y.; Erdakov, I.N.; Mikolajczyk, T.; Soliman, M.S.; El Rayes, M.M. Optimization of cutting conditions using artificial neural networks and the Edgeworth-Pareto method for CNC face-milling operations on high-strength grade-H steel. *Int. J. Adv. Manuf. Technol.* **2019**, *105*, 2151–2165. [CrossRef]

25. Zhang, X.; Yu, T.; Dai, Y.; Qu, S.; Zhao, J. Energy consumption considering tool wear and optimization of cutting parameters in micro milling process. *Int. J. Mech. Sci.* **2020**, *178*, 105628. [CrossRef]

26. Merani, W.; Boujelbene, M.; Salem, S.B.; Singh, H.P. Machining time and quadratic mean roughness optimization in ball end milling of titanium alloy Ti-6Al-4V—Aeronautic field. *Mater. Today Proc.* **2020**, *26 Pt 2*, 2619–2624. [CrossRef]

27. Narita, H. A Study of Automatic Determination of Cutting Conditions to Minimize Machining Cost. *Procedia CIRP* **2013**, *7*, 217–221. [CrossRef]

28. Xing, K.; Liu, X.; Liu, Z.; Mayer, J.R.R.; Achiche, S. Low-Cost Precision Monitoring System of Machine Tools for SMEs. *Procedia CIRP* **2021**, *96*, 347–352. [CrossRef]

29. Colosimo, B.M.; Cavalli, S.; Grasso, M. A cost model for the economic evaluation of in-situ monitoring tools in metal additive manufacturing. *Int. J. Prod. Econ.* **2020**, *223*, 107532. [CrossRef]

30. Bem-Arieh, D.; Li, Q. Web-based cost estimation of machining rotational parts. *Prod. Plan. Control. Manag. Oper.* **2003**, *14*, 778–788. [CrossRef]

31. Xiao, Y.; Jiang, Z.; Gu, Q.; Yan, W.; Wang, R. A novel approach to CNC machining center processing parameters optimization considering energy-saving and low-cost. *J. Manuf. Syst.* **2021**, *59*, 535–548. [CrossRef]

32. Xiao, Y.; Zhang, H.; Jiang, Z.; Gu, Q.; Yan, W. Multiobjective optimization of machining center process route: Tradeoffs between energy and cost. *J. Clean. Prod.* **2021**, *280*, 124171. [CrossRef]

33. Wu, L.; Li, C.; Tang, Y.; Yi, Q. Multi-objective tool sequence optimization in 2.5D pocket CNC milling for minimizing energy consumption and machining cost. *Procedia CIRP* **2017**, *61*, 529–534. [CrossRef]

34. Tlhabadira, I.; Daniyan, I.A.; Masu, L.; Mpofu, K. Development of a model for the optimization of energy consumption during the milling operation of titanium alloy (Ti6Al4V). *Mater. Today Proc.* **2021**, *38*, 614–620. [CrossRef]

35. Kong, M.; Pei, J.; Liu, X.; Lai, P.; Pardalos, P.M. Green manufacturing: Order acceptance and scheduling subject to the budgets of energy consumption and machine launch. *J. Clean. Prod.* **2020**, *248*, 119300. [CrossRef]

36. Wu, Z.; Sun, S. Risk cost estimation of job shop scheduling with random machine breakdowns. *Procedia CIRP* **2019**, *83*, 404–409. [CrossRef]

37. Jadhav, P.; Ekbote, N. Implementation of lean techniques in the packaging machine to optimize the cycle time of the machine. *Mater. Today Proc.* **2021**, *46*, 10275–10281. [CrossRef]

38. Pal, S.; Saini, S.K. Experimental investigation on cycle time in machining of forged crankshaft. *Mater. Today Proc.* **2021**, *44*, 1468–1471. [CrossRef]

39. Timar, S.D.; Farouki, R.T.; Smith, T.S.; Boyadjieff, C.L. Algorithms for time-optimal control of CNC machines along curved tool paths. *Robot. Comput.-Integr. Manuf.* **2005**, *21*, 37–53. [CrossRef]

40. Sadizade, B.; Araee, A.; Oliaei, S.N.B.; Farshi, V.R. Plateau honing of a diesel engine cylinder with special topography and reasonable machining time. *Tribol. Int.* **2020**, *146*, 106204. [CrossRef]

41. Cafieri, S.; Monies, F.; Mongeau, M.; Bes, C. Plunge milling time optimization via mixed-integer nonlinear programming. *Comput. Ind. Eng.* **2016**, *98*, 434–445. [CrossRef]

42. Quintana, G.; Ciurana, J. Cost estimation support tool for vertical high-speed machines based on product characteristics and productivity requirements. *Int. J. Prod. Econ.* **2011**, *134*, 188–195. [CrossRef]

43. Eguia, J.; Lamikiz, A.; Uriarte, L. Error budget and uncertainty analysis of portable machines by mixed experimental and virtual techniques. *Precis. Eng.* **2017**, *47*, 19–32. [CrossRef]

44. Siller, H.; Rodriguez, C.A.; Ahuett, H. Cycle time prediction in high-speed milling operations for sculptured surface finishing. *J. Mater. Process. Technol.* **2006**, *174*, 355–362. [CrossRef]

45. Kumar, S.P.L. Experimental investigations and empirical modeling for optimization of surface roughness and machining time parameters in micro end milling using Genetic Algorithm. *Measurement* **2018**, *124*, 386–394. [CrossRef]

46. Ning, F.; Shi, Y.; Cai, M.; Xu, W.; Zhang, X. Manufacturing cost estimation based on the machining process and deep-learning method. *J. Manuf. Syst.* **2020**, *56*, 11–12. [CrossRef]

47. Engehausen, F.; Lodding, H. Managing sequence-dependent setup times—The target conflict between output rate, WIP and fluctuating throughout times for setup cycles. *Prod. Plan. Control.* **2020**, *33*, 84–100. [CrossRef]

48. Li, Q.; Ben-Arieh, D. Remote cost estimation of machined parts. *IFAC Proc. Vol.* **2001**, *34*, 97–102.

49. Costa, C.; Silva, F.J.G.; Gouveia, R.M.; Martinho, R.P. Development of hydraulic clamping tools for the machining of complex shape mechanical components. *Procedia Manuf.* **2018**, *17*, 563–570. [CrossRef]

50. Kumar, S.R.; Krishnaa, D.; Gowthamaan, K.K.; Mouli, D.C.; Chakravarthi, K.C.; Balasubramanian, T. Development of a Re-engineered fixture to reduce operation time in a machining process. *Mater. Today Proc.* **2021**, *37*, 3179–3183. [CrossRef]

51. Kumar, S.; Campilho, R.D.S.G.; Silva, F.J.G. Rethinking modular jigs' design regarding the optimization of machining times. *Procedia Manuf.* **2019**, *38*, 876–883. [CrossRef]

52. Deng, S.; Yeh, T. Using least squares support vector machines for the airframe structures manufacturing cost estimation. *Int. J. Prod. Econ.* **2011**, *131*, 701–708. [CrossRef]

53. Loyer, J.; Henriques, E.; Fontul, M.; Wiseall, S. Comparison of Machine Learning methods applied to the estimation of manufacturing cost of jet engine components. *Int. J. Prod. Econ.* **2016**, *178*, 109–119. [CrossRef]

54. Yoo, S.; Kang, N. Explainable artificial intelligence for manufacturing cost estimation and machining feature visualization. *Expert Syst. Appl.* **2021**, *183*, 115430. [CrossRef]

*Article*

# Using an Artificial Neural Network Approach to Predict Machining Time

André Rodrigues [1], Francisco J. G. Silva [1,2,*], Vitor F. C. Sousa [1,2], Arnaldo G. Pinto [1], Luís P. Ferreira [1,2] and Teresa Pereira [1,2]

[1] ISEP—School of Engineering, Polytechnic of Porto, Rua Dr. António Bernardino de Almeida 431, 4200-072 Porto, Portugal
[2] INEGI—Institute of Science and Innovation in Mechanical and Industrial Engineering, Rua Dr. Roberto Frias 400, 4200-465 Porto, Portugal
* Correspondence: fgs@isep.ipp.pt

**Abstract:** One of the most critical factors in producing plastic injection molds is the cost estimation of machining services, which significantly affects the final mold price. These services' costs are determined according to the machining time, which is usually a long and expensive operation. If it is considered that the injection mold parts are all different, it can be understood that the correct and quick estimation of machining times is of great importance for a company's success. This article presents a proposal to apply artificial neural networks in machining time estimation for standard injection mold parts. For this purpose, a large set of parts was considered to shape the artificial intelligence model, and machining times were calculated to collect enough data for training the neural networks. The influences of the network architecture, input data, and the variables used in the network's training were studied to find the neural network with greatest prediction accuracy. The application of neural networks in this work proved to be a quick and efficient way to predict cutting times with a percent error of 2.52% in the best case. The present work can strongly contribute to the research in this and similar sectors, as recent research does not usually focus on the direct prediction of machining times relating to overall production cost. This tool can be used in a quick and efficient manner to obtain information on the total machining cost of mold parts, with the possibility of being applied to other industry sectors.

**Keywords:** machining; machining times; machining time prediction; cost estimation; artificial intelligence; artificial neural networks

**Citation:** Rodrigues, A.; Silva, F.J.G.; Sousa, V.F.C.; Pinto, A.G.; Ferreira, L.P.; Pereira, T. Using an Artificial Neural Network Approach to Predict Machining Time. *Metals* **2022**, *12*, 1709. https://doi.org/10.3390/met12101709

Academic Editors: Jorge Salguero and Sergey Konovalov

Received: 4 August 2022
Accepted: 7 October 2022
Published: 12 October 2022

**Publisher's Note:** MDPI stays neutral with regard to jurisdictional claims in published maps and institutional affiliations.

## 1. Introduction

With a wide range of applications, plastics comprise a large and varied group of materials, essentially processed using heat and pressure. Injection molding is the leading plastic transformation process, and it is the most technical and the most widespread. It is used for mass production and has a high production rate of items, with tight tolerances and with little or no need for finishing operations [1]. The surface quality of the molded parts is incomparably superior to that of other competing technologies [2]. The injection-molding process requires an injection machine and a specially manufactured mold that defines the geometry of the final product. Despite having similar elements and structures, mold production is individual, which makes the mold-making industry project-driven. One of the primary sources of risk in managing these projects is the inaccurate prediction of manufacturing costs of the mold, which is usually produced using machining services and can influence, for example, up to 45% of a molded automotive part's price [3].

Machining processes are some of the most relevant processes in the industry. The requirements of some parts, such as injection molds, are only fulfilled using these processes, which makes them irreplaceable [4,5]. Due to their complexity and popularity, many studies aim to understand and improve the different variables involved, not only in traditional but also in modern machining processes [6]. In recent years, part of the evolution of these processes is due, above all, to the progress registered in the performance of cutting tools, with the creation and implementation of coatings suited to different and challenging conditions [7,8]. Thereby, performance studies typically focus on the behavior of these coatings on difficult-to-machine materials [9,10]. All this research aims to identify the best cutting conditions for each situation, vibration being one of the main concerns among users [11]. The analysis of the cutting forces developed throughout the process can make a solid contribution to its stabilization and consequent efficiency [12]. Choosing the appropriate cutting parameters also safeguards the cutting operation and reduces process costs and energy consumption [13]. Some proposed models allow a 7.89% reduction in energy consumption [14] and seek to help operators balance energy consumption and processing costs at the same time [15]. Still, some surveys carried out by both manufacturers and end mill users reveal that the quality and process time is more relevant than energy savings [11], which is shown by some studies that prove that the decrease of processing time has more potential to increase energy efficiency than decreasing the process load [16]. The selection of the right machining strategy and cutting tool also plays a vital role in reducing cutting times, as well as the final cost and quality of the process [17]. Some authors seek to improve the calculation of some common cutting strategies, achieving a reduction in cutting forces and longer tool life [18] or reducing machining time by more than 55% [19]. Other approaches have shown that the sequence of cutting tools used can influence the amount of energy consumed and the cost of the process [20], and that the correct planning of operations allows for the reduction of errors, setup times, and non-productivity times [21]. The machining strategy used also influences the roughness and quality of the surface produced [22,23], as well as the wear of the cutting tools [24]. Trying to cool and lubricate the cutting contact area, the cutting fluids also optimize the performance of the cutting tools and, consequently, the entire process [25]. However, the use of conventional cutting fluids has proved to be a threat to the environment and the health of operators. Therefore, the use of solid lubricants has been the subject of several studies, proving to be an economic and ecological alternative in addition to offering better material removal rates and surface quality [26], as well as increased tool life [27]. Cryogenic cutting has also been analyzed, which leads to a reduction in tool wear, energy consumption, surface roughness, machining costs, and carbon emissions [28]. Both solid lubricants and cryogenic cutting have proved to be better alternatives to traditional machining processes. Often neglected, the part clamping system also plays a critical role in machining processes, guaranteeing easy operation to reduce setup time and the number of assemblies while ensuring the quality of the cutting process [4]. Some studies increased the number of parts machined per day by 32% [29] and reduced the machining time by applying suitable modular jigs [30].

All these variables make the machining process complex and difficult to budget, which is one of the main difficulties of machining services. It is common practice to set an hourly cost for the service and to determine the cost of manufacturing a part as a function of its machining time. Machining times, in turn, are commonly determined either through calculation or through human analysis. In the first case, CAM software simulates the operative sequences and the time needed to manufacture the part. In the second, the machining time is defined by a human evaluator based on a visual analysis of the part and their experience in this type of task. The first method is more accurate but requires more time to calculate machining time. The second one is faster but, as it depends on human analysis, it is less accurate, which can lead to underestimation of budgeting, resulting in losses for the company, or overestimation of budgeting, which will drive away potential customers. Thus, different studies have been developed to improve the cost estimation operation related to injection molds, with some methods showing high accuracy in applying cost drivers to estimate the cost of the manufacturing phases of the mold [31], while other authors proposed an analytic approach to determine the cutting time [32]. Other research focuses on calculating the machining cost according to cutting speed [33].

With a broad spectrum of applications in the industrial world, artificial neural networks (ANN) are mathematical models that, inspired by the functioning of the human brain, seek to understand complex relationships existing in each set of data. Therefore, some studies aim to assist in estimating manufacturing costs for different components [34,35] and to optimize the injection mold manufacturing process by implementing ANN [36,37]. The application of ANN in improving the performance of the machining process has also been studied [38,39], making it possible to predict machined surface roughness [40] and production quality [41], cutting tool wear [42], and cutting forces [43,44], as well as other computational approaches [45]. Regarding cutting tool condition monitoring, this is a very useful manner to evaluate the process's overall stability and productivity [46], contributing to overall process improvement and a reduction in the production costs. There are various opportunities to implement ANN and other computational methods for machining tool monitoring [47], either for milling or for turning [48], contributing to an improvement of these processes, particularly regarding fewer tool exchanges and quality improvements of machined parts [49]. Some authors proposed artificial neural networks for feature recognition [50] to further establish relationships between the number of features and the cost of the part [51]. Other authors presented models using ANN to estimate machining time with average errors of 10.20 min [52] or to estimate injection mold manufacturing time with a percentage error of less than 25% [53]. Artificial intelligence tools to estimate manufacturing costs in the design phase have also been studied [54] as the effectiveness of different types of ANN in estimating machining times [55].

As seen from the studies presented in the previous paragraph, the use of ANNs can be quite useful in predicting the cost of manufacturing a certain part. However, there are some constraints regarding the use of this approach, with some approaches showing high degrees of error or deviation [53]. Despite this, there is potential for application of these ANNs in predicting these costs and thus in aiding budgeting operations for machining services. Thus, the budgeting process is time-consuming, prone to mistakes, and heavily dependent on previously acquired empirical knowledge. It is known that the machining time, especially for mold machining operations, is the most impactful parameter on the overall cost of the tool. Given this, by predicting the machining time, a somewhat accurate prediction of the final part's cost can be made. In the present paper, a methodology for using an ANN to predict machining times for standard injection mold parts is presented by choosing and comparing different architectures, input data, and training variables, and thus finding the most accurate way of predicting machining times. The implementation of this ANN was compared to the previously employed method, in which the time is determined through careful modelling and simulation of the part. This is an expensive and time-consuming way of cost-estimating the machined parts. By employing the ANN presented in this work, companies should have a fast and accurate way of predicting the machining times of their produced parts.

## 2. Methods

The present work aims at creating an artificial neural network (ANN) mode that allows the machining time of standard plastic injection mold parts to be estimated in a quick and effective way. For this purpose, it was considered that these parts usually present groups of similar features, which only diverge in dimensions, such as metric threads, clearance holes, fitting holes, rectangular and circular pockets, among others. Thus, the proposed solution consists of an ANN whose input variables are (see Figure 1):

- The workpiece material;
- Each feature group's total number of elements;
- The volume of removed material from each feature group;
- The surface area machined in each feature group.

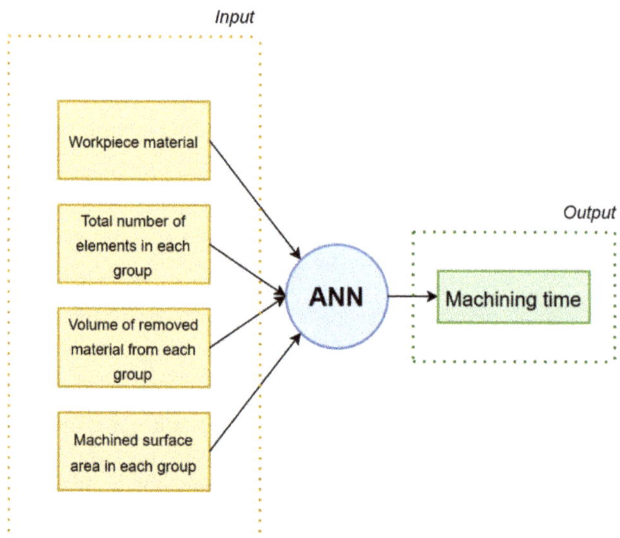


**Figure 1.** Diagram of the input variables of the proposed solution.

## 2.1. Data Preparation

Data preparation was the first step in the development of the ANN. The goal was to simulate a machining process, calculating the machining time of several plastic injection mold plates using CAM programming software (MasterCAM® 2022), and a library of tools created for this purpose. As seen before, the amount of data used for training the ANN influences its accuracy, making it necessary to calculate the machining time of hundreds of different parts to generate enough data. The traditional method for calculating machining times requires a 3D model and the CAM programming of each part, which is time-consuming considering the high number of parts needed; usually this calculated time is then compared to the actual required machining time to produce a said part. Therefore, an empirical-based method was created where the machining time of each element is calculated individually and subsequently assigned to each part at random, allowing the calculation of a multitude of different parts. The workpiece material and the machined geometries are defined based on common requests obtained to produce injection mold parts. These parts are usually standardized. Furthermore, literature research provided insight on the best-suited parameters and strategies to produce said parts. This information is complemented by contacting manufacturers of these standardized components. They usually offer practical and industrially acquired knowledge, which can be very useful when defining these parameters. Then, the machining times of 30 modeled parts are calculated by the traditional (simulation and modeling, offering a comparison of measurements made after component production) and empirical methods to compare the obtained values and validate the devised empirical method. A definition of these methods can be found below:

- Traditional method: To try and predict the required machining time to produce a certain part, these must be modelled in 3D. Subsequently, the CAM program is performed, and the machining time is estimated. Afterwards, these times are validated with those obtained from the part production itself. It is accurate; however, it is also time-consuming and cost inefficient;
- Empirical method: Input and output data are generated for the training of the ANNs, based on the machining time of individual operations (standardized machining operations for mold production). These operations are then compiled and considered for each of the produced part (Figure S1).

### 2.1.1. Features Definition and Sequence of Operations

The features used in this work are presented in Table 1, and the sequence of operations is also defined. This sequence of operations is defined based on the information displayed on the previous section (Section 2.2. Data preparation). In addition to the operations mentioned in Table 1, more machining operations were considered, such as circular and rectangular cavities, guiding holes, and low-depth circular cavities.

**Table 1.** Feature definitions and corresponding sequences of operations.

| Feature Group | Illustration | A (mm) | B (mm) | C (mm) | D (mm) | Sequence of Operations |
|---|---|---|---|---|---|---|
| Metric threads | | M3; M4; M5; M6; M8; M10; M12; M16; M20; M24; M30. | – | – | – | 1st—Drill; 2nd—Chamfer; 3rd—Tap. |
| Clearance holes | | 6.00; 7.00; 9.00; 10.00; 11.00; 11.50; 13.50; 14.00; 15.50; 16.00; 17.50; 20.00; 22.00; 22.50; 27.00. | 46.00; 66.00; 86.00; 96.00; 136.00. | – | – | 1st—Drill; 2nd—Chamfer. |
| Screw clearance holes | | 10.00; 11.00; 14.00; 17.00; 21.00; 23.00; 26.00; 32.00; 37.00. | 6.00; 7.00; 9.00; 11.00; 13.50; 15.50; 17.50; 21.00; 27.00. | 6.00; 7.00; 9.00; 11.00; 13.00; 15.00; 17.00; 21.00; 27.00. | 46.00; 66.00; 86.00; 96.00; 136.00. | 1st—Drill; 2nd—Counter-boring; 3rd—Chamfer. |
| Fitting holes | | 10.00; 12.00; 16.00; 20.00. | 10.00; 16.00; 25.00; 30.00. | – | – | 1st—Drill; 2nd—Bore; 3rd—Chamfer. |

### 2.1.2. Machining Time Calculation

The workpiece material also significantly influences the final machining time by influencing the cutting parameters possible to use with the selected cutting tools. Therefore, two tool steels commonly used in injection mold structures were considered (Table 2).

**Table 2.** Tool steels used to calculate machining time.

| Material No. | AISI | DIN |
|---|---|---|
| 1.1730 | 1045 | C45E |
| 1.2311 | P20 | 40CrMnMo7 |

Based on the features presented in Table 1 and the materials to be machined in Table 2, 30 parts like the one shown in Figure 2 were generated and modeled. Each of the parts was assigned to one of the workpiece materials, as well as the elements in Table 1 in different quantities. Subsequently, and based on the sequences of operations assigned for each feature, a library of cutting tools was created, and machining strategies were defined.

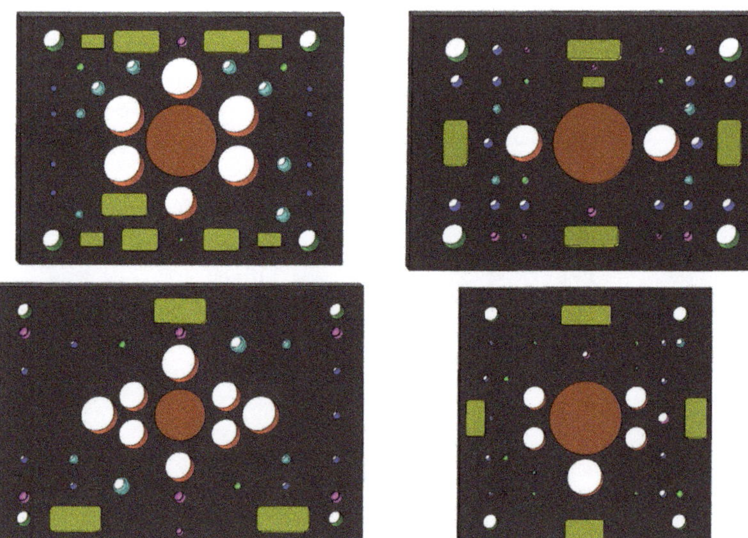

**Figure 2.** Examples of models used to produce the mold parts (Figure S1).

Then, the 30 parts were calculated using traditional and empirical methods, with the machining times obtained being shown and compared in Figure 3.

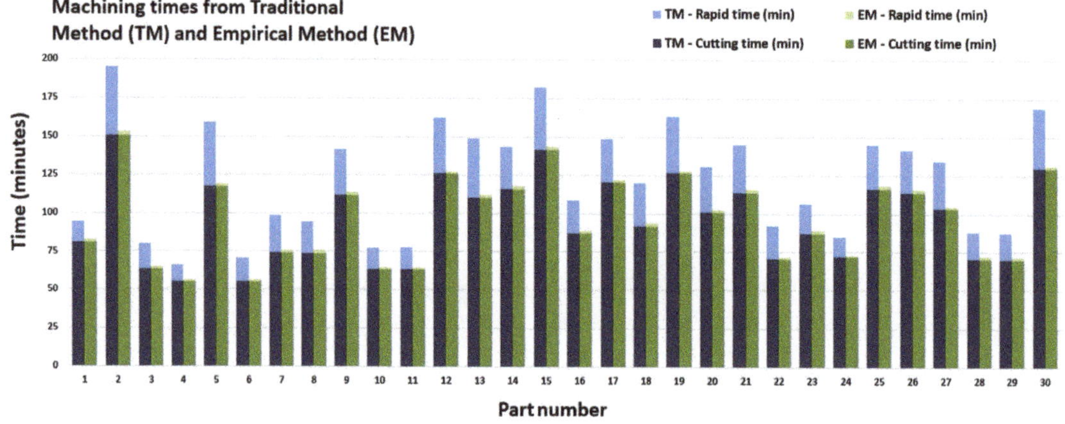


**Figure 3.** Machining times of each of the 30 parts obtained by the traditional method (TM) and the empirical method (EM).

### 2.1.3. Validation of the Empirical Method

Machining times are divided into cutting time (Ct) and rapid travelling (Rt) movement time when the tool approaches the workpiece and moves between elements. These times are shown in Figure 3 and allow one to conclude that the cutting times obtained in the two methods are the same, which is explained by the fact that the same tools were used with the same cutting parameters and sequence of operations. The difference between the values obtained is in the rapid movement times. In the traditional method, the tool displacement time between elements and the tool approach time to the workpiece is calculated, while in the empirical method, due to lack of knowledge about the position of the elements in the workpiece, only the tool approach time is counted. This is the main reason for the difference in times obtained between these two methods. Thus, the empirical method is valid for calculating cutting times but is invalid for calculating rapid movement times. Thus, bearing in mind that they are the primary influence on the final machining time, as seen in Figure 4, only cutting times are used to develop the artificial neural networks.

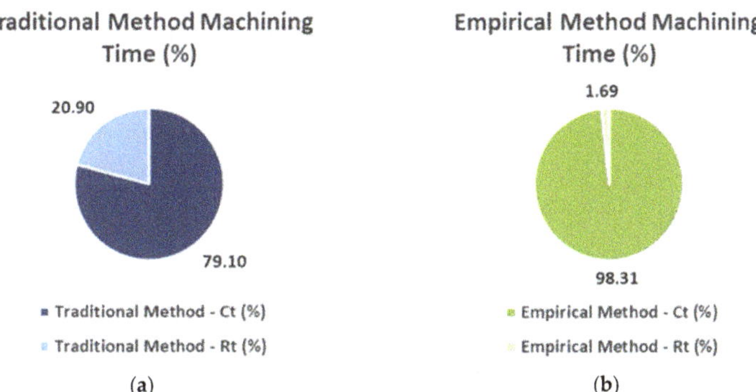


**Figure 4.** Traditional method (**a**) and empirical method (**b**) machining time in percentage (Ct—cutting time; Rt—rapid travelling time).

After the validation of the empirical method, which has the advantage of being able to generate an infinite number of parts, two databases were created based on the features shown in Table 1 for the development of ANNs:

- trainingdata: database with 750 parts for training, testing, and validation of ANNs to develop;
- testdata: database with 100 parts to evaluate the performance of the developed ANNs.

### 2.2. ANN

The back error-propagation algorithm is one of the most used algorithms in engineering prediction models, being mainly used in layered feed forward ANNs. This algorithm is trained considering several series of examples (database), which encompass input arrangements and the desired output ranges. In this kind of algorithm, the input is introduced to the network, and the error between the current and the desired output is propagated backward in the network to calibrate the weights to improve the predictions accuracy. Several ANNs were developed to find the one that best fit the proposed problem. The ANNs were developed using MATLAB®, with the 'feedforwardnet' command, as it is suitable for regression problems where the objective is to estimate a specific numerical result with a single output, and with 80% of the data for training, 10% for validation, and the remaining 10% for testing. The activation function of the hidden layers was the 'tansig' function, and the last layer's activation function, the 'purelin' function. 'Tansig' is a hyperbolic tangent sigmoid activation function generally used in intermediate layers of ANNs, since it is entirely derivable and compatible with most training algorithms used. 'Purelin', in turn, is a linear activation function used in regression problems. The algorithms used in the network training are the default ones for the 'feedforwardnet' command in MATLAB®.

Although ANNs are tested immediately after training, it was decided to test the developed ANNs in estimating the machining times of the 100 parts contained on the database 'testdata'. In this way, the performance of the ANNs are evaluated against the same input data and not against random data. The evaluation of the network's performance is conducted by calculating the mean error (*ME*), the mean absolute error (*MAE*), the mean squared error (*MSE*), and the percentage error (*PE*) between the real and estimated values. The maximum error (EMax) is also analyzed.

The mean error (*ME*) was calculated according to the expression:

$$ME = \frac{\sum_{t=1}^{n} e_t}{n} \tag{1}$$

where $e_t$ is calculated as the difference between the real time ($Z_t$) and the estimated time ($Z_t^e$), according to the expression:

$$e_t = Z_t - Z_t^e \tag{2}$$

The mean absolute error (*MAE*), in turn, is obtained according to:

$$MAE = \frac{\sum_{t=1}^{n} |e_t|}{n} \tag{3}$$

Expression (4) was used to calculate the mean square error (*MSE*):

$$MSE = \frac{\sum_{t=1}^{n} e_t^2}{n} \tag{4}$$

The percentage error was calculated according to the expression:

$$PE = \frac{|Z_t^e - Z_t|}{Z_t} \times 100 \tag{5}$$

### 2.3. Comparative Methods

The first test looked at the network architecture with the best accuracy. The 750 parts of the 'trainingdata' database were used. The input variables of these networks were the quantity and volume of each of the features groups and the material of the workpiece. The second one sought to study the influence of the amount of input data on ANN accuracy. For this purpose, the ANN network architecture with the lowest percentage error of Test 1 was used, differing from the amount of data used in training. Test 3 studied the influence of input variables on the accuracy of ANNs. Once again, and keeping the percentage error as a criterion, the best ANN from the previous tests was selected, and the input variables were studied. Workpiece material was always used as input. To understand the tables referring to Test 3, it is essential to define the following:

- Q—Total number of elements of each feature group;
- A—Machined surface area in each feature group;
- V—Volume of removed material from each feature group;
- Qt—Total amount of elements to machine in the part;
- At—Total surface area to machine in the part;
- Vt—Total volume of removed material in the part.

## 3. Results

In this section, the results obtained in each of the tests are presented and analyzed.

### 3.1. Test 1—Variation of Network Architectures

From the application of ANN of Test 1 to 'testdata', it was observed that the ANNs with the lowest ME were T1_01 and T1_05, the ANN with the lowest MAE was T1_03, and T1_07 presented the lowest MSE. The lowest EMax recorded was 19.54 min, in T1_06. Using the percentage error as a criterion, the ANN with the best performance was T1_07, with a percentage error of 2.52% (please see Table 3).

**Table 3.** Regression and error of the ANNs from Test 1.

| Test | Arch. | R | | | | Error | | | | |
|---|---|---|---|---|---|---|---|---|---|---|
| | | Train | Validation | Test | Total | ME (min) | MAE (min) | MSE (min) | EMax (min) | PE (%) |
| T1_01 | 4-8-1 | 0.98 | 0.97 | 0.98 | 0.98 | 0.16 | 5.00 | 42.22 | 17.14 | 2.63 |
| T1_02 | 4-4-1 | 0.94 | 0.93 | 0.85 | 0.93 | −1.82 | 8.40 | 221.91 | 71.23 | 6.15 |
| T1_03 | 4-9-1 | 0.99 | 0.98 | 0.97 | 0.98 | 0.57 | 4.74 | 38.72 | 23.13 | 2.78 |
| T1_04 | 4-10-1 | 0.98 | 0.98 | 0.96 | 0.98 | 1.95 | 5.31 | 43.74 | 15.30 | 2.86 |
| T1_05 | 4-15-1 | 0.98 | 0.97 | 0.98 | 0.98 | 0.16 | 5.37 | 47.20 | 21.72 | 2.89 |
| T1_06 | 4-20-1 | 0.98 | 0.97 | 0.98 | 0.98 | 1.38 | 5.20 | 45.47 | 19.54 | 2.66 |
| T1_07 | 4-8-8-1 | 0.99 | 0.99 | 0.98 | 0.98 | 0.26 | 4.85 | 41.42 | 25.71 | 2.52 |
| T1_08 | 4-10-10-1 | 0.99 | 0.98 | 0.97 | 0.98 | −0.66 | 5.57 | 62.06 | 37.06 | 2.88 |
| T1_09 | 4-4-4-1 | 0.97 | 0.96 | 0.96 | 0.97 | 2.09 | 6.62 | 75.98 | 34.69 | 3.13 |

Analyzing Table 3, the following conclusions can be drawn:

- The training samples with the best results were T1_03, T1_07, and T1_08, showing an R value of 0.99;
- The T1_07 was the network with the best validation results, having an R value of 0.99;
- T1_01, T1_05, T1_06, and T1_07 showed the second-best results with an R value of 0.98;
- The T1_07 network exhibited the overall best results, while T1_02 showed the worst results;
- Only the R value of T1_02, being 0.93, was lower, showing a 0.85 value for the test samples. The remaining networks exhibited values above 0.93, indicating good training of these networks.

*3.2. Test 2—Influence of the Amount of Input Data*

Test 2, where the amount of data introduced for training was varied, revealed that the ANN with the lowest ME was T2_13, with an average error of −0.08 min. With an MAE of 4.85 min, T2_15 was the ANN with the best result. The ANN with the lowest MSE was T2_14 with 40.18 min. The smallest EMax recorded was 16.73 min, which occurred in T2_09. The lowest PE belonged to T2_15, like T1_07, and with a PE of 2.52%, which ensured the best performance (please see Table 4).

From the training of the networks for Test 2, it was verified that:

- The networks T2_07, T2_09, T2_13, and T2_15 were the ones that exhibited the best performance in the training samples, with a R value of 0.99;
- The T2_15 network was the network with the best performance regarding the validation sample, showing a R value of 0.99;
- With a R value of 0.98, the T2_05, T2_10, T2_11, T2_13, and T2_15 networks were the ones with best performance in the test sample;
- Regarding the training results, the T2_13 network showed the best results (R value equal to 0.99);
- It was observed that the R value would increase with an increase in number of considered parts. It was noted that this value would stabilize at around 250 parts.

The graph in Figure 5 shows the percentage error variation as a function of the number of input data. The error decreases as the number of parts increases. From 250 parts onwards, the percentage error is reduced to a lesser extent.

**Table 4.** Regression and error of the ANNs from Test 2.

| Test | Input Parts | R | | | | Error | | | | |
|------|-------------|-------|------------|------|-------|----------|-----------|-----------|------------|--------|
| | | Train | Validation | Test | Total | ME (min) | MAE (min) | MSE (min) | EMax (min) | PE (%) |
| T2_01 | 50 | 0.63 | 0.86 | 0.90 | 0.67 | 2.45 | 21.64 | 799.52 | 78.15 | 12.82 |
| T2_02 | 100 | 0.85 | 0.74 | 0.66 | 0.82 | 6.74 | 23.99 | 906.34 | 96.35 | 11.29 |
| T2_03 | 150 | 0.97 | 0.88 | 0.93 | 0.96 | 2.69 | 12.50 | 265.21 | 46.08 | 7.48 |
| T2_04 | 200 | 0.97 | 0.97 | 0.93 | 0.97 | −0.32 | 9.47 | 175.14 | 44.10 | 5.01 |
| T2_05 | 250 | 0.98 | 0.96 | 0.98 | 0.98 | 0.43 | 6.94 | 85.75 | 37.75 | 3.95 |
| T2_06 | 300 | 0.98 | 0.97 | 0.97 | 0.97 | −0.29 | 6.95 | 81.23 | 33.91 | 3.78 |
| T2_07 | 350 | 0.99 | 0.95 | 0.97 | 0.98 | −1.16 | 6.27 | 62.82 | 23.14 | 3.55 |
| T2_08 | 400 | 0.98 | 0.97 | 0.96 | 0.98 | −1.35 | 6.43 | 67.74 | 24.18 | 3.49 |
| T2_09 | 450 | 0.99 | 0.98 | 0.97 | 0.98 | −0.96 | 5.85 | 52.05 | 16.73 | 3.26 |
| T2_10 | 500 | 0.98 | 0.97 | 0.98 | 0.98 | 1.42 | 5.59 | 54.26 | 25.16 | 3.15 |
| T2_11 | 550 | 0.98 | 0.98 | 0.98 | 0.98 | 1.80 | 5.72 | 57.80 | 27.27 | 2.99 |
| T2_12 | 600 | 0.99 | 0.97 | 0.97 | 0.98 | 0.62 | 5.27 | 45.28 | 22.02 | 2.98 |
| T2_13 | 650 | 0.99 | 0.98 | 0.98 | 0.99 | −0.08 | 5.49 | 55.31 | 27.80 | 2.98 |
| T2_14 | 700 | 0.98 | 0.98 | 0.97 | 0.98 | 0.27 | 4.88 | 40.18 | 17.09 | 2.57 |
| T2_15 | 750 | 0.99 | 0.99 | 0.98 | 0.98 | 0.26 | 4.85 | 41.42 | 25.71 | 2.52 |

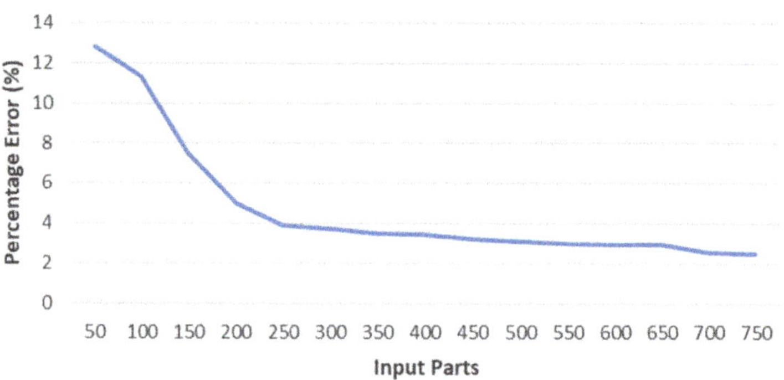


**Figure 5.** Percentage error (PE) variation as a function of the amount of input data.

### 3.3. Test 3—Influence of Input Variables

From the application of the ANNs of Test 3 to the 'testdata' database, where the input variables are studied, the ANN with the lowest ME was T3_07, with an ME of 0.21 min. The ANN with the lowest MAE was T3_01, with an MAE of 4.85 min. With an MSE of 41.42 min, T3_01 was the best ANN in this parameter. The smallest maximum error recorded was 20.20 min and occurred in T3_07. The ANN with the lowest percentage error was T3_01, like T1_07 and T2_15, with a PE of 2.52% (please see Table 5).

**Table 5.** Regression and error of the ANNs from Test 3.

| Test | Input Variables | R | | | | Error | | | | |
|---|---|---|---|---|---|---|---|---|---|---|
| | | Train | Validation | Test | Total | ME (min) | MAE (min) | MSE (min) | EMax (min) | PE (%) |
| T3_01 | Q + V | 0.99 | 0.99 | 0.98 | 0.98 | 0.26 | 4.85 | 41.42 | 25.71 | 2.52 |
| T3_02 | Q + A | 0.96 | 0.95 | 0.95 | 0.95 | 2.13 | 8.90 | 143.17 | 42.02 | 4.58 |
| T3_03 | V + A | 0.97 | 0.95 | 0.97 | 0.97 | −2.76 | 6.65 | 68.63 | 23.18 | 3.21 |
| T3_04 | QT + VT | 0.72 | 0.76 | 0.74 | 0.73 | 1.89 | 18.85 | 600.44 | 67.70 | 10.26 |
| T3_05 | QT + AT | 0.79 | 0.81 | 0.85 | 0.80 | 30.06 | 31.46 | 1373.17 | 91.06 | 14.66 |
| T3_06 | VT + AT | 0.83 | 0.83 | 0.84 | 0.83 | 25.90 | 27.23 | 1154.57 | 105.98 | 12.54 |
| T3_07 | Q + V + A | 0.99 | 0.98 | 0.97 | 0.98 | 0.21 | 5.74 | 52.53 | 20.20 | 2.77 |
| T3_08 | Q + VT | 0.89 | 0.85 | 0.86 | 0.88 | 18.37 | 19.17 | 561.75 | 67.30 | 8.44 |
| T3_09 | Q + AT | 0.93 | 0.93 | 0.94 | 0.93 | 1.80 | 8.98 | 133.10 | 32.55 | 4.43 |

From the training of the networks for Test 3, it can be concluded that:

- The networks T3_01 and T3_07 are the ones that showed the best training sample results, exhibiting an R value of 0.99;
- The best R value obtained for validation samples was the one obtained for T3_01;
- The T3_01 network also showed the best R results for the training sample, with a value of 0.98;
- The networks that showed the best overall results (Total) were the T3_01 and the T3_07 networks, with a R value of 0.98.

## 4. Discussion

The results obtained with Test 2 allowed for concluding something already expected, namely, that by increasing the number of data provided to the ANN, the learning process leads to high accuracy results, and thus the amount of data introduced in the future should be even higher, allowing for more accurate predictions. This directly results from the observations of Table 4 and Figure 5. This trend was previously argued by Tay and Cao [56], namely, that ANN results depend to a large extent on the amount of data provided to the system, being a time-consuming task.

Based on the results previously described, it can be stated that many network architectures have provided very accurate results, with eight architectures showing errors lower than 3.15% (96.85% accuracy), which can be considered an excellent result in terms of machining cost prediction. This result is less than those previously obtained by other researchers [57], who reported average accuracies of about 98.5% using ANNs, which has been shown to be a more accurate method to detect the different blends of fuel than RSM (response surface methodology) methods. Moreover, another study [58] about predictions of AISI 1050 steel machining performance has shown that ANNs presented the most accurate prediction value (92.1%), which is below the accuracy provided by the eight different architectures used in this work. In that work [58], it remains clear that ANN can provide more accurate values than other prediction techniques, such as the adaptive neuro-fuzzy inference system (ANFIS), which provided an accuracy of 73% compared to the 92.1% achieved using ANN. Saric et al. [55] also claimed an error of 2.03% using back-propagation neural networks in the estimation of CNC machining times. However, using self-organizing map neural networks, the results are less accurate, showing errors of about 10.05%. Ning et al. [52] used as a training dataset 21,943 features, and 4338 for validation, obtaining an accuracy of 97.7%, which is very similar to that achieved through this work. Thus, it is expected that a higher number of features used for training of the model can contribute to a better accuracy of the predictions, but too much data cannot contribute to accuracy improvement and can cause higher computing time, which can be unnecessary.

Regarding Table 5, the best results were obtaining using as input variables the volume of removed materials (V) and the total number of elements of each feature (Q). Indeed, the second most accurate result was obtained using the same variables plus the machined surface area (A). However, adding this variable, the results started to degrade. Among the other combinations considered in this work, only the Q + A combination presented acceptable results, together with the combination Q + At. Thus, it is possible to infer that the volume of removed materials plays an important role in terms of keeping the accuracy as high as possible, which can be combined with V for the best results and combined with A for close to the best results.

## 5. Conclusions

This work allowed us to analyze, study, and test the applicability of artificial neural networks in estimating machining times. One of the main contributions of the presented work is the development of an ANN that predicts the machining time of standard parts for plastic injection molds with a percent error of 2.52%.

The application of ANNs in this work proved to be a quick and efficient way to estimate machining times. The introduction of the total quantity and volume of each of the most common features regarding the standard parts proved to be sufficient for this purpose, and it was not necessary to detail individually in the training of the networks the dimensions of each element to be machined. The specification of each element to be machined would make the presented method slower, which would not meet the proposed goals.

The different tests carried out demonstrated that:

- Network architectures had a minor influence on the accuracy of ANNs;
- The amount of data used in network training proved to be of great importance. The ANNs trained with a more significant number of data had a lower percentage of error and better training values. However, excessive amounts of data were time-consuming in terms of computing and did not generate significant gains in terms of accuracy;
- The decrease in the percentage of error of the trained network was less accentuated as the number of data used in training increased;
- The volume of removed material and the total number of elements to be machined proved to be the input variables that provided the lowest percentage error, i.e., the best accuracy of predicted machining costs;
- Contrary to what would be expected, the introduction of the area to the quantity and volume of each group of elements did not represent a decrease in the percentage error;
- The variables total amount of elements to machine, total volume to machine, and the total area to machine did not show good results. This fact indicates that it is necessary to at least group the elements to be machined in similar groups;
- In all tests, a good relationship was confirmed between the regression results of the network training and the percentage error results.

The present work brings new knowledge about the possibility of applying ANNs in the estimation of machining times, which is not very common in the recently analyzed literature. This enables the use of these methods in industry sectors, such as the mold industry, enabling the determination of the machining time and overall production cost of standardized mold parts. However, there are some current limitations of the proposed work. This study was developed for an application regarding the production of mold parts, and although this methodology can be employed in other industry sectors (particularly machining sectors of standardized parts), this would involve calibration and training of new ANNs. Despite this fact, this work highlights the use of ANNs compared to more time-consuming and expensive alternatives when determining operation times and costs.

**Supplementary Materials:** The following supporting information can be downloaded at: https://www.mdpi.com/article/10.3390/met12101709/s1, Figure S1: List of produced parts.

**Author Contributions:** A.R.: investigation, formal analysis and writing—review and editing; F.J.G.S.: conceptualization, methodology, project administration, resources, supervision and writing—review and editing; V.F.C.S.: formal analysis, validation, visualization and writing—review and editing; A.G.P.: formal analysis, validation and writing—review and editing; L.P.F.: formal analysis, validation and writing—review and editing; T.P.: formal analysis, validation and writing—review and editing. All authors have read and agreed to the published version of the manuscript.

**Funding:** This research received no external funding.

**Institutional Review Board Statement:** Not applicable.

**Informed Consent Statement:** Not applicable.

**Data Availability Statement:** No data is made available regarding this work.

**Acknowledgments:** Authors thank INEGI—Institute of Science and Innovation in Mechanical and Industrial Engineering for its support.

**Conflicts of Interest:** The authors declare no conflict of interest.

# References


1.  Matarrese, P.; Fontana, A.; Sorlini, M.; Diviani, L.; Specht, I.; Maggi, A. Estimating energy consumption of injection moulding for environmental-driven mould design. *J. Clean. Prod.* **2017**, *168*, 1505–1512. [CrossRef]
2.  Oropallo, W.; Piegl, L.A. Ten challenges in 3D printing. *Eng. Comput.* **2016**, *32*, 135–148. [CrossRef]
3.  Altan, T.; Lilly, B.; Yen, Y.C. Manufacturing of dies and molds. *CIRP Ann. Manuf. Technol.* **2001**, *50*, 404–422. [CrossRef]
4.  Costa, C.; Silva, F.; Gouveia, R.M.; Martinho, R. Development of hydraulic clamping tools for the machining of complex shape mechanical components. *Procedia Manuf.* **2018**, *17*, 563–570. [CrossRef]
5.  Hällgren, S.; Pejryd, L.; Ekengren, J. Additive Manufacturing and High Speed Machining -cost Comparison of short Lead Time Manufacturing Methods. *Procedia CIRP* **2016**, *50*, 384–389. [CrossRef]
6.  Rao, R.V.; Kalyankar, V. Optimization of modern machining processes using advanced optimization techniques: A review. *Int. J. Adv. Manuf. Technol.* **2014**, *73*, 1159–1188. [CrossRef]
7.  Sousa, V.F.C.; Silva, F.J.G. Recent Advances in Turning Processes Using Coated Tools—A Comprehensive Review. *Metals* **2020**, *10*, 170. [CrossRef]
8.  Sousa, V.F.C.; Silva, F.J.G. Recent Advances on Coated Milling Tool Technology—A Comprehensive Review. *Coatings* **2020**, *10*, 235. [CrossRef]
9.  Silva, F.J.G.; Martinho, R.P.; Martins, C.; Lopes, H.; Gouveia, R.M. Machining GX2CrNiMoN26-7-4 DSS alloy: Wear analysis of TiAlN and TiCN/Al2O3/TiN coated carbide tools behavior in rough end milling operations. *Coatings* **2019**, *9*, 392. [CrossRef]
10. Martinho, R.P.; Silva, F.J.G.; Martins, C.; Lopes, H. Comparative study of PVD and CVD cutting tools performance in milling of duplex stainless steel. *Int. J. Adv. Manuf. Technol.* **2019**, *102*, 2423–2439. [CrossRef]
11. Thorenz, B.; Oßwald, F.; Schötz, S.; Westermann, H.-H.; Döpper, F. Applying and Producing Indexable End Mills: A Comparative Market Study in Context of Resource Efficiency. *Procedia Manuf.* **2020**, *43*, 167–174. [CrossRef]
12. Sousa, V.F.C.; Silva, F.J.G.; Fecheira, J.S.; Lopes, H.M.; Martinho, R.P.; Casais, R.B.; Ferreira, L.P. Cutting forces assessment in cnc machining processes: A criticalreview review. *Sensors* **2020**, *20*, 4536. [CrossRef] [PubMed]
13. Tlhabadira, I.; Daniyan, I.; Masu, L.; Mpofu, K. Development of a model for the optimization of energy consumption during the milling operation of titanium alloy (Ti6Al4V). *Mater. Today Proc.* **2021**, *38*, 614–620. [CrossRef]
14. Zhang, X.; Yu, T.; Dai, Y.; Qu, S.; Zhao, J. Energy consumption considering tool wear and optimization of cutting parameters in micro milling process. *Int. J. Mech. Sci.* **2020**, *178*, 105628. [CrossRef]
15. Xiao, Y.; Jiang, Z.; Gu, Q.; Yan, W.; Wang, R. A novel approach to CNC machining center processing parameters optimization considering energy-saving and low-cost. *J. Manuf. Syst.* **2021**, *59*, 535–548. [CrossRef]
16. Westermann, H.-H.; Kafara, M.; Steinhilper, R. Development of a Reference Part for the Evaluation of Energy Efficiency in Milling Operations. *Procedia CIRP* **2015**, *26*, 521–526. [CrossRef]
17. Gouveia, R.M.; Silva, F.J.G.; Reis, P.; Baptista, A.P.M. Machining Duplex Stainless Steel: Comparative Study Regarding End Mill Coated Tools. *Coatings* **2016**, *6*, 51. [CrossRef]
18. Huang, N.; Jin, Y.; Lu, Y.; Yi, B.; Li, X.; Wu, S. Spiral toolpath generation method for pocket machining. *Comput. Ind. Eng.* **2020**, *139*, 106142. [CrossRef]
19. Cafieri, S.; Monies, F.; Mongeau, M.; Bes, C. Plunge milling time optimization via mixed-integer nonlinear programming. *Comput. Ind. Eng.* **2016**, *98*, 434–445. [CrossRef]
20. Wu, L.; Li, C.; Tang, Y.; Yi, Q. Multi-objective Tool Sequence Optimization in 2.5D Pocket CNC Milling for Minimizing Energy Consumption and Machining Cost. *Procedia CIRP* **2017**, *61*, 529–534. [CrossRef]
21. Silva, J.; Silva, F.J.G.; Campilho, R.D.S.G.; Sá, J.C.; Ferreira, L.P. A Model for Productivity Improvement on Machining of Components for Stamping Dies. *Int. J. Ind. Eng. Manag.* **2021**, *12*, 85–101. [CrossRef]
22. Michalik, P.; Zajac, J.; Hatala, M.; Mital, D.; Fecova, V. Monitoring surface roughness of thin-walled components from steel C45 machining down and up milling. *Measurement* **2014**, *58*, 416–428. [CrossRef]
23. Li, Y.W.; Sun, Y.S.; Zhou, X.G. Theoretical Analysis and Experimental Verification that Influence Factors of Climb and Conventional Milling on Surface Roughness. *Appl. Mech. Mater.* **2013**, *459*, 407–412. [CrossRef]

24. Hadi, M.; Ghani, J.; Haron, C.C.; Kasim, M.S. Comparison between Up-milling and Down-milling Operations on Tool Wear in Milling Inconel 718. *Procedia Eng.* **2013**, *68*, 647–653. [CrossRef]

25. Bouzakis, K.D.; Makrimallakis, S.; Skordaris, G.; Bouzakis, E.; Kombogiannis, S.; Katirtzoglou, G.; Maliaris, G. Coated tools' performance in up and down milling stainless steel, explained by film mechanical and fatigue properties. *Wear* **2013**, *303*, 546–559. [CrossRef]

26. Agarwal, V.; Agarwal, S. Performance profiling of solid lubricant for eco-friendly sustainable manufacturing. *J. Manuf. Process.* **2021**, *64*, 294–305. [CrossRef]

27. Makhesana, M.; Patel, K.; Mawandiya, B. Environmentally conscious machining of Inconel 718 with solid lubricant assisted minimum quantity lubrication. *Met. Powder Rep.* **2020**, *76*, S24–S29. [CrossRef]

28. Agrawal, C.; Wadhwa, J.; Pitroda, A.; Pruncu, C.I.; Sarikaya, M.; Khanna, N. Comprehensive analysis of tool wear, tool life, surface roughness, costing and carbon emissions in turning Ti–6Al–4V titanium alloy: Cryogenic versus wet machining. *Tribol. Int.* **2021**, *153*, 106597. [CrossRef]

29. Kumar, S.; Campilho, R.; Silva, F. Rethinking modular jigs' design regarding the optimization of machining times. *Procedia Manuf.* **2019**, *38*, 876–883. [CrossRef]

30. Kumar, S.R.; Krishnaa, D.; Gowthamaan, K.; Mouli, D.C.; Chakravarthi, K.C.; Balasubramanian, T. Development of a Re-engineered fixture to reduce operation time in a machining process. *Mater. Today Proc.* **2020**, *37*, 3179–3183. [CrossRef]

31. Fiorentino, A. Cost drivers-based method for machining and assembly cost estimations in mould manufacturing. *Int. J. Adv. Manuf. Technol.* **2014**, *70*, 1437–1444. [CrossRef]

32. Bouaziz, Z.; Ben Younes, J.; Zghal, A. Cost estimation system of dies manufacturing based on the complex machining features. *Int. J. Adv. Manuf. Technol.* **2006**, *28*, 262–271. [CrossRef]

33. Narita, H. A Study of Automatic Determination of Cutting Conditions to Minimize Machining Cost. *Procedia CIRP* **2013**, *7*, 217–221. [CrossRef]

34. Deng, S.; Yeh, T.-H. Using least squares support vector machines for the airframe structures manufacturing cost estimation. *Int. J. Prod. Econ.* **2011**, *131*, 701–708. [CrossRef]

35. Loyer, J.-L.; Henriques, E.; Fontul, M.; Wiseall, S. Comparison of Machine Learning methods applied to the estimation of manufacturing cost of jet engine components. *Int. J. Prod. Econ.* **2016**, *178*, 109–119. [CrossRef]

36. Lee, S.; Cho, Y.; Lee, Y.H. Injection Mold Production Sustainable Scheduling Using Deep Reinforcement Learning. *Sustainability* **2020**, *12*, 8718. [CrossRef]

37. Viharos, Z.J.; Mikó, B. Artificial neural network approach for injection mould cost estimation. In Proceedings of the 44th CIRP Conference on Manufacturing Systems, Madison, WI, USA, 1–3 June 2009; pp. 1–6.

38. Tansel, I.; Ozcelik, B.; Bao, W.; Chen, P.; Rincon, D.; Yang, S.; Yenilmez, A. Selection of optimal cutting conditions by using GONNS. *Int. J. Mach. Tools Manuf.* **2006**, *46*, 26–35. [CrossRef]

39. De Filippis, L.A.C.; Serio, L.M.; Facchini, F.; Mummolo, F.F.A.G. ANN Modelling to Optimize Manufacturing Process. In *Advanced Applications for Artificial Neural Networks*; El-Shahat, A., Ed.; Intechopen: London, UK, 2018. [CrossRef]

40. Mundada, V.; Narala, S.K.R. Optimization of Milling Operations Using Artificial Neural Networks (ANN) and Simulated Annealing Algorithm (SAA). *Mater. Today Proc.* **2018**, *5*, 4971–4985. [CrossRef]

41. Chen, Y.; Sun, R.; Gao, Y.; Leopold, J. A nested-ANN prediction model for surface roughness considering the effects of cutting forces and tool vibrations. *Measurement* **2017**, *98*, 25–34. [CrossRef]

42. Hesser, D.F.; Markert, B. Tool wear monitoring of a retrofitted CNC milling machine using artificial neural networks. *Manuf. Lett.* **2019**, *19*, 1–4. [CrossRef]

43. El-Mounayri, H.; Briceno, J.F.; Gadallah, M. A new artificial neural network approach to modeling ball-end milling. *Int. J. Adv. Manuf. Technol.* **2010**, *47*, 527–534. [CrossRef]

44. Al-Abdullah, K.I.A.-L.; Abdi, H.; Lim, C.P.; Yassin, W.A. Force and temperature modelling of bone milling using artificial neural networks. *Measurement* **2018**, *116*, 25–37. [CrossRef]

45. Zain, A.M.; Haron, H.; Sharif, S. Genetic Algorithm and Simulated Annealing to estimate optimal process parameters of the abrasive waterjet machining. *Eng. Comput.* **2011**, *27*, 251–259. [CrossRef]

46. Mohanraj, T.; Shankar, S.; Rajasekar, R.; Sakthivel, N.R.; Pramanik, A. Tool condition monitoring techniques in milling process—A review. *J. Mater. Res. Technol.* **2020**, *9*, 1032–1042. [CrossRef]

47. Serin, G.; Sener, B.; Ozbayoglu, A.M.; Unver, H.O. Review of tool condition monitoring in machining and opportunities for deep learning. *Int. J. Adv. Manuf. Technol.* **2020**, *109*, 953–974. [CrossRef]

48. Kuntoğlu, M.; Aslan, A.; Pimenov, D.Y.; Usca, Ü.A.; Salur, E.; Gupta, M.K.; Mikolajczyk, T.; Giasin, K.; Kapłonek, W.; Sharma, S. A Review of Indirect Tool Condition Monitoring Systems and Decision-Making Methods in Turning: Critical Analysis and Trends. *Sensors* **2021**, *21*, 108. [CrossRef]

49. Pimenov, D.Y.; Bustillo, A.; Wojciechowski, S.; Sharma, V.S.; Gupta, M.K.; Kuntoğlu, M. Artificial intelligence systems for tool condition monitoring in machining: Analysis and critical review. *J. Intell. Manuf.* **2022**, 1–43. [CrossRef]

50. Kataraki, P.S.; Abu Mansor, M.S. Automatic designation of feature faces to recognize interacting and compound volumetric features for prismatic parts. *Eng. Comput.* **2020**, *36*, 1499–1515. [CrossRef]

51. Ning, F.; Shi, Y.; Cai, M.; Xu, W.; Zhang, X. Manufacturing cost estimation based on the machining process and deep-learning method. *J. Manuf. Syst.* **2020**, *56*, 11–22. [CrossRef]

52. Atia, M.R.; Khalil, J.; Mokhtar, M. A Cost estimation model for machining operations.; an ann parametric approach. *J. Al-Azhar Univ. Eng. Sect.* **2017**, *12*, 878–885. [CrossRef]
53. Florjanič, B.; Kuzman, K. Estimation of time for manufacturing of injection moulds using artificial neural networks-based model. *Polim. Časopis Plast. Gumu* **2012**, *33*, 12–21.
54. Yoo, S.; Kang, N. Explainable artificial intelligence for manufacturing cost estimation and machining feature visualization. *Expert Syst. Appl.* **2021**, *183*, 115430. [CrossRef]
55. Saric, T.; Simunovic, G.; Simunovic, K. Estimation of Machining Time for CNC Manufacturing Using Neural Computing. *Int. J. Simul. Model.* **2016**, *15*, 663–675. [CrossRef]
56. Tay, F.E.; Cao, L. Application of support vector machines in financial time series forecasting. *Omega* **2001**, *29*, 309–317. [CrossRef]
57. Mahmodi, K.; Mostafaei, M.; Mirzaee-Ghaleh, E. Detecting the different blends of diesel and biodiesel fuels using electronic nose machine coupled ANN and RSM methods. *Sustain. Energy Technol. Assess.* **2022**, *51*, 101914. [CrossRef]
58. Sada, S.; Ikpeseni, S. Evaluation of ANN and ANFIS modeling ability in the prediction of AISI 1050 steel machining performance. *Heliyon* **2021**, *7*, e06136. [CrossRef]

MDPI
St. Alban-Anlage 66
4052 Basel
Switzerland
www.mdpi.com

*Metals* Editorial Office
E-mail: metals@mdpi.com
www.mdpi.com/journal/metals